Palaeohydrology and Environmental Change

Palaeohydrology and Environmental Change

Edited by

G. Benito
Environmental Sciences Centre, CSIC, Madrid, Spain

V.R. Baker
University of Arizona, Tucson, USA

and

K.J. Gregory
Goldsmiths College, University of London, UK

JOHN WILEY & SONS
Chichester · New York · Weinheim · Brisbane · Singapore · Toronto

Library of Congress Cataloging-in-Publication Data
Palaeohydrology and environmental change / edited by G. Benito,
    V.R. Baker, and K.J. Gregory.
        p. cm.
    Includes bibliographical references (p.   −) and index.
    ISBN 0-471-98465-5 (acid-free paper)
    1. Paleohydrology.   2. Climatic changes.   I. Benito, G. (Gerardo)
    II. Baker, Victor R.   III. Gregory, K.J. (Kenneth John)
    QE39.5.P27P358   1998
    551.48—dc21   98-7027
                                                                  CIP

British Library Cataloguing in Publication Data
A catalogue record for this book is available from the British Library

ISBN 0-471-98465-5
Typeset in 10/12pt Times from the author's disks by Vision Typesetting, Manchester
Printed and bound in Great Britain by Biddles Ltd, Guildford and Kings's Lynn
This book is printed on acid-free paper responsibly manufactured from sustainable forestation,
for which at least two trees are planted for each one used for paper production.

Contents

PART III   Palaeohydrology and Environmental Changes

PART IV   Late Pleistocene Palaeohydrology

PART V   Tropical Zone Palaeohydrology

PART VI   Response of Extreme Events to Climate Change

List of Contributors

P.J. ASHWORTH School of Geography, University of Leeds, Leeds, LS2 9JT, UK

V.R. BAKER Department of Hydrology and Water Resources, J.W. Harshbarger Building, University of Arizona, Tucson, AZ 85721, USA

G. BENITO Environmental Sciences Centre, CSIC, Serrano 115 bis, 28006 Madrid, Spain

P. BISHOP Department of Geography and Topographic Science, University of Glasgow, G12 8QQ, UK

A.G. BROWN Department of Geography and Wetland Research Centre, University of Exeter, Rennes Drive, Exeter, EX4 4RJ, UK

A. DÍEZ-HERRERO Departamento de Medio Ambiente, Facultad de Ciencias, Universidad Europea de Madrid, 28670 Villaviciosa de Odón, Madrid, Spain

J.J. van DIJKE Department of Soil Science and Geology, Wageningen Agricultural University, PO Box 37, 6700 AA Wageningen, The Netherlands

E. FRANZINELLI University of Amazonas, Caixa Postal 885, Centro Manaus AM-69011-970, Brazil

K.J. GREGORY Goldsmiths College, University of London, New Cross, London, SE14 6NW, UK

M.G. GROSSWALD Institute of Geography, Russian Academy of Sciences, 29 Staromonetny Street, 109017 Moscow, Russia

J. HERGET Institute of Geography, Ruhr-Universität Bochum, Universitätsstr. 150, 44801 Bochum, Germany

M. HUISINK Faculty of Earth Sciences, Vrije Universiteit, de Boelelaan 1085, 1081 HV Amsterdam, The Netherlands

T. KALICKI Institute of Geography and Spatial Organisation, Polish Academy of Sciences, Ul. Sw. Jana 22, 31-018 Cracow, Poland

C. KASSE	The Netherlands Centre for Geo-Ecological Research (ICG), Faculty of Earth Sciences, Vrije Universiteit, de Boelelaan 1085, 1081 HV Amsterdam, The Netherlands
K. KELTS	Limnological Research Center, University of Minnesota, 220 Pillsbury Hall, Minneapolis, MN 55455, USA
M.J. KIRKBY	School of Geography, University of Leeds, Leeds, LS2 9JT, UK
A.V. KISLOV	Department of Meteorology and Climatology, Faculty of Geography, Moscow State University, Moscow 119899, Russia
H. KITAGAWA	International Research Center for Japanese Studies, Oeyama-chou, Goryou, Nishikyou-ku, Kyoto 610-11, Japan
L. LAÍN-HUERTA	Instituto Tecnológico GeoMinero de España, Ríos Rosas 23, 28003 Madrid, Spain
E.M. LATRUBESSE	University of Acre, BR 364 Km4, Campus, 69915-900 Rio Branco AC, Brazil
A. LÓPEZ-AVILÉS	School of Geography, University of Leeds, Leeds, LS2 9JT, UK
G.S. MAAS	School of Geography, University of Leeds, Leeds, LS2 9JT, UK
M.J. MACHADO	Environmental Sciences Centre, CSIC, Serrano 115 bis, 28006 Madrid, Spain
M.G. MACKLIN	School of Geography, University of Leeds, Leeds, LS2 9JT, UK
S. MIYAMOTO	Department of Geoecology, Krasnoyarsk State University, Svobodny 79, 660041 Krasnoyarsk, Russia
C.T. OGUCHI	Institute of Geoscience, University of Tsukuba, Ibaraki 305, Japan
T. OGUCHI	Center for Spatial Information Science, University of Tokyo, 7-3-1 Hongo, Bunkyo-Ku, Tokyo 113, Japan
A. PÉREZ-GONZÁLEZ	Departamento de Geodinámica, Facultad de Geología, Universidad Complutense de Madrid, 28040 Madrid, Spain
I. REINFELDS	School of Geosciences, University of Wollongong, NSW 2522, Australia
A. RUDOY	Laboratory of Pleistocene Geology and Palaeogeography, Department of Geography, Tomsk State Pedagogical University, Komsomolosky Av. 75, Tomsk 634041, Russia

A.F. SANKO	Institute of Geological Sciences, National Academy of Sciences, ul. Zodinskaya str. 7, 220141 Minsk, Republic of Belarus
M.L. dos SANTOS	Department of Geography, University of Maringá, 87020-900 Maringá, PR, Brazil
A. SOPEÑA	Instituto de Geología Económica, Facultad de Ciencias Geológicas, Universidad Complutense de Madrid, 28040 Madrid, Spain
L. STARKEL	Institute of Geography and Spatial Organisation, Polish Academy of Sciences, 31-018 Cracow, Ul. Sw. Jana 22, Poland
J.C. STEVAUX	Department of Geography, University of Maringá, 87020-900 Maringá, PR, Brazil
G.V. SURKOVA	Institute of Geography, Laboratory of Climatology, Russian Academy of Sciences, Staromonetny 29, 109017 Moscow, Russia
M.F. THOMAS	Department of Environmental Sciences, University of Stirling, Stirling, FK9 4LA, UK
B. VALERO-GARCÉS	Instituto Pirenaico de Ecología CSIC, Apdo. 202, 50080 Zaragoza, Spain
A. VELDKAMP	Department of Soil Science and Geology, Wageningen Agricultural University, Duivendaal, 6701 AR Wageningen, The Netherlands
A.F. YAMSKIKH	Department of Geoecology, Krasnoyarsk State University, Svobodny 79, 660041 Krasnoyarsk, Russia
Y. YASUDA	International Research Center for Japanese Studies, Oeyama-chou, Goryou, Nishikyou-ku, Kyoto 610-11, Japan
E. ZEROUAL	Limnological Research Center, University of Minnesota, 220 Pillsbury Hall, Minneapolis, MN 55455, USA

Preface

Over the last decade, the scientific understanding of global change has become one of the main issues for the scientific community. The International Geosphere–Biosphere Programme was established to study various aspects of global change. In the context of environmental change, there is a general consensus that future human activity and economic development will be conditioned by water availability. The hydrological effects of global climate change are predicted via the use of general circulation models. These are elegant models of the atmosphere, but they are unable to reconstruct feasible catchment-scale scenarios or to develop realistic simulations of basic hydrological processes. For this reason, the study of the characteristics of past hydrological cycles and their role in past environmental changes is crucial for the prediction of future environmental change. Past hydrological changes provide long-term data which can be used to validate the response of global or regional climate models in different environmental settings.

The Second International Meeting on Global Continental Palaeohydrology (GLOCOPH), held in Toledo, Spain, in September 1996, brought together academic and applied scientists from around the world to review current knowledge and methodological advances on global palaeohydrology for the past 20 000 years. The 34 oral presentations and the 15 posters presented at the conference, including examples from all over the world, stimulated continuous discussion on the contribution made by palaeohydrology to the understanding of past environmental changes. A selection of the papers presented was externally reviewed and, along with additional articles, these form the chapters of this volume.

Palaeohydrology and Environmental Change explores real hydrological scenarios during past environmental changes and covers extensive areas of Europe, America, Africa, Asia and Australia. The review describes the most recent scientific and methodological advances with respect to understanding the compositional reconstruction, distribution and movement of the Earth's main continental water bodies during past hydrological cycles in an ever-changing environment.

The introduction, written by Victor Baker (President of the GLOCOPH Commission of the International Quaternary Association), discusses the relationship of palaeohydrology to the hydrological sciences, underlining the various roles of palaeohydrology providing information regarding the Earth system, as a tool to identify the phenomena to be modelled, and as a source of new hypotheses about the nature of environmental change. This introduction is followed by 23 chapters which cover six major sections. The first section relates the progress and research strategies of particular aspects of palaeohydrology, and the remaining sections include contributions which propose an interdisciplinary approach to the palaeoenvironmental and palaeohydrological reconstruction of Earth's different climatic regions.

Several of the questions examined in the review are felt to be of interest to the scientific community with respect to issues concerning global change. Firstly, there is major emphasis on areas which are extremely sensitive to climatic change such as the Mediterranean, tundra regions, and the tropical parts of Asia, Africa and America. Secondly, new interdisciplinary methods for the reconstruction of palaeohydrological changes are described. These include palaeoecology and pollen analysis, identification of anthropogenic stresses, and changes in climate circulation regimes. In the third instance, several papers describe conceptual and numerical macro- and meso-scale models which reconstruct palaeoenvironments of the Quaternary.

Finally, the editors are very grateful to the authors for their time and effort in preparing contributions and would like particularly to acknowledge the assistance of the colleagues who acted as referees. Maria José Machado at the Environmental Science Centre has assisted in many ways vital for finalizing preparations for the publication of this volume, and Juan García-Vaquero has patiently redrawn some of the figures.

Whilst every effort has been made to trace the owners of copyright material, in a few cases this has proved impossible and we take this opportunity to offer our apologies to any copyright holders whose rights we may have unwittingly infringed.

G. Benito
V.R. Baker
K.J. Gregory

CHAPTER 1

Paleohydrology and the Hydrological Sciences

VICTOR R. BAKER

Department of Hydrology and Water Resources, The University of Arizona,
Tucson, Arizona, USA

INTRODUCTION

Modern hydrology can be viewed alternatively (1) as an applied science and branch of civil/environmental engineering, or (2) as a basic geophysical science (National Research Council, 1991). The goal in (1) is to achieve useful designs, management plans, and/or control works – all in relation to water as a resource or as a hazard. The goal in (2) is considered by many to be the development of explanatory, predictive models for water-related phenomena. Continental water processes and the global water balance are core elements of hydrology as a geophysical science. Cross-disciplinary elements of this science include the physical, chemical, and biological study of aqueous phenomena at the full range of naturally occurring temporal and spatial scales.

The traditional emphases of applied/engineering hydrology are on structural designs, somewhat rigid control works, and other short-term solutions to local, at-site problems. Today, greater emphasis is being placed on long-term management because of concerns over environmental issues, social disruption, and immense cost associated with the older approaches. Environmental sustainability is replacing the 'quick fix' in applied water science and engineering.

The concerns of basic, geophysical hydrology are also changing. Whereas considerable past work emphasized field and laboratory measurements of local processes, newer work is moving the study of land/atmosphere interactions, especially as integrated over continental and global scales, through new generations of climate models. The new modeling tools are making it possible to explore conceptually, a variety of systems interactions, and to evaluate the water resources consequences of future climate change. Computer modeling is becoming the major operating tool in both applied/engineering and basic/geophysical hydrology.

Paleohydrology is the study of past occurrences, distributions, and movements of continental waters (Baker, 1983). It is an interdisciplinary effort linking scientific hydrology to the sciences of Earth history and past environments (Schumm, 1967). The linkage to history is noteworthy because it affords a major distinction from the two branches of modern hydrology described above. Historical sciences emphasize observational experience over theory and experiment (Baker, 1995, 1996a). This is not because these sciences are less rigorous than mathematical/predictive sciences; it is because these sciences do not deal with phenomena amenable to controlled experimentation (Baker, 1996b). They emphasize concrete particular happenings of the past. Mathematical/predictive sciences, in contrast, emphasize conceptual, timeless generalizations. They deal with general principles presumed to apply at all times to all places. The facts of the past are tools for testing such theories. In the historical sciences, theories are the tools for elucidating the facts.

THEORY, PREDICTIVE MODELING, AND TESTING

Chow (1964) described the history of hydrology as one of progressive advancement through successive stages, each emphasizing states of knowledge, successively characterized as (1) empirical, (2) rational, and (3) theoretical. If we were to add a fourth stage, it would probably be that of empowering theory with computational efficiency. Predictive computer modeling has revolutionized the ability of scientists and engineers to consolidate theoretical knowledge into convenient conceptual packages that can be used to simulate the behavior of 'systems' over ranges of conditions and for processes presumed to operate in the real world. Because of the increased ease in performing such modeling, the apparent rigor of the methodology, and utility of the results, predictive computer modeling has become the principal operating tool of modern hydrology.

Perhaps it would be well to define some terms at this point. A hydrological theory is an organized system of statements about things hydrological. The organization involves various assumptions, including rules of logic and axiomatic principles. A theory both applies to a system, and is a system itself. The term 'system' refers to some set of interrelated or interconnected elements. These elements are human abstractions, deemed to be important or essential interrelations or interconnections. The 'true' system of all 'real' interconnections and interactions is an intellectual ideal, the realization of which is either unknowable or untestable.

Models are human constructs that simplify reality. Models embody elements abstracted from the world by their inventors, such that these elements represent essential workings of that world. Thus, the model system is both fully known and testable. The advantage of this model system is that, through its relative ease of manipulation, the modeler can be led to potential understanding about the more complex real world that the model has been presumed to represent. The disadvantage is that the issue of representation is more complex than most modelers realize.

In applied/engineering hydrological modeling, the best available theory is incorporated into models in order to achieve accurate representation (simulation) of the system of interest. In the engineering case, this is a predefined system, not the 'unknowable and untestable' real or true system noted above. This engineering system is predefined by exact criteria for design, management, or control. It is then necessary to obtain facts through direct measurements on this system, facts that we call 'data'. Here, the data are used to make initial estimates of system parameters and to help the selection of appropriate model parameters that reduce uncertainties in the prediction of processes. In other words, data serve to calibrate the models. Thus, engineering models apply to specific, predefined criteria. They are valid only for the problem domain to which those criteria apply. The 'validation' of an engineering hydrological model is a process somewhat circularly confined to the idealized system of a problem definition (Table 1.1).

Table 1.1   Steps in engineering model validation

1. Specific criteria (for model applicability) are established.
2. Model is constructed.
3. Model is tested against criteria established in 1.
4. Model is modified until the criteria are satisfied ('validation').
5. Model can only be applied to the problem domain for which criteria established in 1 apply.

To be scientific, a theory must relate to the empirical world, that is, to facts and observations, measured or sensed from the world of experience. While engineering models are calibrated to perform as practical tools, geophysical hydrological models function to advance scientific understanding, and ultimately to achieve new theories. Philosophers of science have observed that falsification, not calibration, is the essential element in science. The more general notion is that of testing predictions made by the model formulations of theory.

Geophysical hydrology can be viewed as a mathematical/predictive science studying phenomena that can be manipulated through controlled experimentation. Such phenomena are well exemplified in physics, which is '...the science devoted to discovering, developing and refining those aspects of reality that are amenable to mathematical analysis' (Ziman, 1978). Reductionist idealist philosophers believe in the essentially untestable notion that all science will eventually reduce to physics. The approach of this mathematical/predictive science (physics) is conceptual, seeking universal classes of phenomena that can be generalized by means of the underlying physical laws presumed to govern nature. In this form of science, understanding is achieved in the Platonic sense of discovering mathematically expressible functional relationships between abstractly conceived processes and objects. Such science concerns these abstractions rather than the particular things seen in the world about us. The latter may be used as objects for testing or verifying the all-important conceptual schemes, after the prediction of their ideal behavior or properties by those schemes. As noted above, for this to work, there must be a well-defined 'system' amenable to manipulation in controlled experimentation, the object of which is falsification, not verification of model predictions (Popper, 1959, 1969). Regrettably, in hydrology, true controlled experimentation is rarely achieved, and the two worlds of model prediction and verification are all too often hopelessly confused (Klemes, 1987; Oreskes et al., 1994; Baker, 1996c).

THE ROLE OF PALEOHYDROLOGICAL DATA

The major function of paleohydrology in regard to applied/engineering and basic/geophysical hydrology is the supplying of data. For the applied role, this is in regard to calibrating models, say to simulate long-term hydrological behavior in regard to an underground nuclear waste repository or to achieve an estimate of flood risk. The geophysical role of paleohydrological data is in regard to testing models, in a manner that advances understanding. A rather sophisticated version of the latter role is exemplified by the Past Global Changes (PAGES) project of the International Geosphere–Biosphere Programme: A Study of Global Change (IGBP) of the International Council of Scientific Unions. The inferential framework used to test models for future global changes involves two parallel reasoning paths (Oldfield, 1997). The modeling path starts with certain boundary conditions and hypothesized climate forcing. Numerical models of climate, oceans, and biogeochemistry are then used to simulate (predict) some parameter of future and past change. The paleodata path begins with the record of past phenomena, such as a sequence of inferred ancient streamflows or lake levels. This record is then transformed to one of the same parameters simulated by the modeling path. The transformation is generally termed a 'reconstruction', even though it is not clear what is meant by the implied 'construction' of the 'original' parameter.

The PAGES project seeks intercomparisons between (1) the climate change parameters predicted by climate models, and (2) the climate change parameters 'reconstructed' through study of the paleodata (Oldfield, 1997). The ultimate aim is to improve model validity and predictive power. The motivation for this testing, however, is applied, not scientific. As Oldfield (1997, p. 2) observes, this goal '... requires that evidence from the past be refined in order to test the models against reconstructed past conditions. Models form one of the major links between science and decision-making. To be fully effective in simulating future climates, models need testing using climate conditions and boundary conditions unlike those of the present. Unless models can achieve an adequate level of realism in simulating past climate conditions, their performance in predicting future conditions will remain less certain than decision makers would like.'

Note that the logic of the PAGES approach involves a hybrid between the applied/engineering goal of achieving controlled outcomes (predictions with reduced uncertainties) and the basic/geophysical method of testing model predictions against reality. Missing is the recognition that the system framework of the applied/engineering approach (Table 1.1) is completely different from the real or true system implied by the basic/geophysical approach (Baker, 1996c).

The hybrid approach of PAGES and of much current global change science encounters a dilemma that was well stated by Oreskes et al. (1994): '... model verification is only possible in closed systems in which all the components of the system are established independently and known to be correct'. It is not sufficient that model predictions (or retrodictions) match observational data. This can be achieved through multiple means. Indeed, a famous principle of logic, the Duhem–Quine thesis, explicitly disallows such model verification. Simply put, the Duhem–Quine thesis precludes conclusive theory (model) falsification because any predictive clash with observation can always be blamed not on the theory itself, but rather on the maze of assumptions (some not explicitly stated) in the test situation (Oreskes et al., 1994).

For the complex world of environmental processes, models cannot be strictly verified for geophysical science because the real-world systems are not closed, the model predictions are non-unique (Duhem–Quine thesis), model input parameters and assumptions are almost always poorly known, scaling issues for non-additive properties are poorly known, and data themselves pose problems (Oreskes et al., 1994). This latter issue is central to paleohydrology. Can we treat paleohydrological data as exact substitutes for some natural phenomena, or should we consider these data to be inference-laden signifiers of those phenomena? If it is the phenomena that are of primary interest, rather than various theories about them, how do we deal with the limitation of limited access to the phenomena appreciated by understanding data as signifiers of phenomena, rather than as substitutes (proxies) for those phenomena?

PROXIES AND SIGNS

Much current research in paleohydrology is conducted under a research paradigm that emphasizes a conceptual abstraction: climate. Climate is expected weather, or probable trends in the temporal sequence of weather. Climate change, predicted by models of the climate system, is the driving concern of much current international hydrological science.

Paleoclimatology concerns climate prior to instrumental measurement. It focuses on climate-dependent natural phenomena that incorporate into their structure a measure of this dependency (Bradley, 1985). These phenomena are studied as proxy records of climate. The concept of a 'proxy' is that of a substitute, or an agency of one entity acting for another. In this dual role of one thing substituting for another, the entire emphasis is on the perfection or accuracy of that substitution. Thus, much of paleoclimatology consists of a search for the most accurate and detailed substitutes. The various substitutes for climate are all known to contain some climatic signal, but many of these are weakly embedded in considerable extraneous 'noise' arising from various non-climatic influences (Bradley, 1985).

Examples of extremely detailed and (presumably) accurate proxies for paleoclimate include ice cores, deep sea cores, and tree rings. Considerable paleoclimatological research concerns the filtering, processing, and calibration of these records to understand their climate dependence, and, ultimately, to test hypotheses about the causes of climate change (Bradley, 1985). An example of this is illustrated by the PAGES program described above.

The alternative data role to phenomenon substitution (proxies) is that of inference-laden signification. This is a philosophical issue involving our point of view in regard to data. The treatment of data as proxies limits the status of the data to how well they substitute for some conceptual abstraction. The alternative point of view is semiotic, which considers data as part of a complex interpretive structure mediated and sustained by signs (Deely, 1990; Sebeok, 1994). In semiotics, signs stand for something (the object) in relation to something else (the interpretant). The signs of paleohydrology are indexical signs, in which they relate to objects via causation. Thus, in paleoflood hydrology (Baker, 1987), various causal indicators (indexical signs) reflect causal processes as their objects, in this case extreme floods. Although the natural indices found in the field, including various flood-emplaced sediments and landforms, are virtually semiotic in themselves, their interpretant function is realized through the human investigator.

Consider the extreme flood shown in Figure 1.1. After its passage, a slackwater deposit (Figure 1.2) provides an indexical sign of its past activity. The details of such slackwater deposits can be studied for further signs of past flood activity (Figure 1.3). The human investigation transforms signs through their interpretants to new signs, which are similarly transformed, leading to connections that enhance understanding of the past flood processes. Table 1.2 illustrates some of the kinds of indices used in this process of scientific inquiry applied to ancient floods (paleofloods).

The notion of data as indexical signs, rather than as proxies used in theory testing, can be further understood via a simple analogy. Consider a detective at a murder scene. The detective finds a small hole in the wall with an embedded lead particle. This observation

Figure 1.1   Pedernales River in central Texas during Tropical Storm Amelia flooding in August, 1978. Note the high-velocity water in the constriction at the center of the photo. The downstream flow expansion (right center) involves velocity reduction, especially on the right bank

Figure 1.2   Photograph taken shortly after flood recession at the same location as in Figure 1.1. Note the prominent slackwater deposit that developed at the site of velocity reduction in sediment-laden floodwater shown in Figure 1.1. This deposit provides an indexical sign of the causative flooding process

Figure 1.3   Interbedded slackwater sands and tributary stream gravels near Aravaipa Creek, southeastern Arizona. The investigator is sampling an organic-rich zone for radiocarbon dating

Table 1.2   Semiotic structure of paleohydrological data

Indexical sign	Derived paleoflood data	Causal interpretant
Slackwater deposit (SWD)	Elevation of SWD	Minimum paleostage of flood
Scour mark, silt line, or other paleostage indicator (PSI)	Elevation of PSI	Paleostage of flood
Paleochannel	Channel dimensions	Hydraulic geometry of responsible flow regime
Flood-scarred tree	Elevation of flood scar	Minimum paleostage of flood
Largest flood-transported boulders	Boulder sizes	Paleocompetency of flood

might serve as the test of the theory (model) that a gun was used in the crime, firing the bullet, making the hole, etc. Conceptually, the presumed bullet hole serves as a proxy for the gun, a murder weapon that is missing. Of course, other theories can explain these circumstances. Perhaps the gun firing the bullet is not related to the crime. In the semiotic viewpoint, however, the crime theory is not the immediate concern. Instead the detective is a student of bullet holes, much as Sherlock Holmes was a student of exotic tobacco. It is a rather secure notion that the hole and embedded projectile provide an index of a process: gunfire. The detective studies this sign (clue) and combines it with other clues, developing a web of interconnecting clues (signs). Eventually a narrative connecting these clues emerges as the solution of the crime.

One notes in this example that the proxy approach is limited in regard to the particular model or crime theory in mind. The semiotic approach is part of an open process of inquiry that will immediately be recognized by the readers of detective stories. Lest the logic of this analogy seem trivial, it is worthy to note that the logic of detectives is exactly the same as that of abductive/retroductive reasoning in science (Eco and Sebeok, 1988). Eco and Sebeok (1988) develop a full analysis of this 'logic' through reference to detective stories by Edgar Allan Poe and Conan Doyle.

HYDROLOGICAL SEMIOTICS

Are paleohydrological data being most effectively employed in various modeling frameworks? Baker (1995, 1996a) envisions a third role for these data, applicable both to basic and applied science. This is the retroductive function, which serves as the source of creative discovery concerning hydrological processes as they vary over time and space. Here the goal is not to calibrate or test models, but rather to recognize the appropriate real-world phenomena that should be modeled. It is the intellectual challenge of paleohydrologists and the increasingly model-oriented hydrologists to balance these approaches in science in order to achieve the best possible engineering and science of Earth's continental waters.

Retroductive reasoning, or 'abduction' (Von Engelhardt and Zimmermann, 1988; Baker, 1996b), underpins various naturalistic/historical sciences (Baker, 1996a). The focus of these sciences is not on idealized theories that can only be verified under very special laboratory conditions. Rather, their prime concern is with realized phenomena observed in the natural world, uncontrolled by artificial constraints. By not limiting herself to the world amenable to mathematical analysis and/or controlled experimentation, the naturalist/historical scientist takes the world as it is. Instead of general principles of universal application, concrete particulars are the objects of primary concern. Because the richest source of such reality is the various happenings of the past, the naturalistic sciences merge with the historical, constituting an entity that formerly was proclaimed with pride: natural history.

How does one relate predictive computer models to this semiotic scheme? In its semiotic context, as opposed to isolated abstraction, a predictive computer model functions as a system of signs or symbols representing the operation of some 'system' of interest in relation to some subsequent interpretation. Because the system of interest is a symbolic simplification/representation of real processes, a critical issue involves qualitative choices made by the modeler in that simplification/representation. These choices can be hidden in the subsequent interpretation unless questions are raised, not so much about the self-referential quality of the model, as about its referential context. Merely 'testing' the fit of a model to data does not adequately deal with this issue, since such a fit is 'self-referential', i.e. meaningless outside the limited context of the defined symbols. The more complex notion of 'testing a model' is fraught with potential misunderstanding, both for modelers and for those who invest belief in their results (Klemes, 1997).

DISCUSSION AND CONCLUSIONS

The usual goal of hydrology, as a geophysical science, is to achieve theories capable of explaining with satisfactory accuracy the phenomena of interest. Through the rapidly accelerating power and versatility of digital computing technology, theory development and application are immensely facilitated via increasingly sophisticated predictive modeling schemes. The latter have now become the principal operating tools both for applied/engineering hydrology and for basic/geophysical hydrology. The role of paleohydrology in regard to these two hydrological approaches is to supply data. For engineering, the data serve to calibrate and validate models (Table 1.1). For geophysical hydrology, the data serve to test models. However, in the latter role, the models are only tools for guiding further study; their predictions cannot be subjected to proof for fundamental reasons involving logic and the nature of natural systems (Oreskes et al., 1994). This limitation is of considerable concern in modern global change science (Baker, 1996a), where paleohydrological data are commonly treated as proxies for climate (Bradley, 1985).

Galileo Galilei once observed that the book of the universe is written in the symbolic language of mathematics. To read the book of the universe, one must learn this language. Earth's hydrological book also has its mathematical language, but this symbolic language is not the only one worthy of hydrological inquiry. The language of indices, signs directly representing causative processes, constitutes another text for hydrological reading. This paleohydrological text cannot be fully understood in symbolic terms, which involve language imposed by our own conventions. The indexical language of paleohydrology is learned from nature itself. The landforms and sediments emplaced by past hydrological processes constitute the sign language to be interpreted. By adopting a semiotic point of view, the paleohydrologist is led into a process of inquiry that leads to understanding nature's reality prior to the definition of systems appropriate for explanation through model simulation. This process of inquiry is only obliquely addressed by a viewpoint that treats nature's signs as proxies for our conceptualizations, and then uses these proxies to test the validity of the model simulations. Not only is such model validation/verification logically flawed for the real (open) systems of nature, it stifles the spirit of inquiry that is associated with the semiotic viewpoint.

Neither viewpoint of paleohydrological information, proxy or semiotic, is a matter of direct scientific method. Rather, these viewpoints are metaphysical; they are value-laden choices to be made by individual investigators. Obviously, both can be chosen, especially given the different paths along which they lead. The one principle that might apply here is an old one: do not erect any impediments to free scientific inquiry (Baker, 1996d). Paleohydrologists need to value the revolution in their semiotic understanding of nature's reality with enthusiasm equal to that accorded the computational revolution in the explanatory modeling of systems presumed to reflect that reality.

ACKNOWLEDGEMENTS

I thank Gerardo Benito for his facilitation of this essay, based on my oral presentation at the highly successful Second International Conference on Global Continental Paleohydrology (GLOCOPH 96), which Dr. Benito organized in Toledo, Spain. This essay is Contribution Number 54 of the Arizona Laboratory for Paleohydrological and Hydroclimatological Analysis (ALPHA), The University of Arizona.

REFERENCES

Baker, V.R., 1983. Large-scale fluvial palaeohydrology. In K.J. Gregory (ed.), *Background to Palaeohydrology*, Wiley, Chichester, 453–478.

Baker, V.R., 1987. Paleoflood hydrology of extraordinary flood events. *Journal of Hydrology*, **96**, 79–99.

Baker, V.R., 1995. Global paleohydrological change. *Quaestiones Geographicae*, Special Issue **4**, 27–35.

Baker, V.R., 1996a. Discovering Earth's future in its past: Paleohydrology and global environmental change. In J. Branson, A.G. Brown and K.J. Gregory (eds), *Global Continental Changes: The Context of Palaeohydrology*, Geological Society, London, Special Publication 115, 73–83.

Baker, V.R., 1996b. Hypotheses and geomorphological reasoning. In B.L. Rhoads and C.E. Thorn (eds), *The Scientific Nature of Geomorphology*, Wiley, Chichester, 57–86.

Baker, V.R., 1996c. Modeling global change: Why geologists should not let 'system' come between Earth and science. *GSA Today*, **6** (5), 8–11.

Baker, V.R., 1996d. Tablets of stone: Ten commandments or a golden rule? In S.B. McCann and D.C. Ford (eds), *Geomorphology sans Frontieres*, Wiley, Chichester, 59–67.

Bradley, R.S., 1985. *Quaternary Paleoclimatology: Methods of Paleoclimatic Reconstruction*, Allen and Unwin, Boston.

Chow, V.T., 1964. Hydrology and its development. In V.T. Chow (ed.), *Handbook of Applied Hydrology*, McGraw-Hill, New York, 1.1–1.22.

Deely, J., 1990. *Basics of Semiotics*, Indiana University Press, Bloomington.

Eco, U. and Sebeok, T.A., 1988. *The Sign of Three: Dupin, Holmes, Pierce*, Indiana University Press, Bloomington.

Klemes, V., 1987. Hydrological and engineering relevance of flood frequency analysis. In V.P. Singh (ed.), *Hydrologic Frequency Modeling*, Reidel, Dordrecht, 1–18.

Klemes, V., 1997. Of carts and horses in hydrologic modeling. *Journal of Hydrologic Engineering*, **1**, 43–49.

National Research Council, 1991. *Opportunities in the Hydrologic Sciences*, National Academy Press, Washington, DC.

Oldfield, F., 1997. Forward to the past: An update on PAGES. *Global Change Newsletter*, **31**, 1–3.

Oreskes, N., Shrader-Frechette, K. and Berlitz, K., 1994. Verification, validation, and confirmation of numerical models in the Earth sciences. *Science*, **263**, 641–646.

Popper, K., 1959. *The Logic of Scientific Discovery*, Basic Books, NewYork.

Popper, K., 1969. *Conjectures and Refutations*, Routledge and Kegan Paul, London.

Schumm, S.A., 1967. Paleohydrology: Application of modern hydrological data to problems of the ancient past. In *International Hydrology Symposium*, Proceedings Volume 1, Fort Collins, Colorado, 185–193.

Sebeok, T.A., 1994. *Signs: An Introduction to Semiotics*. University of Toronto Press, Toronto.

Von Engelhardt, W. and Zimmerman, J., 1988. *Theory of Earth Science*, Cambridge University Press, Cambridge.

Ziman, J., 1978. *Reliable Knowledge: An Explanation of the Grounds for Belief in Science*, Cambridge University Press, Cambridge.

PART I
Palaeohydrological Research Strategies

CHAPTER 2

Applications of Palaeohydrology

KEN J. GREGORY

Goldsmiths College, University of London, UK

INTRODUCTION

We can now look back over at least four decades of palaeohydrology as suggested in the introduction (Gregory, 1996) to the Geological Survey Special Publication *Global Continental Changes* (Branson et al., 1996). Over those four decades significant changes have taken place and, although some of these were anticipated, others may not have been expected, and for palaeohydrology it may therefore be timely and important to continue to survey where we have come from, where we are now, and how we might progress. Being one of the older members of the GLOCOPH commission may provide an excuse for my being somewhat reflective. In addition I am conscious, as I look back to earlier publications covering other fields such as *Drainage Basin Form and Process* (Gregory and Walling, 1973) or *The Nature of Physical Geography* (Gregory, 1985), it is arguable that, because of general changes in approach in the environmental sciences over the last two decades, such volumes would now require completely new final chapters. Such new concluding chapters could provide necessary bridges between, and extensions of, the original chapters which focused on temporal change and on applications which are no longer so separate. Perhaps in these days of value for money we also should consider how far the research of the GLOCOPH commission is producing all potential products, so as to have answers ready when asked questions about value for money. In attempting to sketch the scope of applications of palaeohydrology it may be useful to have an outline framework which can be augmented and amplified. This will inevitably be preliminary and provisional, influenced by particular themes in temperate areas, but capable of wider extension (see Chapter 1).

INTRODUCTION: CHANGES OVER FOUR DECADES PROVIDE THE CONTEXT

Over the four decades since the definition of palaeohydrology by Leopold and Miller in 1954 at least four major things have happened. *First* environmental awareness has increased very considerably. Although some writers have envisaged an environmental revolution, the beginnings of the green environment approach in the 1960s, followed by more acute environmental realisation in the 1970s, has meant that public awareness of environment is now very different from the situation that existed four decades ago. *Secondly*, and an integral part of this increased environmental awareness, has been the

Palaeohydrology and Environmental Change. Edited by G. Benito, V. R. Baker and K. J. Gregory
© 1998 John Wiley & Sons Ltd.

greater appreciation of environmental change, particularly arising from the development of global change programmes and from the publicity which they have attracted. The most recent Working Group III contribution to the second report of the UN Intergovernmental Panel on Climate Change (IPCC) has provided the latest view of the economic and social dimensions of climate change. The IPCC has given the basis for a number of scientific assessments such as May (1997) drawing on the work of some 3000 scientists. However, although hydrological impacts are mentioned they are not always sufficiently elaborated. Amongst the range of global change programmes, the Human Dimensions of Environmental Change programme continues to develop and most recently has extended to include a research theme on security. *Thirdly* the understanding of the impact of human action has become much more familiar and, not only has the nature and magnitude of human activity been documented, but it has provided a more informed basis for environmental control and management. The general pattern of human activity has been reviewed in a number of books and in papers (Gregory, 1987) and has been most recently reviewed by Goudie (1993) and Gregory (1995a) although in recent research, especially as related to recent environmental change, the difficulty of separating natural from human impacts becomes increasingly evident (Brown, 1997). *Fourthly* it is evident that palaeohydrology itself has changed over 40 years. During this period of four decades it has been suggested (Gregory, 1996) that it is possible to identify as many as seven themes which have contributed to the advance of palaeohydrological research. These include studies of alluvial chronology, the water balance approach, underfit streams and palaeohydrologic methods, river metamorphosis, palaeoecological approaches, lake fluctuations, and integrated approaches such as that employed by IGCP 158 (International Geological Correlation Programme, Project 158 on the *Palaeohydrology of the Temperate Zone in the Last 15 000 Years*). As more research results have been published, greater world coverage has been achieved, although it still remains too uneven (Gregory, 1996) and the number of papers published varies substantially between areas.

Applications of palaeohydrology have been given relatively little attention and are a central theme in less than 10% of the papers in the GLOCOPH database (Branson et al., 1995). In the evolution of palaeohydrology particularly important has been the development of the think back approach (Baker, 1991), the benefits that have derived from a multidisciplinary research endeavour, and the contributions made by several disciplines. A considerable amount of research has now been achieved and, for example, palaeoecological results have recently been published in an impressive volume by Berglund et al. (1996) identifying palaeoecological events in the last 15 000 years in Europe based upon the results of palaeoecological studies of lakes and mires. In parallel, the fluvial research of IGCP 158 was concluded in a volume on the temperate zone published in 1991 (Starkel et al., 1991) and the basic research foundation for GLOCOPH was published in 1995 (Gregory et al., 1995).

It is against this background that we can envisage applications of palaeohydrology and consider whether there are further opportunities for application of research results. Although research is seen by some in the three categories of pure or blue skies, applicable or grey skies, and applied, it is also possible to differentiate between those results which are applicable to other disciplines and often arise from pure or applicable research, and those results which are particularly related to planning and management (Table 2.1). Planning can be thought of as elaboration of a strategy for action to be carried out over some period of time, as distinct from management which is the critical control exerted

Table 2.1 Outline definitions of types of research and research applications

---

Pure research ('blue skies')
  – basic research which is not specifically related to environment problems and not profitable in the current state of knowledge or technological development.

Applicable research ('grey skies')
  – investigations which give results or new facts which may be applicable to environmental problems.

Applied research
  – research where results are related to environmental problems in a specific area.

Planning research
  – elaboration of a strategy for action to be carried out over a specific period of time, involves greater coherence and purpose.

Management research
  – application of appropriate skill and principles in decision making with critical control exerted over people, activities and resources.

Sustainability research
  – good practice in human exploitation of Earth resources whereby resources are capable of being maintained at a certain rate or level.

---

over people, activities and resources. Not only have planning and management figured in relation to applications of scientific research but, especially since the Rio conference, sustainability has become a prominent theme in many disciplines (British Hydrological Society, 1994). The need to keep the link between research and application constantly under review has been stressed by Wolman (1995), and Starkel (1990) has emphasised the importance of knowledge derived from long-term approaches to the study of environment with particular reference to long-term trends of existing water resources as well as those of extreme events.

To address the issue of applications of palaeohydrology it is possible to think of those already achieved, the further applications that are possible, and the further developments and recommendations that may derive.

APPLICATIONS ACHIEVED: TO OTHER DISCIPLINES, TO APPLICABLE AND APPLIED QUESTIONS

In considering the contributions already made to other disciplines and to planning and management issues, the direct and obvious applications relate to certain process components of the hydrological system. In particular, the estimation of discharge, and especially the retrodiction of discharge allowing extension of hydrological records, has been very well researched and developed. Whereas the range of equations available in the 1980s was compiled by Williams (1984) and has subsequently been utilised and further developed in the database compiled for GLOCOPH (Branson et al., 1995), it is from the palaeostage indicator research and the associated research results that Baker and other researchers have been so successful in developing advances in this area. This has been described (Baker, 1995) as a new tool of palaeoflood hydrology, applicable mainly to late Holocene records, which is of considerable promise for understanding individual extreme events. In addition this can lead to a unique opportunity to better understand long-term trends of existing water resources as well as of extreme events. Particularly important is the knowledge gained about the probable maximum flood. This is perhaps the classic way in which the continuous hydrological records can be extended so that the deficiencies of long-term monitoring commented upon by Burt (1994) can be overcome. Direct applications have also arisen from the insights provided by the chronological investigation of particular sites such as the excellent Polish site in the Vistula Valley (Starkel, 1995a), the channel and flood plain sequences such as those investigated by Brown et al. (1994), channel and flood plain responses that have been identified in particular areas (e.g. Rumsby and Macklin, 1994) and the regional studies such as those of the Mississippi (Knox, 1995). In all these ways it has been possible to benefit from a multidisciplinary approach and this has been well exemplified in research in Mediterranean areas (Lewin et al., 1995).

Specific components, influences or means of analysis of hydrological systems have also been the subjects of significant developments. In relation to the vegetation component, for example, it has been shown how this is extremely important in the development of Holocene channel systems (Brown, 1995a,b), and changes of vegetation together with changes in land use obviously have significant influences upon river channel change (Brookes, 1996). Now that we know more about the impact of woody debris in channels it is possible to suggest (Gregory, 1995b) how flood plain vegetation and woody debris affected channel processes in palaeoenvironmental situations. Because much palaeohydrological research has been based upon vegetation history, the dynamics of the vegetation in relation to stream flows is obviously a major element in understanding the operation of palaeohydrological systems. Thus in certain areas it has been demonstrated that present river channel activity is not in accord with flood plain development and that rivers and flood plains are not environmentally equilibrated (Lewin, 1996). It was for these and other reasons that Brookes (1996) argued for further studies of change which consider periods longer than ten or 15 years and which are needed to address the 500 year timescale as the basis for the improved prediction of more recent hydrological impacts and for the development of ecologically sound management tools. In considering applications of hydrology at the end of the recently published volume (Branson et al., 1996), Brown (1996a) also cites the example of the mobilisation of contaminants held in alluvial sediments as an area appropriate for applications, especially as remobilisation of contaminants affected by rate and locations of channel change can significantly affect water quality. In his stimulating chapter entitled 'A bright future for old flows', Baker (1991) identified five major contributions made by that time as including advances in palaeoflow estimation, in geochronology and stratigraphy, in relation to patterns of climate and fluvial regions, in scientific explanation, and in global patterns. These encompass the major contributions noted above and such applications are included in summary form showing relevance to applicable and to applied questions and to other disciplines in Table 2.2.

These developments therefore emphasise the fact that research results need to be analysed within the context of a drainage basin framework and should build upon an improved understanding of river channel processes. Contributions in palaeohydrology have, as Baker suggested in 1991, contributed to other disciplines and in turn there are developments in related disciplines which can be of use in palaeohydrological research. A number of recent contributions illustrate this direction, for example, the analysis of the geomorphological effectiveness of different kinds of floods (Costa and O'Connor, 1995)

Table 2.2   Applications achieved

Type of application	Relevance to other disciplines	Relevance to applicable and applied questions
Palaeoflow estimation	Extend hydrological records	Probable maximum flood
Chronological investigations	Environmental history	Land and flood plain management
Basin changes, e.g. vegetation	Impacts on channel change	Channel management
Contaminants in sedimentation	Derivation of contaminants	Effects on water quality
Global patterns	Global environmental change	Link atmospheric and hydrological models

and identification of spatial patterns of erosion and deposition within a basin (Miller, 1995) offer two particular examples of ways in which such developments may be relevant to palaeohydrology. In particular it is ultimately necessary to decide whether analogues for global change can be identified. Some progress has been made in finding analogues for greenhouse effects, although Starkel (1995b) has shown that we need to give careful consideration to the sequence of changes and thence to pay more attention to periods of rapid change towards warming rather than simply utilising comparison of types of climate. Further analysis of the synchroneity of individual events and climate (Starkel, 1995c) is therefore necessary to make progress in this regard.

It is inevitable that palaeohydrology should seek to contribute to global change (Arnell, 1996) but as suggested by Gregory (1995a) it is notable that global change programmes hitherto have made insufficient reference to palaeohydrological research results. This is somewhat surprising in view of the original approach taken by Schumm (1965) and Dury (1964), both of whom produced schemes for global palaeohydrological analysis that are capable of further refinement and amplification. Such refinement can come about through ideas that have subsequently been proposed, and the use of concepts of thresholds, metamorphosis and complex response as identified by Schumm (1977) can be further developed and reinforced as hinted by Vandenberghe (1995a,b), who argued that reconstructed mean discharge, peak discharge and floods are the principal proxy data from the fluvial environment which allow the direct derivation of palaeoclimatic data. A particularly stimulating approach was provided by Baker (1995) when he posed the question as to whether uniformitarianism can apply to the forward time projection in the way that it applies to the past, and he questioned whether knowledge of the past could apply to the future (Baker, 1995, p. 27). It was for this reason that he then used a retroductive approach to extend to habitability (Baker, 1995, p. 33).

FURTHER POSSIBLE APPLICATIONS

There are several ways in which future research might benefit from greater application of palaeohydrological research results to other disciplines and also to practical problems. These can be thought of *first* in relation to the use of different or modified components of environment whereby we can extend and develop what has already been achieved. For example we now know much about channel processes, channel vegetation and in-channel debris (e.g. Gurnell et al., 1995; Gregory and Davis, 1993) but it is possible to focus upon river corridors as a significant landscape element and Petts (1990), for example, considers them a lost resource. The river corridor is the river and river channel together with their associated wildlife and the adjacent riparian ecosystem; Petts (1990) contends that, according to documentary evidence, the corridors of most large alluvial rivers in northern mid-latitudes were forested in the past and so he argues that landscaping and forest management must be integrated with policies for flow regulation and for maintaining geomorphological and hydrochemical processes in restored river corridors. The classification of river corridors has been addressed (Gurnell et al., 1994) and this can be combined with implications for management of forest aquatic habitats (Gurnell et al., 1995). It is equally imperative that models of Holocene flood plain evolution are developed and clarified, and palaeohydrology can contribute to this in the way that Brown and Keough (1992) identified their stable bed aggrading bank model. It is also important to be able to gain further understanding of valley development and of the significance of episodic influences (Teisseyre, 1995). Palaeoenvironmental analysis of flood plains can illuminate understanding of flood plain processes and flood plain evolution (Brown, 1996b) and so inform the way in which present and future flood plains are managed. Petts (1996) has quoted selected features of the twenty-first century flood plain and these features need to be seen in a holistic way and informed by knowledge of flood plains of the past.

It is *secondly* also important to ensure that certain influences upon palaeohydrological systems are not neglected. Brown and Bradley (1995) have focused attention upon ground water; a greater ecological basis linked to the river basin is still required as implicitly shown by Harper and Ferguson (1995); and the importance of tectonics in certain areas (Starkel, 1990; Lewin et al., 1995) needs to be encompassed. The importance of a global view, as reflected in the GLOCOPH database (Branson et al., 1995), has to be kept in mind and is obviously very pertinent in relation to inputs to global climatic change. *Thirdly*, however, particular aspects which are still especially significant include establishing the links between different spatial and temporal scales, utilising an improved understanding of the mechanics of the hydrological basin model, and proceeding towards a more comprehensive energetically based dynamic approach. Although progress has been made with refining hydrological models (e.g. Kirkby, 1994), Kite (1995) has drawn attention to the fact that general circulation models (GCMs) are still not adequately coupled to the hydrological distributed models; he suggested that this is an outstanding requirement and it is surely one to which palaeohydrological research can contribute. In Kite's words: 'The remaining difficulty is the connection between the grid of the atmospheric model and the watershed'. This will be achieved by a greater understanding of the linkages between the functioning of different parts of the basin as developed in an Australian example by Prosser et al. (1994), who showed in southeastern Australia how the Holocene history can be viewed as a complex response to events in the late Pleistocene. Ideally we should have theoretical basin models developed for each of the major world zones and this could be one of the products of GLOCOPH (e.g. Baker et al., 1995; Kadomura, 1995).

Such potential areas of application of palaeohydrological research emphasise that global forecasts and environment need to reflect greater awareness of the time dimension. In a completely different area, it has been suggested (Spencer, 1996) that simplistic assumptions about the impact of sea level rise on coral reefs should not simply concentrate upon two variables and ignore the lessons of longer term studies. This emphasises the fact that although many models are interpreted as being definitive they really reflect scientific perception at a point in time, and Bianchi (1994) has organised a view of the perception of global change. Global and regional models dealing with hydrological change could therefore encompass an input from palaeohydrological research results. This need has been accepted in the Land–Ocean Interaction Study (LOIS) established to gain an understanding of, and an ability to predict, the nature of environmental change in the coastal zone of the UK. This includes a core programme of the Land–Ocean Evolution Perspective Study (LOEPS) which is designed to yield a detailed history of changes in the coastal study area. New approaches, such as the use of caesium-137 (^{137}Cs) measurements to provide estimates of rates of landform change, can provide new ways of linking landform and process (Quine et al., 1997). However, the time dimension has not yet been as readily absorbed in hydrological studies related to GCMs.

In relation to further developments it is therefore possible to envisage at least three major stages because the changing physical environment and the changing patterns of research do not always exactly correspond (Gregory, 1992). *First* it is necessary to be concerned with landforms, a specific detailed area or sections of valley, because this is particularly relevant to the philosophy of restoration or renovation as a key to working with the river as an ethos in channel management. Whereas recent approaches to river management have assumed that restoration and renovation simply aimed to restore the natural channel, the question of what is natural has been raised by Tapsell (1995). She recognises as a result of recent surveys (e.g. ECON, 1993) that there is a problem in using the term 'river restoration' because we are really concerned with rehabilitating rivers or enhancing them and not restoring them to a pre-disturbance condition as the term suggests. Brookes (1995) concludes that complete restoration to a pre-disturbance state is unattainable. Graf (1996) has reviewed the matter of what is natural in relation to American rivers with particular reference to quality of the environment for the future. Sear (1994) notes that the restoration process has been pioneered primarily by aquatic ecologists and landscape designers (Cairns, 1991) working in conjunction with civil engineers (e.g. Gore, 1985), so that geomorphology has much to offer (Large and Petts, 1996) but has so far concentrated superficially on the scaling and siting of instream fluvial features. In developing a catchment approach to planning river channel restoration, Kondolf and Downs (1996) suggest that 'cookbook' design guidelines, tied to a classification scheme and implemented by staff with only a brief training in geomorphology, will lead to further degradation rather than to channel improvement. It is also necessary to have a palaeohydrological appreciation as well as a geomorphological one. River restoration techniques are now important not only for those channels which are restored after channelisation but also in cases downstream of dams that are deconstructed (Williams, 1997). In managing human impact Newson (1995a) queries whether we have done enough and it may be that a greater long-term contribution from palaeohydrology should figure more prominently. There is certainly a need, and Osborne et al. (1993) demonstrate the value of state-of-the-art knowledge from process engineering, waste-water engineering, agriculture, hydrology and chemistry, hydraulics and ecology; but surely palaeohydrology should be added to this list to provide knowledge of former environmental characteristics and also of the dynamics of change. Sear (1994) has argued that successful river restoration schemes should be underpinned by real geomorphology and it is equally important to have an understanding of palaeohydrology.

Secondly it is important to be aware of what individuals, not only scientists but also the general public, perceive and prefer in rivers and river basins in relation to environmental change. It has been shown (Mosley, 1989; Gregory and Davis, 1993) that preferences for

Table 2.3   Examples of further possible applications

Modified components of environment river corridors flood plain evolution
Influences upon palaeohydrological systems ground water ecological basis tectonics global view
Overall hydrological basin model link spatial and temporal scale greater understanding of temporal change

landscape are often for the humanised rather than the natural, and this may lead to misapprehensions about what should be the objective of management and restoration schemes. It is therefore important to embrace education of others by the results of palaeohydrological research (Table 2.3). A *third* dimension is to remember the need to be holistic, and Petts (1995) has argued for a return to larger rivers, linking geomorphology and ecology, together with applications, as the three main challenges for the future; these are all very appropriate for palaeohydrology because the first two are already characteristic of palaeohydrological research. This would involve the separation of local and other factors from climatically controlled changes as required by Vandenberghe (1995a,b) and it must ultimately lead towards a notion underlying integrated basin management that, at this level, it is possible to utilise the theme of working with the river, not against it, although this also presupposes a background knowledge of palaeohydrological change (Downs et al., 1991). A very significant and well-established book by McHarg (1967) offered an approach to design with nature but this can incorporate a more substantial time dimension, an appreciation of environmental change and also a greater awareness of what nature really is.

Table 2.4 suggests some further developments and recommendations; these are grouped into those that are primarily internal to palaeohydrology and those that are external in that they apply to, and interact with, other disciplines. It is then possible to visualise such developments as relevant to applicable and to applied questions and also to other disciplines. Appreciation of the character of a river system to achieve progress against the objectives of Table 2.4 must be developed against a greater understanding of the river basin within the fluvial system. Therefore it has been argued that integrated basin management can benefit from a knowledge of palaeohydrology (Downs et al., 1991), and Newson (1995b) has emphasised the role of fluvial geomorphology in informing, and in some cases in restraining, headlong moves for further river restoration.

CONCLUSIONS: DEVELOPMENTS AND RECOMMENDATIONS

It is important to avoid being unnecessarily restrictive in setting the boundaries between disciplines and sub-disciplines but it is now clear that there are examples of palaeohydrological research results which can be applied to other disciplines and others that can

Table 2.4 Conclusions: developments and recommendations

	Relevance to other disciplines	Relevance to applicable and applied questions
Primarily internal		
Develop palaeohydrological global models	Employ energetics	Impact of global change scenarios
Adapt drainage basin models, discharge estimation	Basin changes, distributed models	Flow prediction, flood estimation
Incorporate all elements of the drainage basin as appropriate, e.g. ground water, tectonics	Regional palaeohydrological variations	Impact of all factors on system change
Demonstrate how components of basin have changed and over what period, e.g. river corridors, flood plains, valley heads	Refined models of environmental change	Inform landscape management
Time dimension, mechanics of change, sensitivity	Temporal models and thresholds	Landscape stability and sensitivity
Primarily external		
Natural landscape dynamic not static	Landscape has a dynamic history	River management benefits from palaeohydrological research
Education and greater awareness of landscape	Magnitude of human impact	Significance of 'natural' landscape change
Holistic approach	Integrated palaeohydrological systems	Integrated basin management

offer an input to specific applied problems and to landscape management. This is attempted in Table 2.4 which endeavours to distinguish those primarily internal developments from external ones and also suggests applications to other disciplines and to particular applicable and applied questions. In attempting progress in this way we should be holistic and focus on the overall basin and avoid too great an emphasis upon separate particular themes, although this was expedient in the past; we should underline the dynamics of systems especially in relation to the systems components and the ways in which they affect hydrological change; and we should make contributions in the study of temporal change with particular reference to what is natural, what is actual and what is sensitive (Downs and Gregory, 1993, 1995). We have to appreciate that to make such progress will require the education of others (Leopold, 1994), and an associated change in present perceptions. This could come about, for example, by the greater involvement of palaeohydrology in conservation (Gregory, 1997) and Sites of Special Scientific Interest by involving greater emphasis upon the hierarchy, or indeed the uniqueness, of individual locations or of basins (Downs and Gregory, 1994), or by the development of multidisciplinary, palaeohydrologically based research projects as attempted in the UK for the Millennium. Perhaps it is salutary to remember that regional studies or transects are not an end in themselves but merely a vehicle to enable us to make progress in palaeohydrology, and that the progress made by GLOCOPH research can contribute to other disciplines and to practical problems in contemporary river basins. It is imperative that we keep this in mind in relation to the research contributions in this volume and in other important work achieved by GLOCOPH.

REFERENCES

Arnell, N.W., 1996. Palaeohydrology and future climate change. In J. Branson, A.G. Brown and K.J. Gregory (eds), *Global Continental Changes: The Context of Palaeohydrology*, Geological Society, London, Special Publication 115, 19–25.

Baker, V.R., 1991. A bright future for old flows. In L. Starkel, K.J. Gregory and J.B. Thornes (eds), *Temperate Palaeohydrology*, Wiley, Chichester, 497–520.

Baker, V.R., 1995. Global palaeohydrological change. *Quaestiones Geographicae*, Special Issue **4**, 27–36.

Baker, V.R., Bowler, J.M., Enzel, Y. and Lancaster, M., 1995. Late Quaternary Palaeohydrology of arid and semi-arid regions. In K.J. Gregory, L. Starkel and V.R. Baker (eds), *Global Continental Palaeohydrology*, Wiley, Chichester, 202–231.

Berglund, B.J., Birks, H.J.B., Wright, H.E. and Ralska-Jasiewiczowa, M. (eds), 1996. *Palaeoecological Events During the Last 15000 Years – Regional Synthesis of Palaeoecological Studies of Lakes and Rivers in Europe*, Wiley, Chichester.

Bianchi, E. (ed.), 1994. *Global Change Perception*, Guerini Studio Milan.

Branson, J., Clark, M.J. and Gregory, K.J., 1995. A database for global continental palaeohydrology: technology or scientific creativity? In K.J. Gregory, L. Starkel and V.R. Baker (eds), *Global Continental Palaeohydrology*, Wiley, Chichester, 303–321.

Branson, J., Brown, A.G. and Gregory, K.J. (eds), 1996. *Global Continental Changes: the Context of Palaeohydrology*, Geological Society, London, Special Publication 115.

British Hydrological Society, 1994. *Sustainability in a Changing World: The Key Role of Hydrology*, The National Research Strategy of the British Hydrological Society, 1994.

Brookes, A., 1995. River channel restoration: Theory and practice. In A.M. Gurnell and G.E. Petts (eds), *Changing River Channels*, Wiley, Chichester, 369–388.

Brookes, A., 1996. River channel change. In G. Petts and P. Calow (eds), *River Flows and Channel Forms, Selected Extracts from the Rivers Handbook*, Blackwell Science, Oxford, 221–242.

Brown, A.G., 1995a. Vegetation and lake level change. In K.J Gregory, L. Starkel and V.R. Baker (eds), *Global Continental Palaeohydrology*, Wiley, Chichester, 131–150.
Brown, A.G., 1995b. Late Glacial–Holocene Sedimentation in lowland temperate environments: Flood plain metamorphosis and multiple channel systems. In B. Frenzel (ed.), *European River Activity and Climatic Change During the Late Glacial and Early Holocene*, Special Issue: ESF Project European Palaeoclimate and Man 9, European Science Foundation, Strasbourg, 21–35.
Brown, A.G., 1996a. Palaeohydrology: prospects and future advances. In J. Branson, A.G. Brown and K.J. Gregory (eds), *Global Continental Changes: the Context of Palaeohydrology*, Geological Society, London, Special Publication 115, 257–265.
Brown, A.G., 1996b. Flood plain palaeoenvironments. In M.G. Anderson, D.E. Walling and P.D. Bates (eds), *Floodplain Processes*, Wiley, Chichester, 95–138.
Brown, A.G., 1997. *Alluvial Geoarchaeology: Floodplain Archaeology and Environmental Change*, Cambridge Manuals in Archaeology, CUP.
Brown, A.G. and Bradley, C., 1995. Geomorphology and ground water: convergence and diversification. In A.G. Brown (ed.), *Geomorphology and Ground Water*, Wiley, Chichester, 1–20.
Brown, A.G. and Keough, M.K., 1992. Holocene floodplain metamorphosis in the Midlands, United Kingdom. *Geomorphology*, **4**, 433–445.
Brown, A.G., Keough, M.K. and Rice, R.J., 1994. Flood plain evolution in the East Midlands, United Kingdom: the Lateglacial and Flandrian alluvial record from the Soar and Nene Valleys. *Philosophical Transactions of the Royal Society of London, Series A*, **348**, 261–293.
Burt, T.P., 1994. Long term study of the natural environment – perceptive science or mindless monitoring. *Progress in Physical Geography*, **18**, 475–496.
Cairns, J., 1991. The status of the theoretical and applied science of restoration ecology. *The Environmental Professional*, **13**, 186–194.
Costa, J.E. and O'Connor, J.E., 1995. Geomorphically effective floods. In J.E. Costa, A.J. Miller, K.W. Potter and P.R. Wilcock (eds), *Natural and Anthropogenic Influences in Fluvial Geomorphology. The Wolman Volume*, American Geophysical Union, Geophysical Monograph 89, 45–56.
Downs, P.W. and Gregory, K.J., 1993. The sensitivity of river channels in the landscape system. In D.S.G. Thomas and R.J. Alison (eds), *Landscape Sensitivity*, Wiley, Chichester, 15–30.
Downs, P.W. and Gregory, K.J., 1994. Evaluation of river conservation sites: the context for a drainage basin approach. In D. O'Halloran, C. Green, M. Harley, M. Stanley and J. Knill (eds), *Geological and Landscape Conservation*, Geological Society Monograph, 139–143.
Downs, P.W. and Gregory, K.J., 1995. Approaches to river channel sensitivity. *Professional Geographer*, **47**, 168–175.
Downs, P.W., Gregory, K.J. and Brookes, A., 1991. How integrated is river basin management? *Environmental Management*, **15**, 299–309.
Dury, G.H., 1964. *Principles of Underfit Streams*. US Geological Survey, Professional Paper 452A.
ECON (Ecological Consultancy), 1993. *The River Restoration Project Phase 1: The Feasibility Study*, Final report to the River Restoration Project, ECON, University of East Anglia.
Gore, J.A. (ed.), 1985. *The Restoration of Rivers and Streams: Theories and Experience*, Butterworth, Boston.
Goudie, A.S., 1993. *The Human Impact on the Natural Environment*, 4th edn, Blackwell, Oxford.
Graf, W.L., 1996. Geomorphology and policy for restoration of impounded American rivers: what is 'natural'? In B.L. Rhoads and C.E. Thorn (eds), *The Scientific Nature of Geomorphology*, 1996 Binghampton Geomorphology Symposium.
Gregory, K.J., 1985. *The Nature of Physical Geography*, Arnold, London.
Gregory, K.J., 1987. River channels. In K.J. Gregory and D.E. Walling (eds), *Human Activity and Environmental Processes*, Wiley, Chichester, 207–235.
Gregory, K.J., 1992. Changing physical environment and changing physical geography. *Geography*, **77**, 323–335.
Gregory, K.J., 1995a. Human activity in palaeohydrology. In K.J. Gregory, L. Starkel and V.R. Baker (eds), *Global Continental Palaeohydrology*, Wiley, Chichester, 151–172.
Gregory, K.J., 1995b. The increased significance of vegetation on river channel dynamics. *Quaestiones Geographicae*, Special Issue **4**, 117–120.
Gregory, K.J., 1996. Introduction. In J. Branson, A.G. Brown and K.J. Gregory (eds), *Global*

Continental Changes: The Context of Palaeohydrology, Geological Society, London, Special Publication 115, 1–8.

Gregory, K.J. (ed.), 1997. *Fluvial Geomorphology of Great Britain*, Geological Conservation Review Series, Joint Nature Conservation Committee, Chapman and Hall, London.

Gregory, K.J. and Davis, R.J., 1993. The perception of riverscape aesthetics: an example from two Hampshire rivers. *Journal of Environment Management*, **39**, 171–185.

Gregory, K.J. and Walling, D.E., 1973. *Drainage Basin Form and Process*. Arnold, London.

Gregory, K.J., Starkel, L. and Baker, V.R. (eds), 1995. *Global Continental Palaeohydrology*, Wiley, Chichester.

Gurnell, A.M., Angold, P. and Gregory, K.J., 1994. Classification of river corridors: issues to be addressed in developing an operational methodology. *Aquatic Conservation: Marine and Fresh Water Ecosystems*, **4**, 219–231.

Gurnell, A.M., Gregory, K.J. and Petts, G.E., 1995. The role of coarse woody debris in forest aquatic habits: implications for management. *Aquatic Conservation: Marine and Fresh Water Ecosystems*, **5**, 143–166.

Harper, D.M. and Ferguson, A.J.D. (eds), 1995. *The Ecological Basis for River Management*. Wiley, Chichester.

Kadomura, H., 1995. Palaeoecological and palaeohydrological changes in the humid tropics during the last 20,000 years with reference to equatorial Africa. In K.J. Gregory, L. Starkel and V.R. Baker (eds), *Global Continental Palaeohydrology*, Wiley, Chichester, 177–202.

Kirkby, M.J. (ed.), 1994. *Process Models and Theoretical Geomorphology*, Wiley, Chichester.

Kite, G.W. (ed.), 1995. *River Essays by Eminent Hydrologists*, Water Resources Publications LLC, Highlands Ranch, Colorado.

Knox, J.C., 1995. Fluvial systems since 20,000 years BC. In K.J. Gregory, L. Starkel and V.R. Baker (eds), *Global Continental Palaeohydrology*, Wiley, Chichester, 87–108.

Kondolf, G.M. and Downs, P.W., 1996. Catchment approach to planning channel restoration. In A. Brookes and F.D. Shields Jr (eds), *River Channel Restoration: Guiding Principles for Sustainable Projects*, Wiley, Chichester, 129–148.

Large, A.R.G. and Petts, G.E., 1996. Historical channel – flood plain dynamics along the River Trent; implications for river rehabilitation. *Applied Geography*, **16**, 191–210.

Leopold, L.B., 1994. *A View of the River*, Harvard University Press Cambridge, Massachusetts.

Leopold, L.B. and Miller, J.P., 1954. *Postglacial Chronology for Alluvial Valleys in Wyoming*, United States Geological Survey, Water Supply Papers 1261, 61–85.

Lewin, J., 1996. Floodplain construction and erosion. In G. Petts and P. Calow (eds), *River Flows and Channel Forms, Selected Extracts from the Rivers Handbook*, Blackwell Science, Oxford, 203–220.

Lewin, J., Macklin, M.G. and Woodward, J.C., 1995. Mediterranean quaternary river environments – some future research needs. In J. Lewin, M.G. Macklin and J.C. Woodward (eds), *Mediterranean Quaternary River Environments*, Balkema, Rotterdam, 283–284.

May, R., 1997. *Climate Change: a Note by the UK Chief Scientific Advisor*, Office of Science and Technology, Department of Trade and Industry, London, October 1997.

McHarg, I.L., 1967. *Design with Nature*, Wiley, Chichester.

Miller, A.J., 1995. Valley morphology and boundary conditions influencing spatial patterns of flood flow. In J.E. Costa, A.J. Miller, K.W. Potter and P.R. Wilcock (eds), *Natural and Anthropogenic Influences in Fluvial Geomorphology, The Wolman Volume*, American Geophysical Union, Geophysical Monograph, 57–82.

Mosely, M.P., 1989. Perceptions of New Zealand river scenery. *New Zealand Geographer*, **45**, 2–13.

Newson, M. (ed.), 1995a. *Managing the Human Impact on the Natural Environment: Patterns and Processes*, Wiley, Chichester.

Newson, M.D., 1995b. Fluvial geomorphology and environmental design. In A. Gurnell and G. Petts (eds), *Changing River Channels*, Wiley, Chichester, 413–432.

Osborne, L.L., Bayley, P.B., Higler, L.W.G., Statzner, B., Triska, F. and Moth Iversen, T., 1993. Restoration of lowland streams: an introduction. *Freshwater Biology*, **29**, 187–194.

Petts, G., 1990. Forested river corridors: A lost resource. In D. Cosgrove and G. Petts (eds), *Water, Engineering and Landscape, Water Control and Landscape Transformation in the Modern Period*, Belhaven Press, London, 12–34.

Petts, G.E., 1995. Changing river channels: the geographical tradition. In A. Gurnell and G.E. Petts (eds), *Changing River Channels*, Wiley, Chichester, 1–24.

Petts, G.E., 1996. Sustaining the ecological integrity of large floodplain rivers. In M.G. Anderson, D.E Walling and P.D. Bates (eds), *Floodplain Processes*, Wiley, Chichester, 535–551.

Prosser, I.P., Chappell, J. and Gillespie, R., 1994. Holocene valley aggradation and gully erosion in head water catchments, South Eastern highlands of Australia. *Earth Surface Processes and Landforms*, **19**, 465–480.

Quine, T.A., Govers, G., Walling, D.E., Zhang, X., Dermet, P.J.J., Zhang, Y. and Vandaele, K., 1997. Erosion processes and landform evolution in agricultural land – New perspectives from Caesium-137 measurements and topographic based erosion modelling. *Earth Surface Processes and Landforms*, **22**, 799–816.

Rumsby, B.T. and Macklin, M.G., 1994. Channel and flood plain response to recent abrupt climate change: the Tyne Basin, Northern England. *Earth Surface Processes and Landforms*, **19**, 499–515.

Schumm, S.A., 1965. Quaternary Palaeohydrology. In H.E. Wright and D.G. Frey (eds), *The Quaternary of the United States*, Princeton University Press, Princeton, 783–794.

Schumm, S.A., 1977. *The Fluvial System*, Wiley, Chichester.

Sear, D.J., 1994. River restoration and geomorphology. *Aquatic Conservation: Marine and Freshwater Ecosystems*, **4**, 169–177.

Spencer, T., 1996. Paper given on Geomorphic Hazards in Coastal Environments at *28th International Geographical Congress*, The Hague, August 1996.

Starkel, L., 1990. Fluvial environment as an expression of geoecological changes. *Zeitschrift für Geomorphologie*, Supplement Band, **79**, 133–152.

Starkel, L., 1995a. *Evolution of the Vistula river valley during the last 15,000 years*, Polish Academy of Sciences, Institution of Geography and Spatial Organisation, Geographical Studies Special Issues 8. Part V.

Starkel, L., 1995b. Changes of river channels in Europe during the Holocene. In A. Gurnell and G.E Petts (eds), *Changing River Channels*, Wiley, Chichester, 27–42.

Starkel, L., 1995c. Introduction to global palaeohydrological changes. In K.J. Gregory, L. Starkel and V.R. Baker (eds), *Global Continental Palaeohydrology*, Wiley, Chichester, 1–20.

Starkel, L., Gregory, K.J. and Thornes, J.B. (eds), 1991. *Template Palaeohydrology: Fluvial Processes in the Temperate Zone during the last 15,000 years*, Wiley, Chichester.

Tapsell, S.M., 1995. River restoration: What are we restoring to? A case study of the Ravensbourne River in London. *Landscape Research*, **20**, 98–111.

Teisseyre, A.K., 1995. Episodic channels and the development of dry valleys in crop land. *Quaestiones Geographicae*, **17–18**, 65–78.

Vandenberghe, J., 1995a. Postglacial river activity and climate: State of the art and future prospects. In B. Frenzel (ed.), *European River Activity and Climatic Change During the Late Glacial and Early Holocene*, Special Issue: ESF Project European Palaeoclimate and Man 9, European Science Foundation, Strasbourg, 1–9.

Vandenberghe, J., 1995b. The role of rivers in palaeoclimatic reconstruction. In B. Frenzel (ed.), *European River Activity and Climatic Change During the Late Glacial and Early Holocene*, Special Issue: ESF Project European Palaeoclimate and Man 9, European Science Foundation, Strasbourg, 11–19.

Williams, G.P., 1984. Palaeohydrologic equations for rivers: equations and methods. In J.E. Costa and P.J. Fleischer (eds), *Developments and Applications of Geomorphology*, Springer-Verlag, Berlin, 343–367.

Williams, P., 1997. Deconstructing dams. *World Rivers Review*, **12**(4), 2.

Wolman, M.G., 1995. Play: the handmaiden of work. *Earth Surface Processes and Landforms*, **20**, 585–591.

Woodward, J.C., Lewin, J. and Macklin, M.G., 1995. Glaciation, river behaviour and palaeolithic settlement in upland North-West Greece. In J. Lewin, M.G. Macklin and J.C. Woodward (eds), *Mediterranean Quaternary River Environments*, Balkema, Rotterdam, 115–129.

CHAPTER 3

Palaeohydrology, Palaeodischarges and Palaeochannel Dimensions: Research Strategies for Meandering Alluvial Rivers

IVARS REINFELDS

School of Geosciences, University of Wollongong, Australia

AND

PAUL BISHOP

Department of Geography and Topographic Science, University of Glasgow, UK

INTRODUCTION

Palaeohydrology is concerned with the reconstruction of former river flows, be they average flows of low to moderate magnitude and frequency, or infrequent high magnitude events. The latter are commonly reconstructed in bedrock settings using slackwater deposits and other palaeostage indicators (Baker, 1987), and in alluvial settings using a threshold exceedence approach (Helley and LaMarche, 1973; Patton, 1988; Reinfelds, 1995). More commonly, alluvial palaeohydrology has focused on estimating flows of low to moderate magnitude and frequency using palaeochannel dimensions and sedimentological characteristics to reconstruct palaeodischarges, palaeoenvironments and palaeoclimates (Schumm, 1968; Bowler, 1978; Williams, 1984a; Knox, 1985; Page et al., 1991; Fried, 1993; Bishop and Godley, 1994; Wohl and Georgiadi, 1994; Starkel, 1995; Leigh and Feeney, 1995). Such palaeohydrological reconstructions rely on the recognition and measurement of abandoned channels from aerial photographs and in the field, and are generally based on the following two methodological approaches.

(i) *The regime approach or channel geometry method* (Baker, 1994; Goudie et al., 1994; Kolkberg and Howard, 1995; Wharton, 1995). Equations that relate gauged river discharges and channel dimensions are used to provide palaeodischarge estimates based on the dimensions of abandoned channels. The equations may be obtained from the literature or developed explicitly for the study.

Palaeohydrology and Environmental Change. Edited by G. Benito, V. R. Baker and K. J. Gregory
© 1998 John Wiley & Sons Ltd.

(ii) *The sediment transport mechanics approach* (Baker, 1994). Ancient discharges are estimated from flow velocities needed to entrain and transport sediments sampled from the bed of abandoned channels (Maizels, 1983; Bishop and Godley; 1994).

The regime–channel geometry approach is the most commonly used method in palaeohydrological investigations of alluvial channels and is the focus of this chapter. The origins of regime relations between channel width, depth, area, slope, sinuosity, meander wavelength, bend length and radius of curvature can be traced back at least to British observations and design of irrigation canals in India in the 1800s (Blench, 1957; Ackers, 1972; Pickup and Rieger, 1979). Perhaps because of the successful modelling of the dimensions of irrigation canals, the development of similar equations for alluvial rivers became the focus of considerable research (e.g. Leopold and Maddock, 1953; Blench, 1957; Leopold and Wolman, 1957, 1960; Zeller, 1967; Leeder, 1973). Recent compilations of regime equations which relate channel dimensions (width, depth, area) to channel geometric features (meander wavelength, radius of curvature, bend length and sinuosity) have been provided by Williams (1986, 1988).

Other studies related to, and concurrent with, the development of equations between the various channel dimensions focused on the development of equations specifically relating gauged river discharges to various channel dimensions. This work has included studies of at-a-station hydraulic geometry (Goudie et al., 1994), studies of relationships between river dimensions and discharges (Carlston, 1965; Osterkamp and Hedman, 1982; Pickup and Warner, 1984; Wharton et al., 1989, Wharton, 1995), and palaeohydrological studies which have also developed channel dimension–discharge equations (Dury, 1976; Williams, 1984a, 1988; Knox, 1985; Leigh and Feeney, 1995). Williams (1984b, 1988) and Wharton (1995) provide compilations of published equations, and good examples of their derivation are given by Williams (1984a), Knox (1985), Leigh and Feeney (1995) and Wharton (1995).

Reservations concerning the accuracy of the regime approach to palaeohydrology have generally centred on: the magnitude of the standard errors for the commonly used palaeohydrological equations (Williams, 1984a; Dury, 1985); misunderstanding of the definitions of variables in existing equations (Williams, 1988); the inapplicability of commonly used palaeohydrological equations across different environmental settings (Williams, 1988); the misidentification of palaeochannel boundaries (Dury, 1976; Fried, 1993); the use of equations outside the ranges of input values for which they were developed (Williams, 1988); and the compounding of errors when a value estimated by one equation is used as input data for another equation (Williams, 1988; e.g. Wohl and Georgiadi, 1994). Despite these problems, the simplest application of the regime approach to palaeohydrology requires only that values for channel dimensions be entered into existing regime and channel geometry equations (e.g. Bishop and Godley, 1994; Wohl and Georgiadi, 1994). A general concern we hold is that the relative ease with which palaeodischarge estimates can be generated from available equations may lead to excessive emphasis being placed on such estimates. Such emphasis (plus an excessive and unjustified precision in the reporting of palaeodischarge estimates; see Bishop and Godley 1994) may divert attention from other, perhaps more definitive, palaeoenvironmental evidence.

In this chapter, we focus on several aspects of the regime approach to palaeohydrology with the aim of improving the confidence that can be placed in Pleistocene and Holocene palaeohydrological reconstructions of abandoned alluvial channels. We begin with a discussion of measurement strategies for abandoned and active channels, and then narrow our discussion to the accuracy of discharge estimates generated from such measurements. Pre-empting our most important conclusion, we suggest that the best way forward for palaeohydrological reconstructions of Quaternary variations in low to moderate magnitude and frequency discharges is to establish regionally applicable regime–channel geometry equations for gauged rivers in the catchments of the palaeochannels under investigation (e.g. Carlston, 1965; Williams, 1984a; Knox, 1985; Leigh and Feeney, 1995; Wharton, 1995). The observations and suggestions we present in this chapter are based primarily on a review of readily available regime, channel geometry and palaeohydrological literature, supplemented with our own experiences in the investigation of abandoned and active channels of the Yom River in northern Thailand, and the Hunter and Latrobe Rivers, southeastern Australia. Further details of these field areas are given by Bishop and Godley (1994), Erskine et al. (1992), Reinfelds et al. (1995) and Reinfelds (1997).

MEASUREMENT OF CHANNEL DIMENSIONS FOR THE DEVELOPMENT OF REGIONALLY APPLICABLE DISCHARGE PREDICTION EQUATIONS

The most commonly used channel dimensions to predict river discharge are channel bankfull width (W_b), bankfull depth (D_{max}), bankfull mean or hydraulic mean depth (D_b), bankfull cross-sectional area (A_b), radius of curvature (R_c), meander wavelength (L_m) and bend length (L_b). Definitions and symbolic notation for these dimensions follow those of Williams (1984a, 1986, 1988). In this section we provide suggestions as to the number of measurements necessary to determine an 'average' value for any channel dimension, and discuss measurement techniques which are applicable to both palaeochannels and active rivers to aid consistent data compilation and the development of regionally applicable regime–channel geometry equations.

Number of measurements to determine 'average' channel dimensions

There are few guidelines in the literature regarding the appropriate number of measurements to determine an 'average' value for any channel dimension. Williams (1984a) used four values of R_c to characterise the oldest meander cutoffs he studied, and eight to ten measurements to characterise the width of modern meandering rivers in Sweden. Bishop and Godley (1994) used five measurements of bankfull width for abandoned channels in Thailand and Wharton (1995) suggested using three cross-sections to determine an average cross-sectional area for the development of discharge predictive equations.

The sample size (n) necessary to obtain the average of a normally distributed population to a given degree of accuracy is dependent on the standard deviation of the population (S), the required confidence level (in this case 95%) and the acceptable error (L, as a percentage of the mean) such that (Snedecor and Cochran, 1980):

$$n = 4S^2/L^2 \tag{3.1}$$

Given the uncertainties manifest in regime theory (Goudie et al., 1994), we suggest that an error of $\pm\,10\%$ of the true mean would be acceptable for most cases. The relation thus simplifies to:

$$n = 0.04S^2 \qquad (3.2)$$

Put simply, this means that the number of measurements necessary for 95% confidence that the value obtained will be within 10% of the true mean is 4% of the square of the standard deviation of the mean (the standard deviation being expressed as a percentage of the mean). For general application of this formula, however, there are few published data regarding the standard deviations of river channel dimensions. Reinfelds (1997) presented an example of the application of this formula to bankfull width data determined from aerial photography and surveyed cross-sections for the Latrobe River, Australia. In the study, more than 240 bankfull width measurements from aerial photography were compared with 30 bankfull width values determined from surveyed cross-sections. It was found that 30 cross-sections were more than sufficient to determine an average width for the 80 km length of the Latrobe River under investigation, and that 17 cross-sections would probably have been sufficient. However, the study also demonstrated that the degree of variation in bankfull width (expressed as the standard deviation) for the entire length of the study area was the same as for shorter river reaches exhibiting a similar range of width values. In this regard, splitting the 30 cross-section data set into shorter river reaches resulted in insufficient cross-sections in some reaches to investigate changes in channel width between the dates of the surveyed cross-sections and the aerial photography (Reinfelds, 1997). Clearly, therefore, the number of measurements necessary to determine an average value for any channel dimension is dependent not only on the spatial variability of the dimensions under investigation, but also on the questions that are being addressed.

Preservation and measurement of bankfull width

Bankfull width is a fundamental measurement in fluvial geomorphology and the consistent measurement of bankfull width is an important issue for the development of regime–channel geometry equations. The most consistent method of determining a bankfull cross-section (and hence bankfull width) for suspended mixed load meandering rivers requires surveyed cross-sections and measurement of bankfull width at that stage coinciding with the minimum width-to-depth (W/D) ratio (Riley, 1972; Williams, 1978). Many studies do not explicitly state, however, whether D_{max} or D_b is used to determine W/D ratios, and for simplicity we recommend that D_{max} be used in preference to D_b. Indeed, if bankfull mean depth (D_b) is used to determine a W/D ratio, then the W/D_b relation can be simplified to W^2/A_b. For the low gradient, low W/D ratio Latrobe River, the minimum W/D_{max} ratio for all cross-sections corresponds to the 'active channel width' as defined by Osterkamp and Hedman (1982) (Reinfelds, 1997). 'Active channel width' has since been redefined as 'bankfull width' by Wharton (1992, 1995) (Figure 3.1).

Although surveyed cross-sections offer the most objective way to determine the bankfull width of river channels, they are not always the most practicable method. Direct measurements of bankfull width from aerial photography are an appropriate alternative and have been successfully used to investigate sediment storage, bedload transport, bank

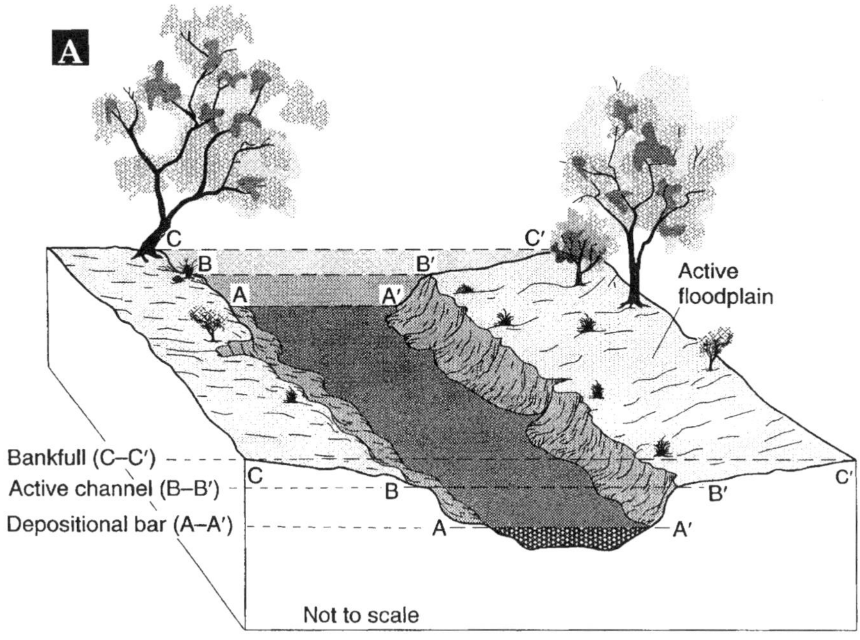

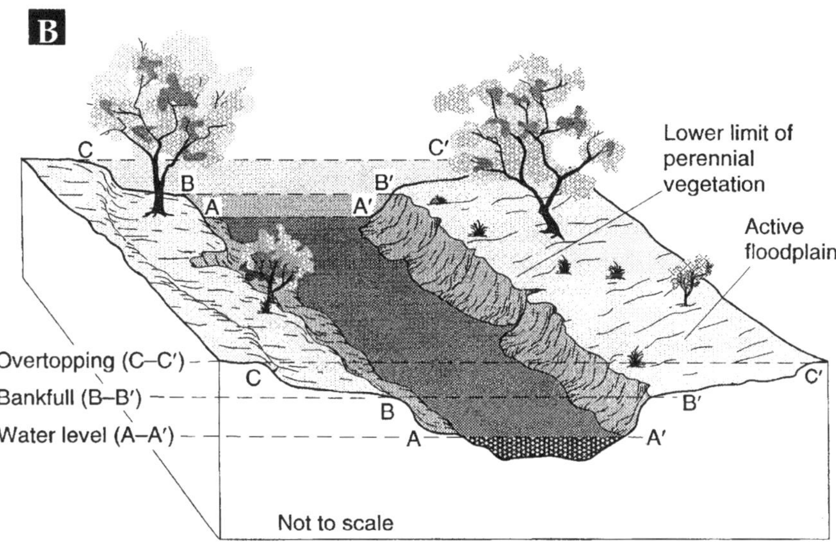

Figure 3.1   Regime–channel geometry reference levels for bankfull discharge: (A) reference levels commonly used by the USGS Water Resources Division (after Osterkamp and Hedman, 1982); (B) reference levels used to develop channel geometry equations in Britain (after Wharton, 1992, 1995). Note the redefinition of 'active channel width' as bankfull width in (B)

stability and channel width changes over time for perennial meandering, wandering and braided rivers (Reinfelds, 1997). The effective use of measurements of bankfull width from aerial photography for a wide variety of purposes suggests that aerial photography is an important but under-utilised data source in the development of regime–channel geometry equations. The elevated moisture levels in palaeochannel infill sediments often mean that the contact between the palaeochannel perimeter and the palaeochannel's post-abandonment infill can be clearly identified on aerial photography, and infra-red photography may facilitate this task even further (e.g. Leigh and Feeney's (1995) use of infra-red aerial photography to delineate banklines of palaeochannels dated to 30 ka). Further guidelines as to the measurement of river bankfull widths from aerial photography have been provided by Reinfelds (1997).

For the development of regime–channel geometry equations, Wharton (1995) recommended that cross-sections should be surveyed, or widths measured, at or near points of inflection so as to reduce variations caused by pool–riffle sequences. The same approach should be adopted for measurement of bankfull widths of palaeochannels. However, we have observed that for old and highly infilled palaeochannels, banklines become increasingly difficult to distinguish away from bend apices. Where banklines of palaeochannels can only be clearly distinguished at bend apices, radius of curvature (R_c) may be more appropriate than bankfull width (W_b) for comparisons between modern and ancient river dimensions (Williams, 1984a).

Palaeohydrologists have evidently routinely assumed that palaeochannels faithfully record bankfull widths of the pre-abandonment active channel, thereby enabling regime-based reconstructions of former discharges. A range of post-abandonment modifications of the palaeochannel, reflecting various combinations of erosion and deposition, however, would seem to be possible (Figure 3.2). Our observations and radiocarbon dating of Latrobe River meander cutoffs suggest that banklines of meander cutoffs along low W/D, suspended load rivers can be readily identified at the floodplain surface for about 2 ka. By about 4 ka, however, Latrobe River meander cutoffs are infilled to such an extent that banklines are obscured and bankfull widths cannot be measured at the floodplain surface. The relatively short duration for which bankfull widths are preserved at the Latrobe floodplain surface suggests, therefore, that preservation of palaeochannel bankfull width at the floodplain surface cannot be assumed automatically. Such preservation should be checked by careful aerial photograph and field examination of the characteristics of the palaeochannel banklines.

Measurement of bankfull depth

Two measurements of bankfull depth are used in palaeohydrological studies and in the development of regime–channel geometry equations: bankfull mean depth (D_b) and bankfull depth (D_{max}). Bankfull mean depth (D_b) is also termed hydraulic mean depth and is defined as bankfull cross-sectional area (A_b) divided by bankfull width (W_b) (Williams, 1988). This measurement is more suited to studies of hydraulic geometry at river gauging sites than alluvial palaeohydrology because it is difficult to determine in abandoned channels (as opposed to gauging stations) unless large numbers of boreholes are successfully drilled to the channel perimeter (e.g. Rotnicki, 1983; Knox, 1985). Bankfull depth (D_{max}), defined as the depth of the channel between the thalweg and the top of the lower of the two floodplain banks where the W/D_{max} ratio is at a minimum

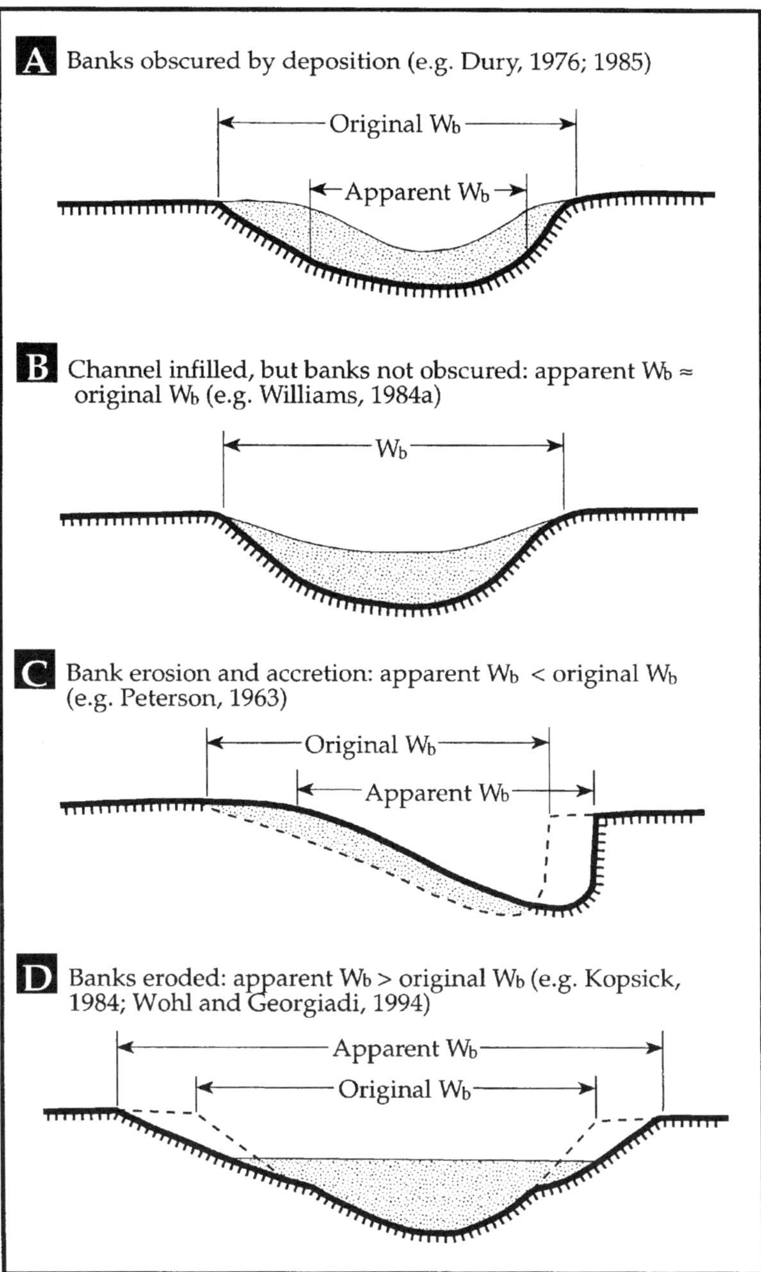

Figure 3.2 Examples of erosion and deposition in meander cutoffs as reported in the literature (see text)

(Riley, 1972; Williams, 1988), is more suitable for palaeohydrological studies and the development of regime–channel geometry equations and is the focus of this section.

Bankfull depth (D_{max}) is easily measured from surveyed cross-sections of active rivers through the application of the minimum W/D_{max} principle. However, D_{max} is more difficult to reconstruct in infilled palaeochannels because of the necessity to drill sufficient boreholes to be confident that the former channel bed has been successfully reached and identified. Success in reaching and identifying the former channel bed (and therefore accurately reconstructing D_{max}) depends to a large degree on the nature of the sediments infilling the abandoned channel, and the contrast between the infilling sediments and the former bedload sediments (Rotnicki and Borowka, 1985). Erskine et al. (1992) were successful in reaching the Hunter River's old channel bed in 55% of auger boreholes because the deeper (older) units in the Hunter River cutoff infills were uniformly fine-grained muds and easily retained by the auger head. Considerable difficulty may be experienced in reaching the former channel bed in cutoffs dominated by a sandy post-abandonment infill, especially when augering below the water table. Comparisons between historical pre-cutoff channel survey data and boreholes in the Latrobe River cutoffs indicate that hand augering is usually unsuccessful in reaching the former bed in sand-filled 1950–60s cutoffs, whereas drilling in clay-filled cutoffs is more successful (Figure 3.3). Our lack of success in reaching the bed in sand-filled cutoffs was largely due to the poor cohesion of the sandy sediments below the water table. More importantly, without detailed historical data on pre-cutoff thalweg elevations, it would have been impossible to know that our drilling had not intersected the channel bed in these sand-filled cutoffs and that our reconstructed bankfull depths were less than the true bankfull (Figure 3.3). The converse problem – over-estimation of bankfull depth – may arise, as we have already noted, in situations where there is little textural difference between the sediments infilling an abandoned channel and the sediments into which the channel is cut (Rotnicki, 1983; Fried, 1993).

For consistency in the development of regime–channel geometry equations, bankfull depth should be reconstructed at or near points of inflection between bend apices in order to minimise variations caused by pool–riffle sequences.

Measurement of bankfull cross-sectional area

Accurate reconstruction of bankfull cross-sectional area for active rivers requires surveyed cross-sections, and in the case of palaeochannels, detailed drilling to the channel perimeter (Rotnicki, 1983; Knox, 1985). Detailed drilling may be possible where channel fills are relatively shallow and the bed clearly defined, such as for higher W/D_{max} ratio gravel-bed rivers, but it is often not practicable for palaeochannels of low W/D_{max} ratio, suspended load rivers. Bishop and Godley (1994) used an alternative approach for the Yom River, Thailand, by estimating two extreme channel cross-sectional areas, rectangular and triangular, based on reconstructions of bankfull width (W_b) and depth (D_{max}). In the rectangular cross-sectional area, bed width equals W_b, and the triangular has zero bed width. These rectangular and triangular cross-sectional area estimates must bracket the true cross-sectional area, and an average of the two estimates is likely to yield a reasonable approximation of the true cross-sectional area of the channel. As for bankfull width and depth, cross-sectional areas should be reconstructed at or near points of inflection to minimise variations caused by pool–riffle sequences.

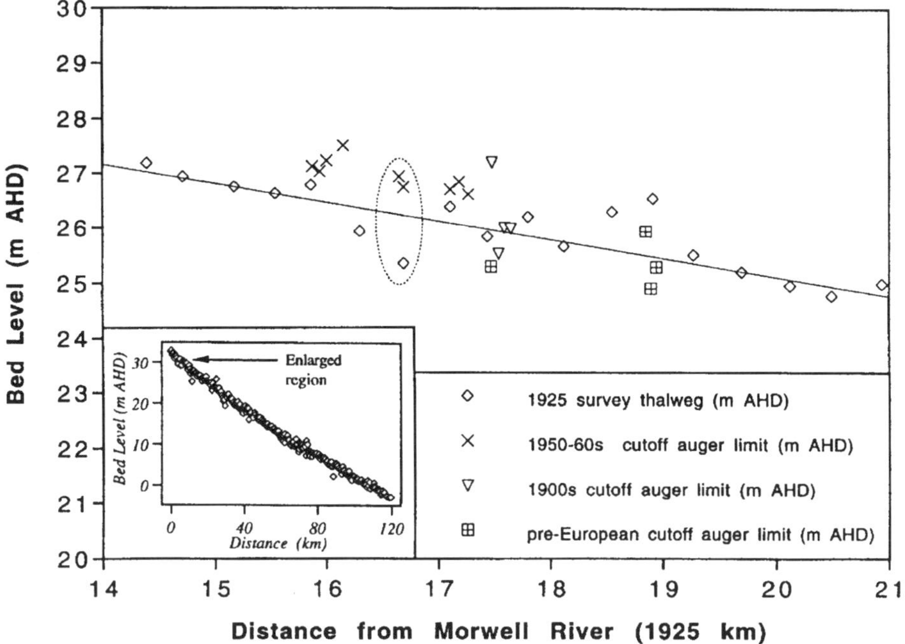

Figure 3.3   Reconstructions of bankfull depth (D_{max}) to Australian Height Datum (AHD) in clay-filled (pre-European and 1900s) and sand-filled (1950–60s) meander cutoffs in reach 1 of the Latrobe River. The inset diagram shows survey data from 1925 (283 cross-sections) which was used to model an 'average' thalweg elevation by fitting a second-order polynomial curve ($r^2 = 0.995$) to the data. This 'average' thalweg elevation appears on the main diagram as the solid line. Auger holes drilled in the 1950–60s sand-filled cutoffs fall short of the 'average' thalweg elevation by approximately 0.5–1.0 m. The ellipse highlights the shortfall between the maximum depth reached by hand augering in deep pools of sand-filled meander cutoffs. Conversely, auger holes in clay-filled pre-European and 1900s cutoffs were generally successful in reaching the bed, as indicated by the coincidence of the square and triangular symbols and the modelled 'average' thalweg elevation

Wharton (1995, p. 656) recommended the use of bankfull cross-sectional area in preference to bankfull width for the development of discharge predictive equations because 'channel area more accurately describes river size than a single width value'. Although this is certainly true, it appears from the data that Wharton (1995, p. 652) presented that bankfull width displays less scatter than cross-sectional area for the prediction of river discharge. For reconnaissance investigations, bankfull width has other advantages over cross-sectional area in that it is relatively simple to measure directly from aerial photography and in the field, and that a large number of measurements can be obtained in a relatively short period of time.

Radius of curvature, meander wavelength and bend length

In many floodplain settings, the oldest meander cutoffs are found at a considerable distance from the present river (e.g. Handy, 1972), and the surface expression of these old cutoffs is best preserved at bends. This greater preservation of the surface expression of meander cutoffs at some distance from the active channel is due largely to the higher sedimentation rates that characterise the entrance and exit reaches of meander cutoffs (Shields and Abt, 1989) and to the greater bankfull depths at bends relative to inflections. The popularity of R_c and L_m in palaeohydrological studies (e.g. Dury, 1976; Williams, 1984a; Wohl and Georgiadi, 1994) is thus justified because of the high preservation potential of these dimensions which can therefore be reconstructed even if only the outer bankline of palaeochannels can be distinguished. This popularity, however, belies problems associated with conventional approaches to the measurement of R_c and L_m to which we now turn.

Of the various channel dimensions commonly used in palaeohydrological studies and regime–channel geometry equations, radius of curvature (R_c) and meander wavelength (L_m) appear to be the parameters that are the most susceptible to variations arising from operator subjectivity. Williams (1984a) has shown that independent measurements of R_c for the same set of palaeomeanders may vary by up to about 50%. Such discrepancies may be partly due to a general misrepresentation of meandering rivers as being characterised by opposing and regular meander loops (Figure 3.4A; e.g. Rotnicki, 1983; Williams, 1986), whereas irregular, non-opposing, compound and sometimes asymmetrical meanders (Figure 3.4B) are characteristic of many meandering rivers (Brice, 1974, 1984; Ferguson, 1975; Carson and Lapointe, 1983). Conventional methods of determining R_c, by assigning a single R_c arc to an entire meander loop, are inappropriate in such rivers (e.g. Figure 3.4A). The method of assigning multiple arcs to the centreline of the channel (Figure 3.4B) provides reasonable correlations with discharge (e.g. Brice, 1974; Williams, 1984a) and is therefore more likely to generate accurate estimates of discharge in both palaeochannels and active rivers.

The measurement of L_m as the straight-line distance between alternate bend apices or inflections (Figure 3.4A; e.g. Rotnicki, 1983; Williams, 1986) is likewise inapplicable to many rivers (Williams, 1984a). Ferguson's (1975) suggestion that meander wavelength be determined as twice the mean inflection spacing measured along the thalweg (Figure 3.4B) eliminates many of the difficulties of determining an average wavelength for rivers with irregular meanders, although we suggest that measurement along the channel centreline is more consistent with Brice's (1974) and Williams' (1984a) methods. Ferguson's (1975) alternative to meander wavelength has since been termed 'bend length' (L_b) (Williams, 1986), and it has been shown that bend length (L_b) has a slightly better correlation with bankfull width than does meander wavelength (L_m) (Hey and Thorne, 1986; Williams, 1986, 1988). Unless the meanders under investigation are clearly opposing and regular (e.g. Figure 3.4A), the use and development of formulae involving meander wavelength (L_m) should be avoided.

PALAEODISCHARGE ESTIMATES

The accuracy of the discharge estimates yielded by a regime–channel geometry equation (in other words, the magnitude of the equation's standard error of estimate) is a critical issue in palaeohydrology, given that one of the fundamental goals of alluvial palaeohydrology is to reconstruct catchment hydrological changes via comparisons of dimensions and discharge estimates in palaeochannels and active rivers. Bishop and Godley (1994) suggested that the discharge estimates be used mainly to highlight *relativities* of discharge magnitudes in ancient and modern channels but the question remains of whether differen-

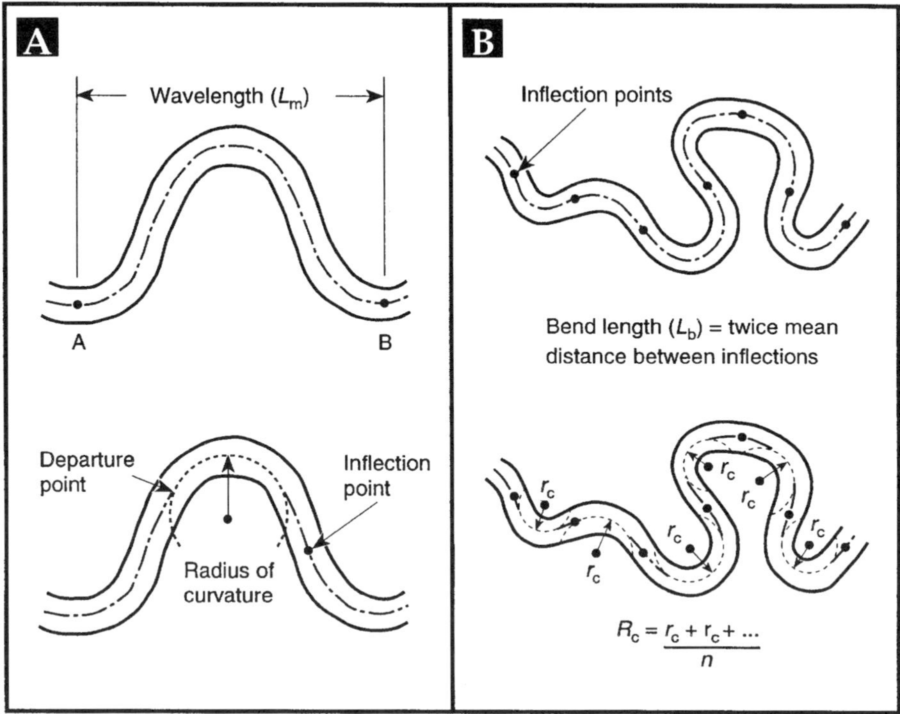

Figure 3.4 Conventional measurement techniques for meander wavelength (L_m) and radius of curvature (R_c) are applicable only to regular, opposing sinusoidal meanders. Measurement techniques for these parameters are illustrated in (A). (B) illustrates alternative measurement techniques for bend length (L_b) and R_c which are more easily applicable to irregular and non-opposing meanders. For further details on these measurement techniques see discussions by Brice (1974, 1984), Ferguson (1975) and Williams (1984a)

ces between discharge estimates are 'real' or not. Inferential statistical techniques are inappropriate for such comparisons because of the small data sets that can be realistically generated in palaeohydrological studies. Bishop and Godley (1994) used a type of sensitivity analysis based on the triangular and rectangular channel cross-sectional areas to assess whether discharge estimates for palaeochannels and the Yom River were 'significantly' different. If a smaller channel's maximum discharge estimate (based on the rectangular cross-sectional area) was less than a larger channel's minimum discharge estimate (triangular cross-sectional area), it was concluded that the difference between the discharge estimates was real. With the application of this technique to the regime approach, a 100% difference in cross-sectional area (i.e. one channel has twice the cross-sectional area of the other) is thus considered 'significant'.

The level of rigour in this type of sensitivity analysis may seem excessive, but is justified by the large standard errors of many established relationships between channel dimensions and discharge (Dury, 1985; Williams, 1988). Indeed, with the exception of the data of Bowler (1978), Williams (1984a), Page et al. (1991) and Leigh and Feeney (1995), few published palaeochannel dimensions are probably sufficiently different from either the present channel or other palaeochannels in the study areas to justify, without other supporting evidence, inferences about palaeohydrological changes based on discharge estimates from available equations. Such evidence could include regional palynological, limnological or other sedimentological data (e.g. Knox, 1985; Starkel, 1995) or, at least, changes in several channel dimensions.

Development of regional versus use of existing palaeohydrological equations

The simplest approach to alluvial palaeohydrology requires only that measured channel dimensions be entered into existing equations to provide discharge estimates. An alternative to the use of existing palaeohydrological equations is to develop regionally applicable regime–channel geometry equations (e.g. Williams, 1984a; Knox, 1985; Leigh and Feeney, 1995; Wharton, 1995). The development of regional equations has several advantages over the application of existing equations available to palaeohydrologists. First, many of the existing formulae involving width, depth and cross-sectional area are based on hydraulic geometry relations at a single or small number of gauging sites for an individual river (Dury, 1976; Osterkamp and Hedman, 1982; Williams, 1986, 1988). Gauging stations are often located at the most stable sites within a river reach (Osterkamp and Hedman, 1982) and may be unrepresentative of the river in general (Knox, 1985). This problem can be eliminated for regional equations if field surveys are conducted in the vicinity of the gauge site to determine average channel dimensions (Wharton, 1995). Second, the magnitude of the standard error for regionally derived equations is sometimes less than for the commonly applied palaeohydrological equations (e.g. Williams, 1984a; Knox, 1985; Leigh and Feeney, 1995). Hydrological changes that are reflected as relatively minor changes in channel dimensions can thus be investigated with greater confidence (e.g. Knox, 1985). Finally, development of regional equations has the additional advantage that the equation which yields the best correlation between discharge and present channel dimensions can be applied to the palaeochannels under investigation.

Bankfull channel dimensions and channel-forming discharges

Channel-forming discharge is commonly associated with bankfull discharge but the identification of a single channel-forming discharge may be an over-simplification of the behaviour of river systems (Pickup and Warner, 1976; Richards, 1982). Furthermore, bankfull channel dimensions, and therefore bankfull discharges, are known to vary as a result of intrinsic channel adjustments that are unrelated to changes in catchment hydrology. Such changes may relate to floodplain–channel evolution (Nanson, 1986; Brizga and Finlayson, 1990) but are more commonly the result of human impacts, for example catchment clearing and river engineering (Knox, 1977, 1985; Reinfelds et al., 1995). In short, the degree to which bankfull channel dimensions, geometries and discharges are products of processes unrelated to catchment climatic and hydrological conditions is a central issue which must be considered in palaeohydrological analyses (Knox, 1985), particularly when bankfull dimensions and discharges of modern fluvial systems are compared with the bankfull characteristics of ancient systems.

An alternative approach to the use of bankfull discharge is to relate discharges with a range of recurrence intervals to bankfull channel dimensions. No single *frequency-defined* discharge (as opposed to *dimensionally defined* bankfull discharge) is less valid than any other for palaeohydrological applications. This is because the magnitude of hydrological shifts producing measurable channel changes would be likely to produce a change in the entire discharge magnitude–frequency relationship. We recommend, therefore, the use of frequency-defined discharges in the development of regional regime–channel geometry equations for palaeohydrological applications. It also seems intuitive to suggest, in the context of palaeohydrological investigations concerned with climatic change, that mean annual discharge (also termed average daily discharge; Williams, 1984a) is an appropriate discharge for estimating past catchment rainfall.

CONCLUSION

We suggest that the best way forward for regime-style alluvial palaeohydrology lies in the development of regional equations which relate channel dimensions to frequency-defined discharges, using the consistent measurement techniques and cautionary approaches outlined in readily available literature and reiterated in this chapter. The development of such equations will not only improve confidence in palaeohydrological interpretations of moderate Quaternary and Holocene climatic fluctuations, but the associated data may also provide opportunities for improvement of regime–channel geometry relationships for rivers in general. If the development of such equations is beyond the scope of palaeohydrological investigations, existing equations should be applied with caution. We have noted that even some of the most obvious problems relating to the application of existing equations, such as the use of equations outside of the data ranges for which they were developed and the compounding of errors where a value estimated by one equation is used as input data for another equation, are sometimes overlooked. Less obvious problems which must be considered in palaeohydrological investigations include comparisons between unlike measurement positions between palaeochannels and modern rivers, especially where the dimensions of modern rivers are derived solely from gauging station sites. These general concerns and the issues raised in this chapter should be considered in the application of the regime–channel geometry approach to palaeohydrological studies. Every effort should also be made to investigate other palaeoenvironmental indicators that are independent of the palaeohydrological techniques.

ACKNOWLEDGEMENTS

We thank the reviewers and Ian Rutherfurd for comments which resulted in considerable improvements to this chapter. This work was supported by an Australian Postgraduate Award to I.R., an Australian Research Council grant to P.B., and Monash University. We also thank land owners of the Yom, Latrobe and Hunter Rivers for access to field sites; Martin Freiverts, Jacky Woodfull, Duang and Goh for assistance in the field; and Natasha Velleley, Gary Swinton and David Tooth for logistical assistance.

REFERENCES

Ackers, P., 1972. River regime: research and application. *Journal of the Institution of Water Engineers*, **26**, 257–281.
Baker, V.R., 1987. Palaeoflood hydrology and extraordinary flood events. *Journal of Hydrology*, **96**, 79–99.

Baker, V.R., 1994. Glacial to modern changes in global river fluxes. In *Material Fluxes on the Surface of the Earth*, National Academy Press, Washington, DC, 86–98.

Bishop, P. and Godley, D., 1994. Holocene palaeochannels at SiSatchanalai, north-central Thailand: ages, significance and palaeoenvironmental indications. *The Holocene*, **4**, 32–41.

Blench, T., 1957. *Regime Behaviour of Canals and Rivers*, Butterworths Scientific Publications, London.

Bowler, J.M., 1978. Quaternary climate and tectonics in the evolution of the Riverine Plain, Southeastern Australia. In J.L. Davies and M.A.J. Williams (eds), *Landform Evolution in Australasia*, ANU Press, Canberra, 70–112.

Brice, J.C., 1974. Evolution of meander loops. *Geological Society of America Bulletin*, **85**, 581–586.

Brice, J.C., 1984. Planform properties of meandering rivers. In J. Elliot (ed.), *River Meandering*, American Society of Civil Engineers, New York, 1–15.

Brizga, S.O. and Finlayson, B.L., 1990. Channel avulsion and river metamorphosis: the case of the Thomson River, Victoria, Australia. *Earth Surface Processes and Landforms*, **15**, 391–404.

Carlston, C.W., 1965. The relation of free meander geometry to stream discharge and its geomorphic implications. *American Journal of Science*, **263**, 864–885.

Carson, M.A. and Lapointe, M.F., 1983. The inherent asymmetry of river meander planform. *Journal of Geology*, **91**, 41–55.

Dury, G.H., 1976. Discharge prediction, present and former, from channel dimensions. *Journal of Hydrology*, **30**, 219–245.

Dury, G.H., 1985. Attainable standards of accuracy in the retrodiction of palaeodischarge from channel dimensions. *Earth Surface Processes and Landforms*, **10**, 205–213.

Erskine, W., McFadden, C. and Bishop, P., 1992. Alluvial cutoffs as indicators of former channel conditions. *Earth Surface Processes and Landforms*, **17**, 23–37.

Ferguson, R.I., 1975. Meander irregularity and wavelength estimation. *Journal of Hydrology*, 26, 315–333.

Fried, A.W., 1993. Late Pleistocene river morphological change, southeastern Australia: the conundrum of sinuous channels during the last glacial maximum. *Palaeogeography, Palaeoclimatology, Palaeoecology*, **101**, 305–316.

Goudie, A., Atkinson, B.W., Gregory, K.J., Simmons, I.G., Stoddart, D.R. and Sugden, D., 1994. *The Encyclopedic Dictionary of Physical Geography*, Blackwell Reference Publications, London.

Handy, R.L. 1972. *Alluvial cutoff dating from subsequent growth of a meander*, Geological Society of America Bulletin, 83, 475–480.

Helley, E.J. and LaMarche, C., 1973. *Historic flood information for northern California streams from geological and botanical evidence*, United States Geological Survey, Professional Paper 465-E.

Hey, R.D. and Thorne, C.D., 1986. Stable channels with mobile gravel beds. *Journal of Hydraulic Engineering*, **112**, 671–689.

Knox, J.C., 1977. Human impacts on Wisconsin stream channels. *Annals of the Association of American Geographers*, **67**, 323–342.

Knox, J.C., 1985. Responses of floods to Holocene climatic change in the upper Mississippi Valley. *Quaternary Research*, **23**, 287–300.

Kolkberg, F.J. and Howard, A.D., 1995. Active channel geometry and discharge relations of U.S. Piedmont and Midwestern streams: the variable exponent model revisited. *Water Resources Research*, **31**(9), 2353–2365.

Kopsick, P.R., 1984. Dimensions of modern and relict meander loops of selected rivers in Kansas and Nebraska. In J. Elliot (ed.), *River Meandering*, American Society of Civil Engineers, New York, 138–146.

Leeder, M.R., 1973. Fluviatile fining-upwards cycles and the magnitude of palaeochannels. *Geological Magazine*, **110**, 265–276.

Leigh, D.S. and Feeney, T.P., 1995. Paleochannels indicating wet climate and lack of response to lower sea level, southeast Georgia. *Geology*, **23**, 687–690.

Leopold, L.B. and Maddock, T., 1953. *The hydraulic geometry of stream channels and some physiographic implications*, United States Geological Survey, Professional Paper 252, 1–57.

Leopold, L.B. and Wolman, M.G., 1957. *River channel patterns – braided, meandering and straight*, United States Geological Survey, Professional Paper 282-B.

Leopold, L.B. and Wolman, M.G., 1960. River meanders. *Bulletin of the Geological Society of America*, **71**, 769–794.

Maizels, J.K., 1983. Palaeovelocity and palaeodischarge determination for coarse gravel deposits. In K.J. Gregory (ed.), *Background to Palaeohydrology*, Wiley, Chichester, 101–139.

Nanson, G.C., 1986. Episodes of vertical accretion and catastrophic stripping: a model of disequilibrium floodplain development. *Bulletin Geological Society of America*, **97**, 1467–1485.

Osterkamp, W.R. and Hedman, E.R., 1982. *Perennial-streamflow characteristics related to channel geometry and sediment in Missouri River Basin*, United States Geological Survey Professional Paper 1242.

Page, K.J., Nanson, G.C. and Price, D.M., 1991. Thermoluminescence chronology of late Quaternary deposition on the Riverine Plain of southeastern Australia. *Australian Geographer*, **22**, 14–23.

Patton, P.C., 1988. Geomorphic response of streams to floods in the glaciated terrain of southern New England. In V.R. Baker, R.C. Kochel and P.C. Patton (eds), *Flood Geomorphology*, Wiley, New York, 261–277.

Peterson, M.S., 1963. Hydraulic aspects of Arkansas River stabilisation. *American Society of Civil Engineers, Journal of the Waterways and Harbours Division*, **WW4**, 29–65.

Pickup, G. and Rieger, W.A., 1979. A conceptual model of the relationship between channel characteristics and discharge. *Earth Surface Processes*, **4**, 42.

Pickup, G. and Warner, R.F., 1976. Effects of hydrologic regime on magnitude and frequency of dominant discharge. *Journal of Hydrology*, **29**, 51–75.

Pickup, G. and Warner, R.F., 1984. Channel adjustment to sediment load and discharge in the Fly and lower Purari, Papua New Guinea. *Catena Supplement*, **5**, 9–41.

Reinfelds, I., 1995. Evidence for high magnitude floods along Waimakariri River, New Zealand. *Journal of Hydrology (New Zealand)*, **34**, 95–110.

Reinfelds, I., 1997. Reconstruction of changes in bankfull width: a comparison of surveyed cross-sections and aerial photography. *Applied Geography*, **17**, 203–213.

Reinfelds, I., Rutherfurd, I. and Bishop, P., 1995. History and effects of channelisation along the Latrobe River, Victoria. *Australian Geographical Studies*, **33**, 60–76.

Richards, K., 1982. *Rivers: Form and Process in Alluvial Channels*, Methuen, London.

Riley, S.J., 1972. A comparison of morphometric measures of bankfull. *Journal of Hydrology*, **17**, 23–31.

Rotnicki, K., 1983. Modelling past discharges of meandering rivers. In K.J. Gregory (ed.), *Background to Palaeohydrology*, Wiley, New York, 321–354.

Rotnicki, K. and Borowka, R.K., 1985. Definition of subfossil meandering palaeochannels. *Earth Surface Processes and Landforms*, **10**, 215–225.

Schumm, S.A., 1968. *River adjustment to altered hydrologic regimen – Murrumbidgee River and palaeochannels, Australia*, United States Geological Survey, Professional Paper 598.

Shields, F.D. and Abt, S.R., 1989. Sediment deposition in cutoff meander bends and implications for effective management. *Regulated Rivers*, **4**, 381–396.

Snedecor, G.W. and Cochran, W.G., 1980. *Statistical Methods*, Iowa State University Press, Ames.

Starkel, L., 1995. Reconstruction of hydrological changes between 7000 and 3000 BP in the upper and middle Vistula River basin, Poland. *The Holocene*, **5**, 34–42.

Wharton, G., 1992. Flood estimation from channel size: guidelines for using the channel-geometry method. *Applied Geography*, **12**, 339–359.

Wharton, G., 1995. The channel-geometry method: guidelines and applications. *Earth Surface Processes and Landforms*, **20**, 649–660.

Wharton, G., Arnell, N.W., Gregory, K.J. and Gurnell, A.M., 1989. River discharge estimated from channel dimensions. *Journal of Hydrology*, **106**, 365–376.

Williams, G.P., 1978. Bankfull discharge of rivers. *Water Resources Research*, **14**, 1141–1154.

Williams, G.P., 1984a. Palaeohydrological methods and some examples from Swedish fluvial environments. II. River meanders. *Geografiska Annaler*, **66A**, 89–102.

Williams, G.P., 1984b. Paleohydrologic equations for rivers. In J.E. Costa and P.J. Fleisher (eds), *Developments and Applications of Geomorphology*, Springer-Verlag, Berlin, 343–367.

Williams, G.P., 1986. River meanders and channel size. *Journal of Hydrology*, **88**, 147–164.

Williams, G.P., 1988. Paleofluvial estimates from dimensions of former channels and meanders. In

V.R. Baker, R.C. Kochel and P.C. Patton (eds), *Flood Geomorphology*, Wiley, New York, 321–334.

Wohl, E.E. and Georgiadi, A.G., 1994. Holocene paleomeanders along the Sejm River, Russia. *Zeitschrift für Geomorphologie*, **38**, 299–309.

Zeller, J., 1967. *Meandering channels in Switzerland*, International Association of Scientific Hydrology, Publication 75, 174–186.

CHAPTER 4

Fluvial Evidence of the Medieval Warm Period and the Late Medieval Climatic Deterioration in Europe

A. G. BROWN

Department of Geography and Wetland Research Centre, University of Exeter, UK

MEDIEVAL CLIMATIC TRENDS

This chapter evaluates the fluvial record for the Late Medieval period (*c.* AD 1000 to 1500) using both recent work in Midland England and wider records from northwest Europe. The Medieval Warm Period (MWP) or Little Climatic Optimum is generally regarded to have lasted from AD 900–1000 to AD 1200–1300 (Hughes and Diaz, 1994) and ended with a climatic deterioration and instability, which was then replaced in the 1600s by the relative climatic stability of the Little Ice Age (LIA, dated to AD 1600 to AD 1800–1900; Grove, 1988). There is, however, considerable variation in the dates quoted for these climatic periods owing to their diachronous nature and the inherent difficulty of sub-dividing continuously varying phenomena. The evidence of the MWP has come from many sources including Alpine and Scandinavian limits of valley glaciers (Grove and Switsur, 1994), Asian mountains (Serebryanny and Solomina, 1996), ice core data (Thompson, 1991), dendroclimatology (Guiot et al. 1988), settlement history and crop limits (Parry, 1978; Zhang and Crowley, 1989) and documents (Lamb, 1982). The evidence points to a higher mean annual temperature of *c.* 1 °C with hotter summers for Europe north of the Alps. However, the MWP is not a global event (Hughes and Diaz, 1994) and where it is manifest it is diachronous even at the continental scale. The record south of the Alps differs from that north of the Alps with warming not occurring until the 14th century. There is some indication of a NE to SW shift with early MWP dates in Russia and the North Atlantic (including Iceland) and later manifestation in middle Europe, and south of the Alps.

The synoptic cause is believed to be northern hemisphere sub-tropical cyclones displaced to the north with weaker mid-latitude (40–60°N) westerlies and more frequent zonally extended anticyclones. Although the fundamental cause is not certain it is probably increased solar output and resultant higher sea surface temperatures in the North Atlantic. The result north of the Alps was hotter and drier summers but probably little difference in winters. The summers, with less cyclonic activity and more convective activity, probably had a higher incidence of convective storms with lightning and rainfall events of high intensity but relatively short duration. South of the Alps the more frequent cyclonic activity may have resulted in a less pronounced summer drought. However, the relationship of hydrology to these postulated climatic trends is complex. Higher temperatures would have produced higher evapotranspiration, lower annual discharges, lower groundwater recharge and possibly less snowmelt contribution. But higher sea surface temperatures (1–5 °C in inshore waters) could have produced greater convective activity and anticyclonic rainfall. This suggests that although mean annual rainfall was lower, storms in late summer may have been of higher intensity but shorter duration with downstream effects depending upon antecedent conditions and infiltration capacities. This may be the cause of the marked increase in wet autumns in the late 1300s and into the early 1400s (Lamb, 1982). The bog record at Bolton Fell Moss also indicates wet shifts between AD 900 and 1100 and 1300 and 1500 (Barber et al. 1994). The later shift marks the relatively clear climate change at the end of the MWP in northwestern Europe. Between AD 1200 and 1400 the circumpolar vortex had been extended but subsequently retreated with increased strength and prevalence of westerlies. The result is a downward curve in both summer and winter temperatures, increased year to year variability with a possible 50 year cycle associated with blocking type circulation patterns (anticyclonal activity). This includes what Wanner and Siegenthaler (1995) refer to as the cold relapses and sea level pressure reversals of the Late Maunder Minimum between AD 1675 and 1704. This change in typical meteorological conditions is illustrated in the North Sea by an increase in storm surges recorded by Gotteschalke (1971–74) which resulted in coastal flooding and increased mobility of sand dunes. A typical result of this was the burial of a small farmstead on the northern shore of Lindisfarne sometime between AD 1400 and 1600 (Walsh et al., 1995).

EVIDENCE OF LATE MEDIEVAL FLUVIAL CHANGE

In Belo-Russia studies of the Dnieper and Berezina valleys have revealed a distinct change in sedimentation of the upper floodplains from silty muds to sandy silts between c. AD 1000 to 1100 interpreted as resulting from an increased frequency of large floods (Kalicki and Krapiec, 1991). Studies of the Danube have suggested an increased rate of overbank sedimentation from the 1300s onwards, associated by Buch (1989) with deforestation (for iron making) and the onset of the LIA. Other reaches of the Danube studied by Becker and Schirmer (1977) also show increased fluvial activity in the High Middle Ages (1300s and 1400s) as revealed by increased deposition of ranen (black oaks). In central and eastern Germany the Ilme, Weser and Werra valleys all contain 'older meadow loams' which date to the 1300s (Rother, 1989) and in Saxony, northern Germany, work on headwater catchments by Bork (1989) has shown dramatic local fluvial responses to severe storms in the 1300s. The best known was the storm of 1342 which destroyed over half the agricultural land of one village and was experienced over a wide area of northern/central Europe (Pfister, 1988). The geomorphological record of this event is important as it illustrates the results of the coincidence of intense storms with intensive arable land use which included few physical land use barriers (i.e. strip farming).

In Spain the Atlantic rivers show high flood peaks between AD 1150 and 1290, 1400 and 1500 and later between 1850 and 1910 and it can be assumed that these were caused by westerly frontal systems (Benito et al., 1996). There is also no discernible winter warming until the mid-14th century. Between AD 1290 and 1400 there is an increase in flooding in easterly Spanish catchments which is associated by Benito et al. (1996) with a higher frequency of cold-pool cells, probably related to a decrease of the zonal circulation caused by an increase in blocking conditions. This record suggests that higher flood frequency was related both to higher sea temperatures in the North Atlantic associated with the MWP and climatic transitions at the end of the MWP (Late Medieval Climatic Deterioration) and towards the end of the LIA.

There is even some indication of a wetter period in the eastern USA with floods of greater magnitude and frequency on the River Washington between AD 1020 and 1390 than immediately before or since (Chatters and Hoover, 1986).

MIDLAND ENGLAND

One of the few well-dated records of Late Medieval fluvial activity from England comes from the Middle Trent River in the Midlands (Figure 4.1). This record is based on work at Colwick by Salisbury and others in the 1970s (Salisbury et al., 1984) and more recent work at Hemington by Salisbury and Brown (Brown, 1996a, in prep.). In 1993 the remains of three Medieval bridges were found during aggregate quarrying. The site at Hemington Fields is located in the Middle Trent valley upstream of Nottingham at the junction of the Trent and the Derwent (which drains from the southern Pennines). The bridges were excavated and have been dated using both dendrochronology and radiocarbon dating. They date from the late 11th century (Bridge 1, AD 1096–), the early 13th century (Bridge 2, AD 1214–) and the mid-13th century (Bridge 3, AD 1240–). Upstream but in the same reach there had previously been found a Norman mill which was destroyed c. AD 1140 (Clay and Salisbury, 1990). The three bridges had been destroyed by floods and overwhelmed by channel gravels. The flood stratigraphy indicates at least three large events with a bipartite cut-and-fill stratigraphy which can partly be correlated with historical evidence of channel change elsewhere in the Trent. Floods are tentatively dated to c. AD 1140, the 1210s and 1403 (Table 4.1). Palaeohydraulic modelling of the bridges (using the HEC-2 step-backwater modelling) and associated channels has suggested a progressive increase in the channel-forming discharge through the channels under each bridge and this has been explained as a result of a changing channel pattern with flow switching from a northwesterly to southeasterly channel in a braided system before a single channel system was re-established by the 18th century. The magnitude of channel change and gravel reworking and the destruction of the bridges suggests that two or three floods of over 1000 m^3s^{-1} occurred in a period of 200 years. Between the 9th century and the 15th or 16th centuries the river in this reach went through a metamorphosis from a single channel/meandering system to a braided and then anastomosing system before a return to single channel/meandering form (Figure 4.2). Since a change in the reach slope is unlikely, the most obvious cause is an increase in flood magnitudes causing increased channel erosion and gravel bar deposition. By the 17th century the channel had stabilised (the Old Trent) and a flood of over 1000 m^3s^{-1} in 1795 did not restimulate braiding and left little or no discernible sedimentary record. Other archaeological remains also suggest a change in flood regime as numerous fish weirs dated to this period suggest relatively shallow water (at least seasonally) facilitating construction and maintenance. There are reasons to believe that the Middle Trent is rather more sensitive to climate than most other lowland rivers in Britain. This is because within a short length of approximately 40 km it receives a large input from two tributaries which drain uplands, the Derwent and the Dove, both of which carry significant snowmelt discharges and are prone to rain-on-snow floods.

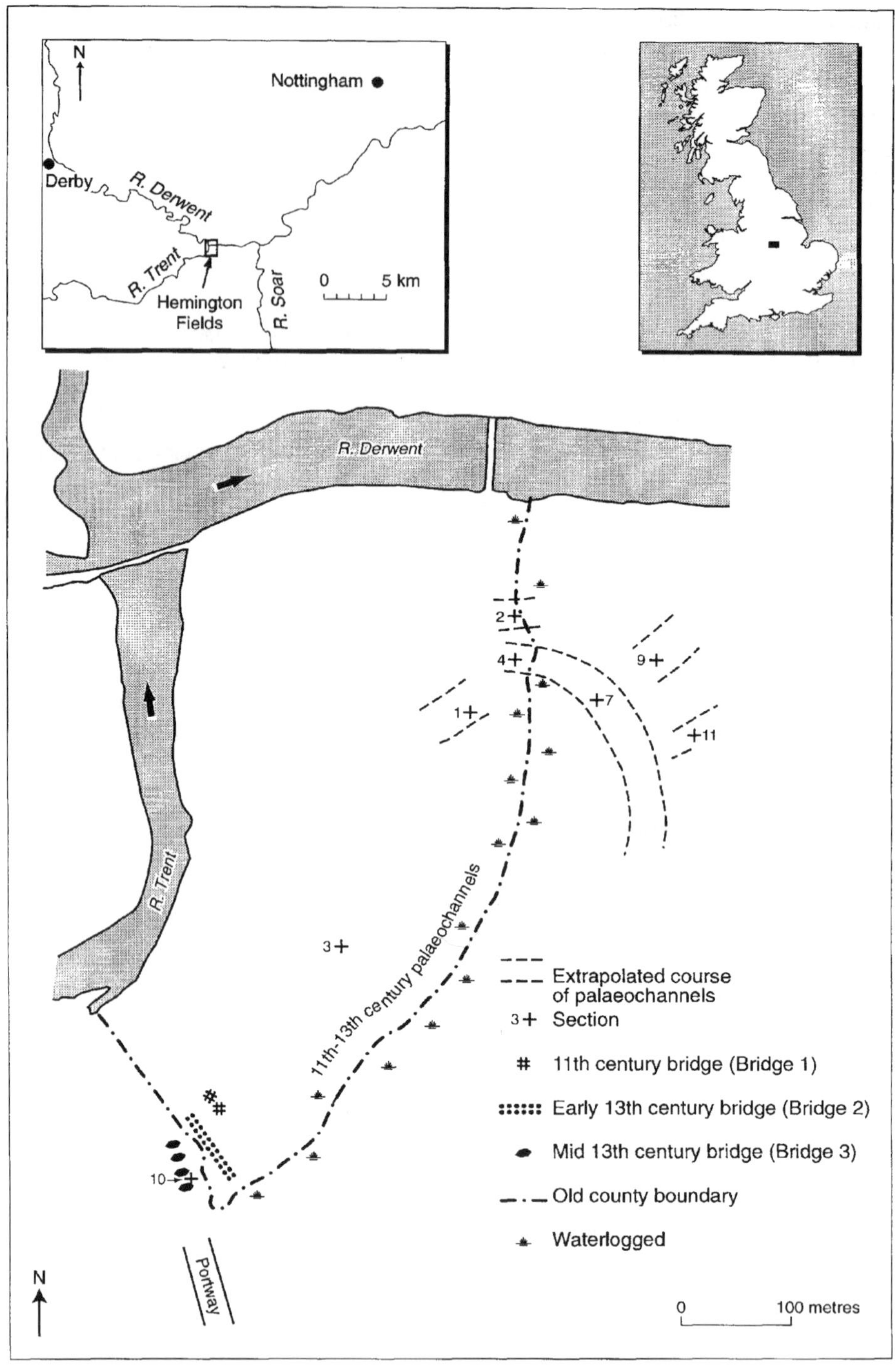

Figure 4.1   Location map of the Middle Trent and the Hemington bridges site

Table 4.1   Floods that occurred in the 11th–14th centuries on the Trent (from Potter, 1964) with the floods which most likely destroyed the bridges at Hemington indicated in bold. The Spalford Bank is a flood embankment which separates the Trent from the floodplain of the Witham, and can only be breached by the largest Trent floods (in excess of 1000 m^{3}s^{-1})

Date	Comments
1141 (Bridge 1)	**First recorded flood on the Trent, breached the Spalford Bank. A**
(2 Feb.)	**rain-on-snow flood similar to 1795 and 1946**
1205	Severe winter, snow and river frozen
1216/17 (Bridge 2)	**Severe winter, snow and river frozen**
1255 July	A summer flood exacerbated by flood debris
1305–1306 Dec–Jan	Severe winter freeze ended by three days of rain
1309/1310 (Bridge 3?)	**Severe winter floods destroyed several bridges and damaged Hethbeth Bridge, Nottingham (pontage issued for its repair)**
1315	Rainfall floods (pontage again issued for Hethbeth bridge)
1322	Severe winter floods
1310–1330	A period of severe winters with damage to bridges almost every winter
1403	**A severe flood which breached the Spalford Bank and caused channel change at Wilne**

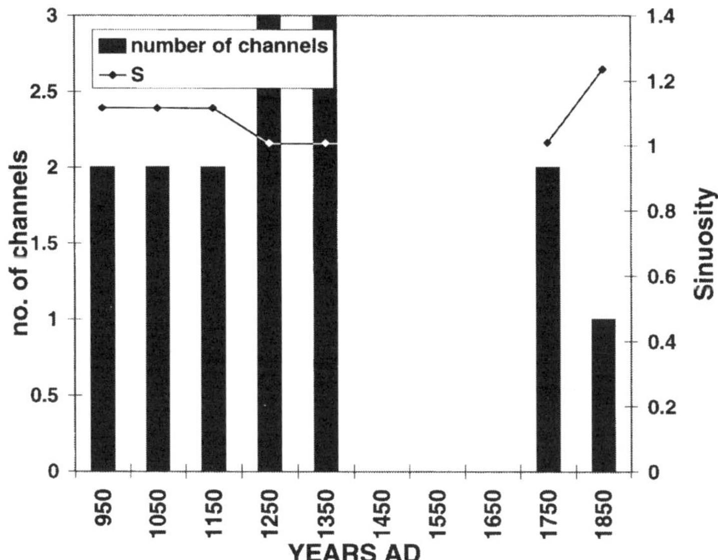

Figure 4.2   Sinuosity (of the main channel) and number of channels of the Hemington Fields reach of the Middle Trent during the Late Medieval period and historical period

There is other evidence of floods during this period from the British Isles, especially in the uplands. There was terrace formation in Glen Feshie *c.* AD 900 and again *c.* AD 1400 (Robertson-Rintoul, 1986) and the Rheidol in Wales underwent a phase of enhanced alluviation between AD 1300–1349 and again at the end of the LIA (Macklin et al., 1992; Rumsby and Macklin, 1996). The middle river Tyne in northern England exhibits increased alluviation between AD 1350–1399 at one reach and AD 1400–1449 at a downstream reach (Rumsby and Macklin, 1996), although this may be a reflection of downstream movement of a single sediment slug rather than two phases of enhanced flood frequency. Although Rumsby and Macklin (1996) regard the period AD 1250–1550 as one of 'enhanced fluvial activity' associated with climatic transition after the Medieval Optimum, the floods recorded here in Midland England and several other areas such as northwest Spain clearly fall during the period of enhanced mean temperatures of the MWP; a possible mechanism may be increased westerly rainfall associated with higher sea surface temperatures.

In one of the western UK piedmont zones the fluvial record for the River Dane shows aggradation and vertical accretion producing a middle terrace during the Middle Ages prior to incision and the formation of a lower terrace since the 18th century (Hooke et al., 1990). In the West Midlands of England, the monks at Bordesley Abbey in Warwickshire had to empty their millpond of sediment with increased regularity during the 10th–13th centuries to the point that it affected the financial viability of the mill (Astill, 1993). Even in Ireland, the part of the British Isles with the least flashy hydrological regimes, there were major hydrogeomorphological changes between AD 1100 and 1200 in the Brosna River (part of the Shannon system), with the beginning of extensive overbank deposition and channel anastomosis (Aalbersberg, 1994).

EVIDENCE FROM ITALY

South of the Alps the Medieval Warm Period is less well marked but there are better records of changes in river flows. In Italy long records exist for the Tiber (400 BC onwards), the Arno and the Po. Camuffo and Enzi (1994) have compared the Tiber and Po flood records and have shown there to be major inhomogeneities in the series with spectral analysis indicating only one common anomaly in the second half of the 15th century (Figure 4.3). This also correlates with other evidence from the Po (Braga and Gervasoni, 1989; Pavese et al., 1994) and possibly with rivers east of the Apennines (Coltorti et al., 1991). In response to a need for multiple catchment studies from within a homogeneous climatic area, five adjacent catchments have been studied from the region just to the north of Rome (Brown and Ellis, 1995). The study has used a variety of dating methods including radiocarbon, luminescence and palaeomagnetics. The region also has excellent lake records from the flooded calderas of Tertiary and Quaternary volcanic activity. The largest lake in the region, Lake Bolsena had low lake levels from AD 1200 to 1500 with a subsequent rise after 1500; however, there is considerable diachrony in lake levels even within central Italy (Dragoni, 1996). The floodplain of the Fiori, the most northerly river, has basal gravels which are generally of Roman/Antic age and exhibits a change in channel pattern between the 14th and 16th centuries (Figure 4.4). The Treia also displays similar Roman/Antic basal gravels and textural changes in flood deposits in the 14th century. Overall the sequences show the following features.

(1) Braided, unstable/laterally mobile channels throughout the Roman period and into the Antic period, with aggradation and lateral erosion that was a persistent engineering problem, examples being the repeated constructions required at Mérida and Salamanca (O'Connor, 1993).

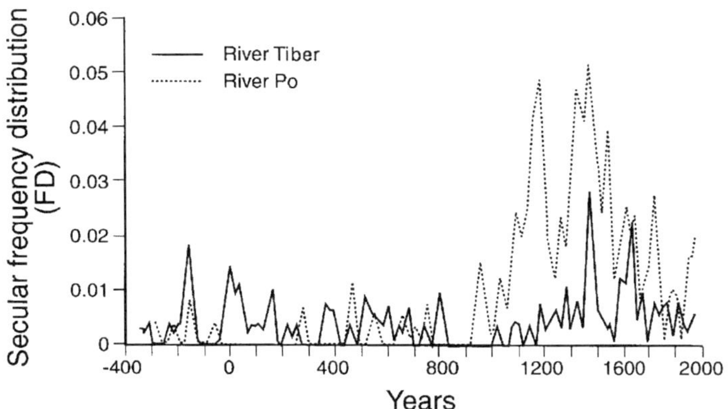

Figure 4.3 Secular frequency distributions of the floods of the rivers Po and Tiber, Italy (after Camuffo and Enzi, 1994)

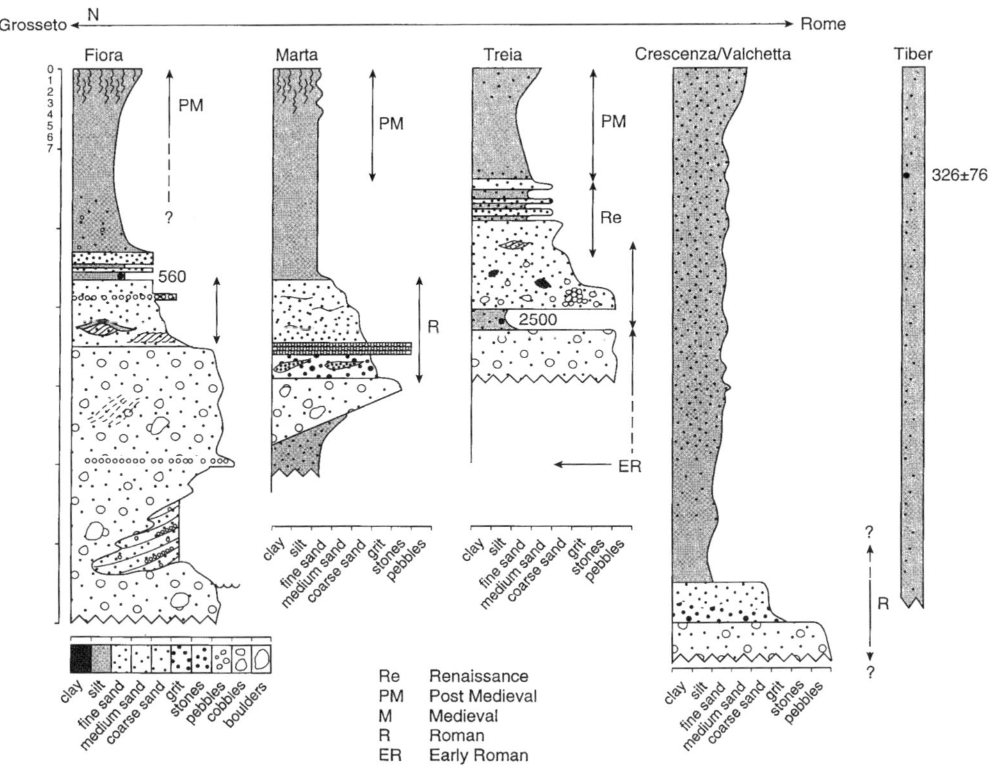

Figure 4.4 Summary stratigraphy of four adjacent catchments in central Italy. Dating is by archaeology and optically stimulated luminescence

(2) A general change in river behaviour in the study area north to south with braiding continuing in post-Roman times in the north (Albagne and Fiori) but replaced by overbank deposition of sand and silt in the southerly basins (Marta and a small tributary of the Tiber).

(3) Changes in river activity in the 14th–16th centuries with increased channel migration in the north and a large increase in overbank deposition rates in the south which completely buried the old braid-plain. The most likely cause of this difference is the intensive arable land use in the southern basins associated with Rome and its economic hinterland and the growth of Rome in the Renaissance created an increased local demand for horticultural goods.

There remain two problems. Firstly it is not clear whether the lack of evidence of fluvial activity in the post-Roman pre-Medieval period (Dark Ages) is entirely real or partly a function of dating problems. Secondly, given the later peak in fluvial activity south of the Alps, it is not clear whether it is a Late Medieval Deterioration phenomenon or an early Little Ice Age phenomenon, although the identification of broadly synchronous cooling and increased flood frequency in the 18th–19th centuries would suggest the former interpretation. A further theoretical problem is that the climatic record and land use effects are not unrelated, so an increase in summer temperatures could increase flood peaks and flood frequency both directly through meso-scale convective activity and indirectly through the expansion and intensification of arable cultivation, thus magnifying the effect of the initial cause.

CONCLUSIONS

The Late Medieval Warm Period and Late Medieval Deterioration are both represented in alluvial stratigraphies in Midland England. This probably reflects little change in winter temperatures with the largest floods still being caused by the coincidence of heavy rain and snowmelt. The record also shows that the previously largest recorded flood was of a return period of no greater than 1:200 years (previous estimate was 1:400 years), although the value of return periods larger than the periodicity of climatic changes is very doubtful (Brown, 1996b). North of the Alps winters may even have been wetter during the Late Medieval Warm Period and the Late Medieval Deterioration. South of the Alps the records from central Italy suggest fluvial change during the 15th–17th centuries probably associated with increased cyclonic activity. This is broadly in line with Vita-Finzi's (1976) explanation for the Younger Fill; however, archaeological evidence and the variation of the fluvial response would suggest that this was exacerbated by intensive arable land use in the Rome region. Another possible cause of European regional diachrony is volcanic activity in the North Atlantic. It is suggested here that only detailed within-region studies can hope to resolve problems of the partial correlations of possible causative factors and the allocation of causality for palaeohydrological change in the Late Holocene.

ACKNOWLEDGEMENTS

The author must thank all who have worked on the Hemington Bridges project and English Heritage for funding it, as well as the British Academy, British School at Rome and Royal Society for funding the Italian work. In particular thanks must go to H. R. Potter, C. Clay, C. Salisbury, G. Barker and T. Potter.

REFERENCES

Aalbersberg, G., 1994. *The Little Brosna River Valley*, Research Report, Faculty of Earth Sciences, Free University, Amsterdam.

Astill, G.G., 1993. *A Medieval Industrial Complex and its Landscape: The Metalworking watermills and Workshops of Bordesley Abbey*. Council for British Archaeology, Research Report 92.

Barber K.E., Chambers, F.M., Maddy, D., Stoneman, R. and Brew, J.S., 1994. A sensitive high-resolution record of late-Holocene climatic change from a raised bog in northern England. *The Holocene*, **4**, 198–205.

Becker, B. and Schirmer, W., 1977. Palaeoecological study on the Holocene valley development of the river Main, Southern Germany. *Boreas*, **6**, 303–321.

Benito, G., Machado, M.J. and Pérez-González, A., 1996. Climate change and flood sensitivity in Spain. In J. Branson, A.G. Brown and K.J. Gregory (eds), *Global Continental Changes: the Context of Palaeohydrology*, Geological Society, London, Special Publication 115, 85–98.

Bork, H-F., 1989. Soil erosion during the past millennium in Central Europe and its significance within the geomorphodynamics of the Holocene. In F. Ahnert (ed.), *Landforms and Landform Evolution in West Germany, Catena Supplement*, **15**, 121–132.

Braga, G. and Gervasoni, S., 1989. Evolution of the Po river: an example of the application of historic maps. In G.E. Petts (ed.), *Historical Change of Large Alluvial Rivers: Western Europe*, Wiley, Chichester, 113–126.

Brown, A.G., 1996a. *Palaeohydrology and archaeology with special reference to the Middle Trent*, British Hydrological Society, Occasional Paper 7, 55–62.

Brown, A.G., 1996b. Human dimensions of palaeohydrological change. In J. Branson, A.G. Brown and K.J. Gregory (eds), *Global Continental Changes: the Context of Palaeohydrology*, Geological Society, London, Special Publication 115, 57–72.

Brown, A.G., in prep. The geomorphological context of the Hemington Bridges. In L. Cooper and S. Ripper (eds), *The Hemington Bridges*, University of Leicester, Archaeological Services Report 97/27.

Brown, A.G. and Ellis, C., 1995. People, climate and alluviation: theory, research design and new sedimentological and stratigraphic data from Etruria, Italy. *Papers of the British School in Rome*, **63**, 45–73.

Buch, M.W., 1989. Late Pleistocene and Holocene development of the Danube valley east of Regensburg. *Catena Supplement*, **15**, 279–287.

Camuffo, D. and Enzi, S., 1994. Reconstructing the climate of northern Italy from archival sources. In R.S. Bradley and P.D. Jones (eds), *Climate Since A.D. 1500*, Routledge, London, 143–154.

Chatters, J.C. and Hoover, K.A., 1986. Changing Late Holocene flooding frequencies on the Columbia river, Washington. *Quaternary Research*, **26**, 309–320.

Clay, P. and Salisbury, C.R., 1990. A Norman mill dam and other sites at Hemington fields, Castle Donington, Leicestershire. *The Archaeological Journal*, **147**, 276–307.

Coltorti, M., Consoli, M., Dramis, F., Gentili, B. and Pambianchi, G., 1991. Evoluzione geomorfologica delle piane alluvionale deele Marche Centro-Meridionale. *Geografia Fisica e Dinamica Quatermaria* **14**, 87–100.

Dragoni, W., 1996. Response of some hydrological systems in Central Italy to climatic variations. In A.N. Angelakis and A.S. Issar (eds), *Diachronic Impacts on Water Resources*, NATO ASI Series, Vol. I 36, Springer-Verlag, Berlin, 193–229.

Gotteschalke, M.K.E., 1971–74. *Stormvloeden en Rivieroverstromingen in Nederland*. Assen, Netherlands (3 volumes).

Grove, J.M., 1988. *The Little Ice Age*. Methuen, London.

Grove, J.M. and Switsur, R., 1994. Glacial geological evidence for the Medieval Warm Period. In M.K. Hughes and H.F. Diaz (eds), *The Medieval Warm Period*, Kluwer, Dordrecht, 143–170.

Guiot, J., Tessier, L., Serre-Bachet, F., Guibal, F. and Gadbin, C., 1988. Annual temperature changes reconstructed in W. Europe and N. W. Africa back to A.D. 100. *Annals Geophysica* **85**

(Special Issue, XIII General Assembly of CGS, Bologna).

Hooke, J.M., Harvey, A.M., Miller, S.Y. and Redmond, C.E., 1990. The chronology and stratigraphy of the alluvial terraces of the River Dane valley, Cheshire. *Earth Surface Processes and Landforms*, **15**, 717–737.

Hughes, M.K. and Diaz, H.F. (eds), 1994. *The Medieval Warm Period*, Kluwer, Dordrecht.

Kalicki, T. and Krapiec, M., 1991. Subboreal 'black oaks' identified from the Vistula alluvia at Grabie near Cracow (south Poland). *Geologia*, **17**, 155–169.

Lamb, H.H., 1982. *Climate, History and the Modern World*, Methuen, London.

Macklin, M.G., Rumsby, B.T. and Newson, M.D., 1992. Historical floods and vertical accretion of fine-grained alluvium in the Lower Tyne valley, Northeast England. In P. Billi, R.D. Hey, C.R. Thorne and P. Tacconi (eds), *Dynamics of Gravel-bed Rivers*, Wiley, Chichester, 564–580.

O'Connor, C., 1993. *Roman Bridges*, Cambridge University Press, Cambridge.

Parry, M.L., 1978. *Climatic Change, Agriculture and Settlement*, Dawson, Folkestone.

Pavese, M.P., Banzon, V., Colacino, M., Gregori, G.P. and Pasqua, M., 1994. Three historical data series on floods and anomalous climatic events in Italy. In R.S. Bradley and P.D. Jones (eds), *Climate Since A.D. 1500*, Routledge, London, 155–170.

Pfister, C., 1988. Variations in the spring–summer climate of central Europe from the High Middle Ages to 1850. In H. Wanner and U. Siegenthaler (eds), *Long and Short Term Variability of Climate*, Springer-Verlag, Berlin, 57–82.

Potter, H.R., 1964. Introduction to the History of the Floods and Droughts of the Trent Basin. Unpublished manuscript of a conference held at Loughborough University, October 1964.

Robertson-Rintoul, M.S.E., 1986. A quantitative soil stratigraphic approach to the correlation and dating of post-glacial river terraces in Glen Feshie, Western Cairngorms. *Earth Surface Processes and Landforms*, **11**, 605–617.

Rother, N., 1989. Holozäne erosion und akkumulation im Ilmetal südniedersachen. *Bayreuth Geowissenschaftliche Arbieten Band*, **14**, 87–94.

Rumsby, B. and Macklin, M.G., 1996. River response to the last neoglacial (the 'Little Ice Age') in northern, western and central Europe. In J. Branson, A.G. Brown and K.J. Gregory (eds), *Global Continental Changes: the Context of Palaeohydrology*, Geological Society, London, Special Publication 115, 217–234.

Salisbury, C.R., Whitley, P.J., Litton, C.D. and Fox, J.L., 1984. Flandrian courses of the river Trent at Colwick, Nottingham. *Mercian Geologist*, **9**, 189–207.

Serebryanny, L.R. and Solomina, O., 1996. Glaciers and climate of the mountains of the former USSR during the neoglacial. *Mountain Research and Development*, **16**, 157–166.

Thompson, L., 1991. Ice-core records with emphasis on the global record for the last 2000 years. In Bradley, R. (ed.), *Global Changes of the Past*, UCAR/Office for Interdisciplinary Earth Studies, Boulder, Colorado, 201–224.

Vita-Finzi, C., 1976. Diachronism in Old World alluvial sequences. *Nature*, **263**, 218–219.

Walsh, K., O'Sullivan, D., Young, R., Crane, S. and Brown, A.G., 1995. Medieval landuse, agriculture and environmental change on Holy Island, Northumbria. In R. Butlin and N. Roberts (eds), *Ecological Relations in Historical Times*, IBG Special Publication, Blackwell, 101–121.

Wanner, H. and Siegenthaler, U. (eds), 1995. *Long and Short Term Variability of Climate*, Springer-Verlag, Berlin.

Zhang, J. and Crowley, T.J., 1989. Historical climate records in China and reconstruction of past centuries. *Journal of Climatology*, **2**, 833–849.

CHAPTER 5

Modelling Long-term Erosion and Sedimentation Processes in Fluvial Systems: A Case Study for the Allier/Loire System

A. VELDKAMP AND J.J. VAN DIJKE

Department of Soil Science and Geology, Wageningen Agricultural University, The Netherlands

INTRODUCTION

A general aim of geomorphological and palaeohydrological modelling is that of simulating climate-related Quaternary dynamics in fluvial basins. Despite this general aim, only relatively few attempts have been made to relate process data and evidence of climatic fluctuations to fluvial basins (Puvaneswaran and Conacher, 1983). Most available models are focused on specific spatial and temporal scales, and according to Anderson (1988), contemporary geomorphological modelling generally emphasises the shorter time base. Although it is often argued that such approaches may eventually provide sub-model components for a longer term model, scale effects will certainly hamper straightforward applications (Rosswall et al., 1988; Anderson, 1988; Beven, 1995). The scale effect is one of the main reasons that many existing models are incompatible. Furthermore, models should also be compatible with the available data. One may consider using an actual hydrological model for a palaeohydrological investigation but this would require accurate and detailed hydrological data which are certainly not available.

On a basin scale, fluvial systems are subject to changes due to erosional and depositional processes which are directly or indirectly driven by climate, tectonics and base level. Neither endogenetic processes such as tectonic uplift, nor exogenic processes such as weathering and erosion can be considered constant, especially in the Quaternary, the most important period for understanding present-day relief development in fluvial basins. Tectonic activity has accelerated over the last 10 Ma, while fluctuations in tectonic rates in time scales between 0.1 and 1000 ka are still largely unknown (Zijerveld et al., 1992; Van den Berg, 1996). Quaternary climate-changes at similar time scales have greatly influenced the rate of weathering, type and rate of fluvial processes, and the position of base level (sea level).

Despite the fact that it is generally recognised that the fluvial system is a strongly dynamic system, limited attempts have been made to capture this behaviour in a model

Palaeohydrology and Environmental Change. Edited by G. Benito, V. R. Baker and K. J. Gregory
© 1998 John Wiley & Sons Ltd.

for Quaternary time spans. This is probably due to fragmentary and mostly incomplete Quaternary terrestrial records. The fluvial record usually consists of stacked truncated sediment bodies with poor age control. A combined erosional and depositional record can be found in fluvial terraces. Fluvial terraces are commonly interpreted as a mainly climate-controlled system (Gibbard, 1988). Although this interpretation may be valid it is important to keep an open mind for alternative explanations. The roles of crustal movements (glacio-isostasy) and base level which have experienced synchronous changes with climate could be severely underestimated. This underestimation is probably related to the fact that these factors are not always directly reconstructable from field data. But the fact that, contrary to prevailing climate, one cannot directly reconstruct base level position or uplift rate from a sediment record does not mean that their role was irrelevant.

A possible solution to prevent such biased interpretations lies in an open and well balanced basin-scale analysis of the investigated system. The system description used in the analysis should be incorporated in a numerical conceptual model which can be applied to test and falsify hypotheses about the system dynamics. Such a modelling exercise is able to demonstrate consequences of assumptions and interpretations.

FLUVER1 is a conceptual long-term (time steps of 2 ka) macro-scale (finite elements of 100 m) model based on such an open system description to simulate scenarios of the development of fluvial terraces during the Quaternary (Veldkamp and Vermeulen, 1989; Veldkamp and van Dijke, 1994). The model has climate, base level and tectonic inputs which all interact and determine fluvial terrace morphology and stratigraphy. This model has been successfully applied for several NW European case studies (e.g. Veldkamp, 1992; Veldkamp and van den Berg, 1993).

A major disadvantage of the model is that it simulates local valley morphology and stratigraphy which requires quantitative descriptions of the dynamics of sediment fluxes through time. These data are usually rare and were derived from estimates based on bulk geochemical sediment research of the alluvial stratigraphy (Veldkamp, 1991). To extend the model to basin scale it is necessary to model sediment flux dynamics along a longitudinal profile with data about base level changes, crustal movements and climatic variations. This fluvial activity is modelled with equations relating sediment transport capacity of rivers to discharge, topography and sediment fluxes from slope processes (Van Dijke and Veldkamp, 1996).

An advantage of the new model version is that it requires a calibration based on obtainable data: net relief changes of the longitudinal profile in time and the sediment budget of the studied system. To test this new model version to a real-world system we used a relatively well documented case where a previous model version was also applied. Therefore we reinvestigated the Allier/Loire system with FLUVER2.

The Allier/Loire system has a relatively well documented record of fluvial terraces along its present course of approximately 600 km. Modelling exercises with FLUVER1 for a segment of the Allier demonstrated the importance of uplift and sediment flux dynamics in explaining the observed terrace altitude distribution and stratigraphy (Veldkamp, 1992). In this paper we will elaborate this research by reconstructing longitudinal profile dynamics for the whole Allier/Loire system. Subsequently, the results will be evaluated at valley level by simulating local valley terrace morphology.

STUDY AREA

The study area is situated in central and western France and comprises the Allier/Loire drainage basin (Figure 5.1). The Allier and Loire drain two separate rift valleys in the Hercynian crystalline Massif Central. Previous investigations on the Allier and Loire terraces are summarised by Macaire (1986) and Pastre (1987) and the general geological setting by Autran and Peterlongo (1980). The geology of the study area, including the terraces, is excellently mapped at a 1:50000 scale by the Bureau de Recherches Géologiques et Minières (BRGM). The Allier/Loire basin covers an area of approximately 110000 km^2 and comprises Hercynian crystalline and metamorphic rocks in the Massif Central and the Amorican Massif along the coast. Marine Mesozoic rocks are exposed in the southern and northwestern parts of the Paris basin, and Tertiary sediments in the rift valleys and in the centre of the Paris basin (Figure 5.1).

Both rivers have eight main terrace levels, numbered from Z to S (present flood plain to oldest terrace level). Larue (1979) made the first thorough investigation of the chronostratigraphy of these terraces based on sediment composition and Pastre (1987) related sand mineralogy with the mineralogy of dated eruptions to correlate and date the different terrace deposits. The Quaternary sediments along the lower reach of the Loire were investigated by Macaire (1986). Veldkamp (1991) investigated the geochemistry of terrace sediments and used ^{14}C and Th/U series for age dating.

The typical sequence of terraces present in the Allier/Loire valley shows that accumulation and vertical erosion alternated repeatedly during the general valley incision. Very

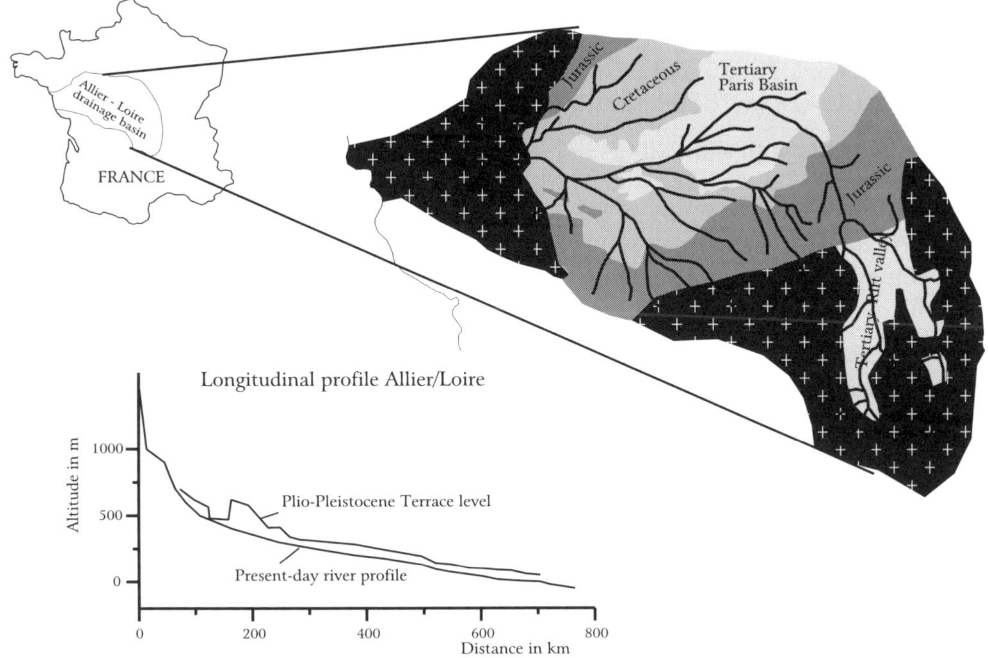

Figure 5.1 Geographical and geological setting of the Allier/Loire basin. The current longitudinal profile is shown with reconstructed position of uplifted Plio-Pleistocene sediments

similar terrace sequences exist along the Meuse, Rhine and Thames (McGregor and Green, 1978; Brunnacker and Boenigk, 1983; Andres, 1989; Van den Berg, 1996). The mechanism of terrace formation in these systems is usually thought to be controlled by climate, tectonism and base level. The knowledge about the role of these three variables for the Allier/Loire river system is limited.

1. *Climate.* Terrace sediments indicate that most of the terrace deposits accumulated during and at the end of glacials. The alternations between accumulation and incision are therefore ascribed primarily to climatic causes (Raynal, 1984). For NW Europe most investigators agree on deposition of most terrace sediments during generally cold-climate conditions but their interpretations of the exact system dynamics involved in creating such deposits differ strongly (Starkel, 1983; Dawson and Gardiner, 1987; Gibbard, 1988; Vandenberghe et al., 1994).
2. *Tectonism.* The contribution of tectonism to terrace formation is obvious from the tendency towards valley deepening for the entire Allier/Loire system. The average uplift for the whole basin is approximately 100 m during the last 1 Ma. Apart from this uplift, which still continues (Giot et al., 1978), longitudinal profiles of terraces indicate numerous displacements by faults (Larue, 1979).
3. *Base level.* A direct or indirect influence of sea level variations is not visible in the Loire longitudinal profile. However, near the actual coast the current estuary shows the effect of a recent sea level rise on the erosion and sedimentation processes. But, as in other fluvial systems, an upstream effect of base level changes can be expected (Schumm, 1993).

MODELLING THE QUATERNARY LONGITUDINAL PROFILE DYNAMICS OF THE ALLIER/LOIRE

FLUVER2 starts with simulating longitudinal profile dynamics as a function of precipitation, tectonics and sea level. After the overall longitudinal profile calculations the local valley morphology is calculated for selected stretches of 10 km applying the original FLUVER model. The calculated longitudinal changes are used as inputs in the valley morphology calculations. The model is valid at basin scale only with a spatial resolution of 3 km and a temporal resolution of 200 years.

Mathematical concepts for numerical modelling of longitudinal fluvial dynamics

The model describes the fluvial system dynamics in a two-dimensional way, the longitudinal profile dynamics. The modelled system allows for feedbacks and inputs related to climate (precipitation and temperature), and changes in relative altitude and relief (tectonics and sea level position). The model describes the fluvial processes at the valley level in terms of sediment transport. Furthermore, slope processes are simulated as an independent process able to supply additional sediments to the fluvial system.

Fluvial processes

Fluvial processes erode, transport and deposit sediments. These actions will change the relief (δz) in time (δt) as well as the shape of the longitudinal profile (δx). Longitudinal profile changes are therefore a function of sediment flux changes (δfs) which are caused by erosion (E) and sedimentation (S_d):

$$\frac{\delta z}{\delta t} = \frac{\delta fs}{\delta dx} = E - S_d \tag{5.1}$$

The changes in sediment flux (δfs) within the longitudinal profile are a function of the amount of sediment in the fluvial system which can be transported and the amount which is available in the system. The maximum amount of sediment which can be transported maximally (transport capacity) is a function of the amount of water and the gradient of the fluvial system.

The maximum amount of sediment (Qs) a river can transport can be described by an empirical relationship which assumes a dependence on discharge (Qw) and slope of the river (S). This relationship was proposed by Kirkby (1971):

$$Qs = k\, Qw\, S^n \tag{5.2}$$

The factor k and the exponent n are constant. The factor k is related to the sediment properties. A relatively high value indicates easily transportable sediments such as clay and silt whereas a small value corresponds to resistant gravel bed and bed rock.

Equation 5.2 is solved by assuming equilibrium conditions in the system. This assumption is sensitive for the chosen time span (time step). Under equilibrium conditions the continuity principle applies and we therefore applied a continuity equation for sediment transport which was solved with the finite difference method.

The altitude H at time t is expressed as a function of the independent variables and H at time $t - 1$:

$$H_t = H_{t-1} + (Qs_{x-1} - Qs_x)\frac{dt}{dx} \tag{5.3}$$

A similar approach is also used by Willgoose et al. (1991) and Howard et al. (1994).

Slope processes

Apart from the amount of sediment available within the fluvial system due to fluvial erosion processes, slope processes also supply sediments to the fluvial system. Slope processes are best described as a diffusion or dispersion process. These processes have a relief-smoothing effect and they are only dependent on the slope (S) and a diffusion constant (D):

$$\frac{\delta}{\delta x} = \frac{\delta}{\delta x} D \frac{\delta z}{\delta x} = \frac{\delta}{\delta x}(D\, S) \tag{5.4}$$

Slope processes were related to permafrost conditions by a link between altitude and temperature. The diffusion constant therefore changes with the altitude of the watershed, effectively causing an increase in slope processes with altitude.

With Equations 5.3 and 5.4 it is now possible to calculate the changes in altitude for every position along the longitudinal profile by taking into account the transport capacity of the available water with respect to the available sediments by both fluvial erosion and slope processes.

Climate-related discharge and sediment supply relationships

The Quaternary has known many astronomically controlled worldwide changes in climate, which can be very satisfactorily explained by the Milankovich theory (Imbrie et al., 1984; Shackleton et al., 1990). Although the relationship between fluvial dynamics and climate behaviour depends on the nature of climatic change and the effects of such changes on discharge and sediment load (Lowe and Walker, 1984; Bradley, 1985), a simple linear relation was assumed between changes in caloric insolation (Berger, 1978) and mean discharge and sediment supply dynamics. Discharge is calculated from precipitation/evaporation ratio and catchment size. The non-fluvial sediment supply derived from slope processes was simulated as a function of slope and altitude (see Equation 5.4). During glacial episodes a large amount of water was stored in glaciers whereas in the continental areas bordering the major ice sheets, cold and dry conditions prevailed causing a decrease in mean river discharges (Starkel, 1983). Owing to this drier and colder glacial environment the vegetation cover decreases (Guiot et al., 1989) causing an increase in sediment supply from the slopes into fluvial systems. Interglacials yield the opposite scenario: an increase in mean discharge and a complementary decrease in the sediment supply.

Mean precipitation and resulting discharge are simulated as the sum of three sinusoidal functions with the periodicities of the precession (23 000 years), obliquity (41 000 years) and eccentricity (96 000 years). The sediment supply to the fluvial system by mass and slope movements is simulated as the sum of the sinusoidal functions of the same three periodicities. Bulk geochemical research (Veldkamp, 1991) demonstrated that sediment fluxes are somewhat out of phase (back-lagging) with the discharge curves and are therefore simulated as cosinus functions. Although these curves do not exactly match the climatic curves derived from deep sea cores, loess sediments and pollen records indicate that they sufficiently describe climatic changes during the Quaternary for our conceptual long-term modelling purposes (Figure 5.2) (Van den berg, 1996).

Tectonism

Two different components of tectonism are incorporated in FLUVER: a component of gradual uplift or flexure of the whole simulated landscape, and an uplift component describing the difference in uplift rates between different tectonic units separated by faults. A simplified longitudinal profile of the actual Allier/Loire and a reconstructed Plio-Pleistocene level using the terrace profile (Larue, 1979; Le Griel, 1983) along the Allier/Loire served as a starting point for the simulations (Figure 5.1). From these profiles the net uplift and incision could be reconstructed. For the initial profile, the current profile minus the mean reconstructed uplift was used. The uplift rates were based on previous reconstructions which indicate a gradual, relatively slow uplift (0.04 m ka^{-1}) during the Early Pleistocene and increased uplift rates with active fault displacements (ranging from 0.08–0.2 m ka^{-1}; Giot et al., 1978) during the last 0.8 Ma (Veldkamp, 1991; Van den Berg, 1996), a trend commonly observed in western Europe (Lilienberg, 1985).

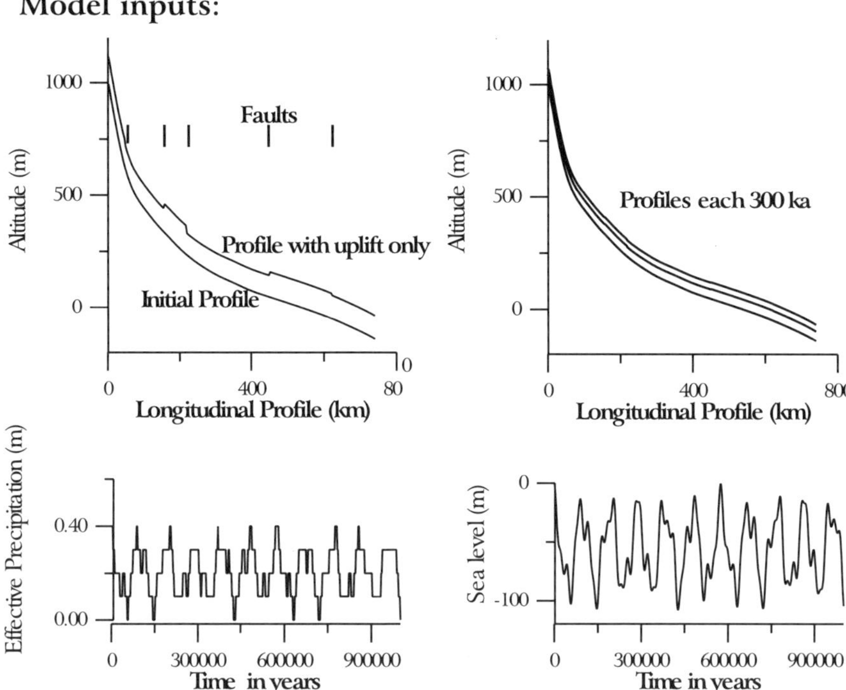

Figure 5.2   Initial simulation conditions, total uplift inputs, changes in the simulated longitudinal profile during the simulated 1 Ma, and changes in effective precipitation and sea level during the simulation

The total uplift relative to the initial profile and fault boundaries used in the simulation are indicated in Figure 5.2.

Sea level

Sea level is also a function of the climate-controlled water balance and is consequently simulated by the same set of periodicities as used for precipitation simulations with an amplitude of 60 m (Pirazzoli, 1991; Figure 5.2).

Calibration

The model simulation was calibrated on two aspects of profile development character-istics. The simulated profile should (1) qualitatively match the current longitudinal profile after the simulated interval of 1 Ma, and (2) the changes in the longitudinal profile should display alternating erosion and sedimentation episodes as reconstructed from terrace geomorphology and stratigraphy. The inputs of changing climate, sea level and uplift rates were not used for calibration purposes. Only the magnitudes of changes in water and sediment supply were used as tuning variables to obtain alternating erosion/deposi-tion conditions. Because the simulated sediment fluxes are the result of a tuning exercise, their magnitudes have only semi-quantitative values, and must be viewed as such.

MODELLING THE LOCAL FLUVIAL TERRACE DEVELOPMENT ALONG THE ALLIER/LOIRE

The longitudinal vertical dynamics, expressed as dh/dt, can be used to derive the erosion/sedimentation conditions along a fluvial system in time. Positive dh/dt values indicate net deposition during a time step and negative erosion values. When these dh/dt profiles are plotted with time, a so-called Profile Evolution Map (PEM) is derived (Van Dijke and Veldkamp, 1996; Figure 5.3). This PEM gives a clear overview of where and when conditions in the fluvial system change. Although the simulated Allier/Loire system has known net erosion through time due to the continuing uplift, phases of deposition alternate with erosional phases throughout the whole profile. For the upstream part the

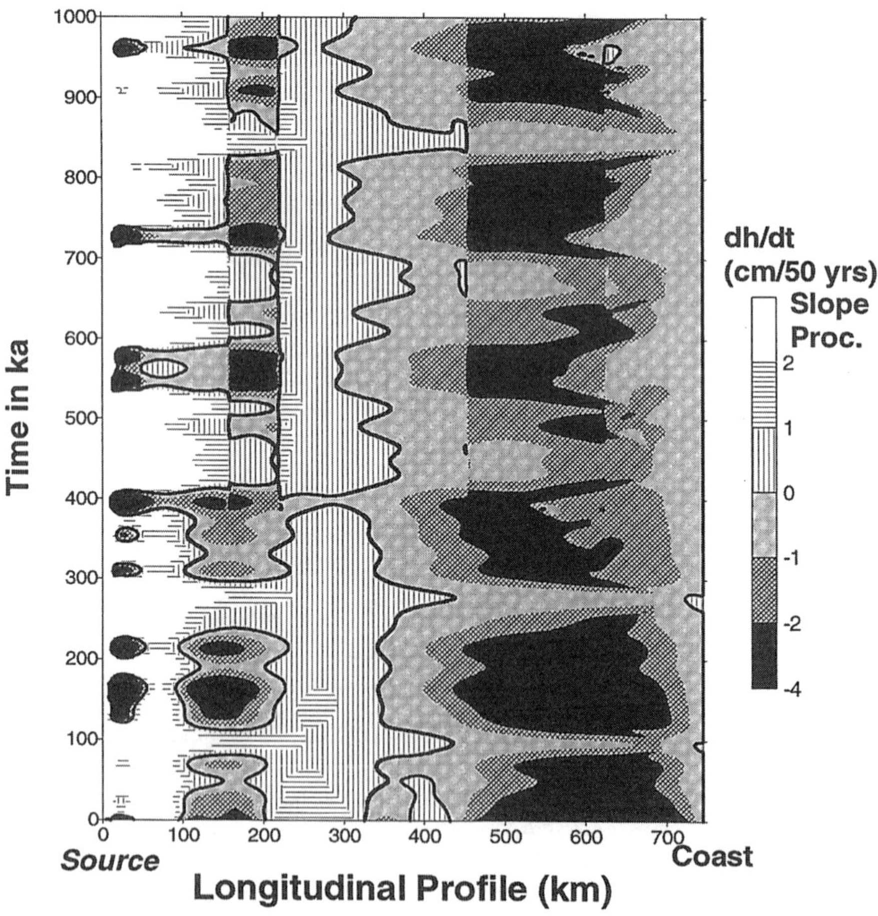

Figure 5.3   The Profile Evolution Map (PEM) of the Allier/Loire longitudinal profile during 1 Ma

erosional phases coincide with interglacials while glacial periods are characterised by net deposition. The more downstream reach has opposite characteristics: glacial erosion and interglacial deposition. Although the simulated longitudinal profile shows no obvious changes in its shape and characteristics (Figure 5.2), the different tectonic units clearly show up on the PEM (Figure 5.3) after 200 ka when the fault displacement is initiated as a result of accelerated uplift. The faults sometimes act as a threshold for the process, confining erosion or sedimentation to one profile segment only.

To simulate in detail the vertical erosion/deposition dynamics, local valley development was simulated for two different reaches (Figures 5.4 and 5.5). Apart from the different basin-wide dynamics, the two selected reaches have known a slightly different tectonic history. The valley reach at 215 km downstream of the water divide (Figure 5.4) has a net uplift of 128 m in 1 Ma (during the first 200 ka the uplift rate was $0.04 \, \text{m ka}^{-1}$ followed by acceleration to a rate of $0.15 \, \text{m ka}^{-1}$ during the remaining 800 ka) while the more downstream reach (450 km from the water divide) has a net uplift of 96 m (during the first 200 ka the uplift rate was $0.04 \, \text{m ka}^{-1}$ followed by a rate of $0.11 \, \text{m ka}^{-1}$ during the remaining 800 ka).

The simulation results at valley scale reflect the differences in uplift rate by the difference in valley depth. The relative position of the simulated valley is illustrated by its size: the valley further downstream (Figure 5.5) is much wider. The combined effect of valley position along the profile and of local uplift regime is expressed by the vertical and horizontal position and the number of terraces. The most upstream valley (Figure 5.4) has four distinct terrace levels, while the downstream valley (Figure 5.5) displays only two terraces. The valley development diagrams (Figures 5.4 and 5.5) clearly illustrate that numerous terraces were formed during the simulation but due to the relatively low uplift rate hardly any terrace is preserved

DISCUSSION

Our simulation of the Allier/Loire system illustrates that although there is a continuous change in erosion and sedimentation along the whole profile, their timing is not always synchronous and, therefore, they may not always be directly related to climatic changes as proposed by Raynal (1984) and Pastre (1987). The system shows shifts in the position of erosional and depositional zones during the simulation. This basin-wide dynamics was found for all systems investigated with FLUVER (Van Dijke and Veldkamp, 1996) and may explain why many researchers have different interpretations of the causes involved in the Quaternary terrace formation for similar river systems (Starkel, 1983; Buch, 1987; Dawson and Gardiner, 1987; Gibbard, 1988; Vandenberghe et al., 1994). This modelling illustrates that even within one fluvial system a basin-wide analysis is required to correlate and match terraces and alluvial deposits along a longitudinal profile. A similar insight was also offered by Bull (1991) who argued that simplistic temporal reconstructions from the longitudinal profile can be misleading.

In the most downstream reach of the Allier/Loire the effect of changing base level dominates the longitudinal profile by causing sedimentation zones during interglacials and erosional phases during glacials. These dynamics have also been described by Schumm (1993). However, due to the general uplift trend of the entire basin this base level effect is limited to the lowest 150 km reach. Similar observations were described for flume experiments without uplift (Schumm et al., 1987). The effects of climate-related

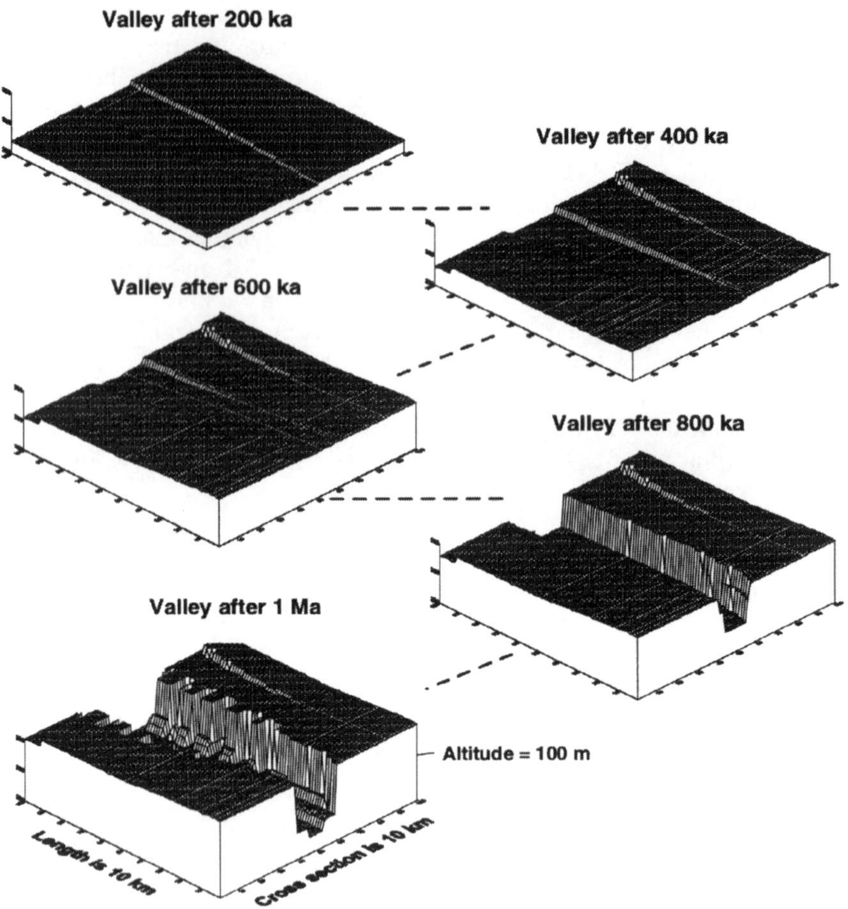

Figure 5.4 The local valley terrace morphology development at 215 km downstream of the water divide

changes in discharge and sediment fluxes dominate the overall characteristics of the Allier/Loire geomorphology. Furthermore, the role of tectonics is illustrated in the PEM of Figure 5.3 where certain erosional or depositional stages are confined to certain tectonic zones. It is this tectonic component which caused the non-synchronous Allier/Loire terraces. Apart from this non-synchroneity of terrace formation, the preservation of a stream terrace greatly depends on the local tectonic regime. In areas with relatively slow or rapid uplift the terrace preservation is strongly reduced. In rapidly uplifting areas the strong vertical incision left only narrow terrace remnants which are easily eroded, while in slowly uplifting areas the frequent lateral shifts of the channel may cause the destruction of existing terraces.

The type of simulations presented here can be applied in other long-term geomor-

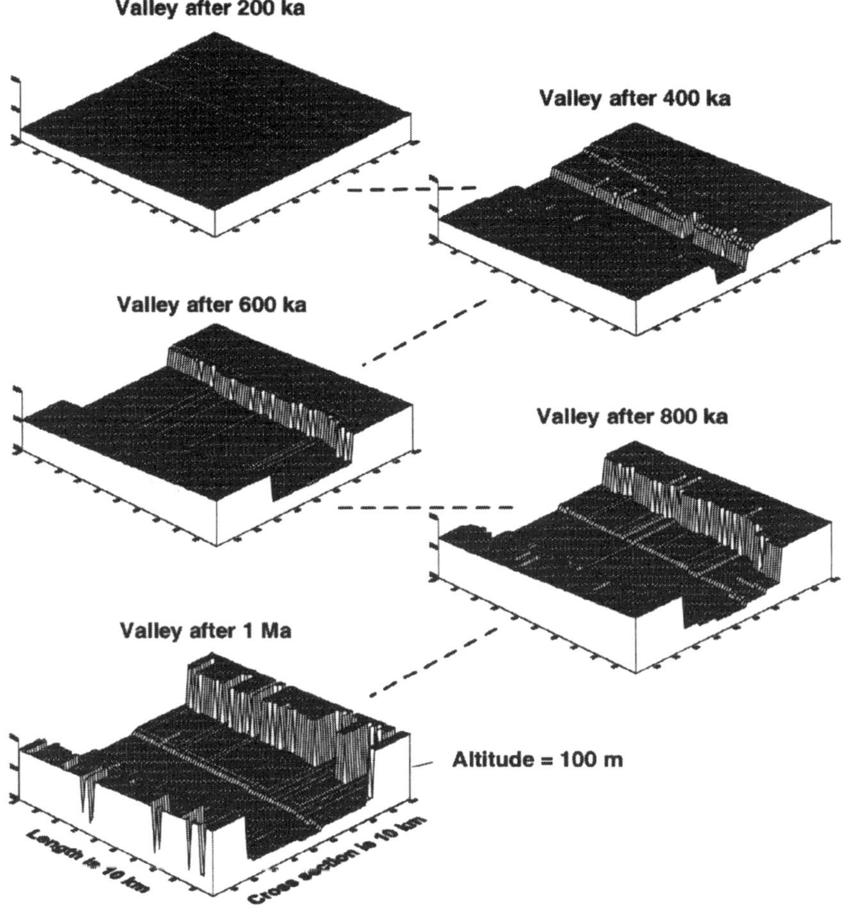

Figure 5.5　The local valley terrace morphology development at 450 km downstream of the water divide

phological and palaeohydrological studies to discern the main steering variables in fluvial system development. The Allier/Loire system contains terraces along a reach of approximately 500 km. Very often researchers have correlated and even 'dated' these terraces using only their relative elevation above present-day thalweg (Larue, 1979; Raynal, 1984). The results of our simulation demonstrate that even in an uplifting system like the Allier/Loire it is rather unlikely that one terrace level can be considered synchronous throughout the entire system. Even within those segments where the overall erosional/depositional dynamics are synchronous, local tectonic effects can strongly affect the preservation of old alluvium as a terrace.

There are several ways to improve our model. In general, more insight into fluvial dynamics during the Quaternary as a whole is needed. Most of the reconstructions are based on the last 40–50 ka (the dating range of ^{14}C). This period can hardly be considered as representative of the Quaternary given the considerable changes in topography, climate and tectonics since the dawn of the Quaternary. Finally, the process descriptions used can be further improved and refined.

CONCLUSIONS

FLUVER2 simulations demonstrate a basin-wide effect of uplift, climate and base level changes during the Quaternary on fluvial dynamics. The effects of these parameters change considerably along the longitudinal profile causing alternating shifts in erosional and depositional zones throughout the system in time. These effects indicate that even within one fluvial system a basin-wide analysis is required to correlate and match terraces and alluvial deposits along a longitudinal profile. Given these longitudinal profile dynamics it is likely that synchronous terrace formation cannot occur in a fluvial system. Simulations of valley development indicated that, as well as longitudinal profile dynamics, local tectonic conditions strongly affect the preservation of terraces.

REFERENCES

Anderson, M.G., 1988. *Modelling Geomorphological Systems*, Wiley, London.

Andres, W., 1989. The Central German upland. *Catena Supplement*, **15**, 25–44.

Autran, A. and Peterlongo, J.M., 1980. Le Massif Central. In *Géologie des Pays Européens, France, Belgique, Luxembourg*, Dunod, Paris, 3–133.

Berger, A.L., 1978. Long-term variations of caloric insolation resulting from the Earth's Orbital elements. *Quaternary Research*, **9**, 139–167.

Beven, K., 1995. Linking parameters across scales: subgrid parameterizations and scale dependent hydrological models. *Hydrological Processes*, **9**, 507–525.

Bradley, R.S., 1985. *Quaternary Paleoclimatology, Methods of Paleoclimatic Reconstruction*, Allen & Unwin, Boston.

Brunnacker, K. and Boenigk, W., 1983. The Rhine valley between the Neuwied basin and the Lower Rhenish Embayment. In K. Fuchs et al. (eds), *Plateau Uplift*, Springer Verlag, Berlin, 62–72.

Buch M.W., 1987. Spätpleistozäne und holozäne fluviale Geomorphodynamik im Donautal östlich von Regenburg – ein Sonderfall unter den mitteleuropäischen Flusssystemen?, *Zeitschrift für Geomorphologie Neue Folge, Supplement-Band*, **66**, 95–111.

Bull, W.B., 1991. *Geomorphic Responses to Climatic Change*, Oxford University Press, New York.

Dawson, M.R. and Gardiner, V., 1987. River terraces, the general model and palaeohydrological and sedimentological interpretation of the terraces of the Lower Severn. In K.J. Gregory, J. Lewin and J. B. Thornes (eds), *Palaeohydrology in Practice*, Wiley, London.

Gibbard, P.L., 1988. The history of the great northwest European rivers during the past three million years. *Philosophical Transactions of the Royal Society of London*, **B318**, 559–602.

Giot, D., Clozier, L. and Fleury, R., 1978. Manifestions tectoniques quaternaires en Limagne d'Allier. *Bulletin du BRGM*, section I, **2**, 150–155.

Guiot J., Pons, A., De Beaulieu, L.J. and Reille, M., 1989. A 140,000 years continental climate reconstruction from two European pollen records. *Nature*, **338**, 309–313.

Howard, A.D., Dietrich, W.E. and Seidl, M.A., 1994. Modeling fluvial erosion on regional and continental scales. *Journal of Geophysical Research*, **99**, 12244–12258.

Imbrie, J., Hays, J., Martinson, D., McIntyre, A., Mix, A., Morley, J., Pisias, N., Prell, W. and Shackleton, N.J., 1984. The orbital theory of Pleistocene climate; support from a revised chronology of the marine δ^{18}O record. In A. Berger, J. Imbrie, J. Hays, G. Kukla and B. Saltzman (eds), *Milankovich and Climate, Understanding the Response to Astronomical Forcing*, Reidel, Dordrecht, 269–305.

Kirkby, M.J., 1971. Hillslope process–response models based on the continuity equation. In D. Brunsden (ed.), *Slopes, Forms and Processes*, Institute of British Geographers, Special Publication 3, 15–30.

Larue, J.P., 1979. *Les nappes alluviales de la Loire et de ses affluents dans le Massif Central et dans le Sud de bassin Parisien: étude géomorphologique*, Thèse Etat Géographie, Clermont University.

Le Griel, A., 1983. Age et principales étapes du dépot des Sables et argiles du Bourbonnais. *Revue de Géologie Dynamique et de Géographie Physique*, **24**, 425–433.

Lilienberg, D.A., 1985. Recent geodynamics of the Alpine Orogenous Belts of southern Europe. *Geomorfologiya*, **4**, 16–29 (in Russian).

Lowe, J.J. and Walker, M.J.C., 1984. *Reconstructing Quaternary Environments*, Longman, New York.

Macaire, J.J., 1986. Apport de l'altération superficielle a la stratigraphie – example des formations alluviales et eoliennes Plio-Quaternaires de Touraine (France). *Bulletin de l'Association Française pour l'Etude du Quaternaire*, **3–4**, 233–245.

McGregor, D.F. and Green, C.P., 1978. Gravels of the River Thames as a guide to Pleistocene catchment changes. *Boreas*, **7**, 197–203.

Pastre, J.F., 1987. *Les formations Plio-Quaternaires du bassin de l'Allier et le volcanisme regional (Massif Central, France)*, Doctoral thesis, University of Paris.

Pirazzoli, P.A. 1991. Quaternary sea-level changes and their impacts: a review of recent advances (1987–1991). Abstract *INQUA XIII*, Beijing, 282.

Puvaneswaran, P. and Conacher, A.J., 1983. Extrapolation of short-term process data to long term landform development: a case study from southwestern Australia. *Catena*, **10**, 321–337.

Raynal, J.-P., 1984. Chronologie des basses terrasses de l'Allier en grande Limagne (Puy-de -Dôme, France). *Bulletin de l'Association Française pour l'Étude du Quaternaire*, **1–3**, 79–84.

Rosswall, T., Woodmansee, R.G. and Risser, P.G. (eds), 1988. *Scales and Global Change. Spatial and Temporal Variability in Biospheric and Geospheric Processes*, Wiley, Chichester.

Schumm, S.A., 1993. River response to base level change: implications for sequences stratigraphy. *Journal of Geology*, **101**, 279–294.

Schumm, S.A., Mosley, M.P. and Weaver, W.E., 1987. *Experimental Fluvial Geomorphology*, Wiley, New York.

Shackleton, N.J., Berger, A. and Peltier, W.R., 1990. An alternative astronomical calibration of the Lower Pleistocene time scale on ODP site 677. *Transactions of the Royal Society of Edinburgh, E, Science*, **81**, 251–261.

Starkel, L., 1983. The reflection of hydrologic changes in the fluvial environment of the temperate zone during the last 15,000 years. In K.J. Gregory (ed.), *Background to Palaeohydrology*, Wiley, London.

Van den Berg, M.W., 1996. *Fluvial sequences of the Maas. A 10 Ma record of neotectonics and climate change at various time-scales*, PhD Thesis, Wageningen Agricultural University.

Vandenberghe, J., Kasse, C., Bohncke, S. and Kozarski, S., 1994. Climate-related river activity at the Weichselian–Holocene transition: a comparative study of the Warta and Maas rivers. *Terra Nova*, **6**, 476–485.

Van Dijke, J.J. and Veldkamp, A., 1996. Long term dynamics of river systems draining volcanic mountain slopes in Costa Rica, assessed by computer simulations. In G. Benito, A. Pérez-González, M.J. Machado and S. de Alba (eds), *Palaeohydrology and Modelling of Environmental Change*, Toledo, 25 (abstract).

Veldkamp, A., 1991. *Quaternary river terrace formation in the Allier basin, France. A reconstruction based on sand bulk geochemistry and 3-D modelling*, Doctoral Thesis, Wageningen Agricultural University.

Veldkamp, A., 1992. A 3-D model of fluvial terrace development in the Allier basin (Limagne, France). *Earth Surface Processes and Landforms*, **17**, 487–500.

Veldkamp, A. and van den Berg, M.W., 1993. Three-dimensional modelling of Quaternary fluvial dynamics in a climo-tectonic dependent system. A case study of the Maas record (Maastricht, the Netherlands). *Global and Planetary Change*, **8**, 203–218.

Veldkamp, A. and van Dijke, J.J., 1994. Modelling of potential effects of long-term fluvial dynamics on possible geological storage facilities of nuclear waste in the Netherlands. *Geologie en Mijnbouw*, **72**, 237–249.

Veldkamp, A. and Vermeulen, S.E.J.W., 1989. River terrace formation, modelling, and 3-D graphical simulation. *Earth Surface Processes and Landforms*, **14**, 641–654.

Willgoose, G., Bras, R.L. and Rodriguez-Iturbe, I., 1991. Results from a new model of river basin evolution. *Earth Surface Processes and Landforms*, **16**, 237–254.

Zijerveld, L., Stephenson, R., Cloetingh, S., Duin, E. and van den Berg, M.W., 1992. Subsidence analysis and modelling of the Roer Valley Graben (S.E. Netherlands). *Tectonophysics*, **208**, 159–171.

CHAPTER 6

Arid Phases in the Western Mediterranean Region During the Last Glacial Cycle Reconstructed from Lacustrine Records

BLAS VALERO-GARCÉS

Instituto Pirenaico de Ecología CSIC, Zaragoza, Spain

ESSAID ZEROUAL AND KERRY KELTS

Limnological Research Center, University of Minnesota, Minneapolis, USA

INTRODUCTION

Despite abundant paleoclimatic evidence in the form of pollen diagrams, lake level records and geomorphological studies from Europe and northern Africa, the Late Quaternary, and especially Holocene, climate history of the western Mediterranean region is still relatively poorly understood (Harrison et al., 1992; Huntley and Prentice, 1993; Zeroual, 1995). Pollen-based paleoclimatic reconstructions are frequently hampered by the absence of modern analogues for many of the pollen assemblages (Huntley and Prentice, 1993) and the slow response of vegetation to abrupt climatic events (Seret et al., 1992). On the other hand, lake level studies are in a preliminary state (Guiot et al., 1993; Harrison et al., 1993; Yu and Harrison, 1995), and the reconstructions are based on the reinterpretation of a very limited number of published records and not on detailed sedimentological studies of selected sites.

Although the same general sequence applies to the deglaciation history in Mediterranean regions as in northern Europe, there are clear divergences about the timing and nature of the events in lower latitudes (Overpeck et al., 1989; Gasse et al., 1990; Zeroual, 1995). The western Mediterranean paleorecords have great potential for deciphering the contribution of North Atlantic and tropical processes during the deglaciation because of their geographic location at the southern limit of influence of the polar front, and in the tropical-sensitive Mediterranean climate zone. In this chapter a number of paleohydrological-sensitive lacustrine indicators (sediment, chemical, and stable isotope stratigraphies) from two sites are studied: the Spanish Pyrenees (Lake Banyoles) and the Moroccan High Atlas Mountains (Lake Isli). With these multiproxy-based

---

Palaeohydrology and Environmental Change. Edited by G. Benito, V. R. Baker and K. J. Gregory
© 1998 John Wiley & Sons Ltd.

paleolimnological reconstructions the relative intensity and nature of arid phases in the western Mediterranean region during the last glacial cycle are analysed.

METHODS

The Banyoles lacustrine sequence was dated by Pérez-Obiol and Julià (1994) using two calibrated ^{14}C dates of the overlying peat and nine U-series analyses of the lacustrine carbonates. An additional Accelerator Mass Spectrometer (AMS) ^{14}C date has been obtained from a pollen concentrate (19.30 m depth, 22 890 $\pm$ 310 ^{14}C yr BP; Table 6.1). A detailed pollen stratigraphy was reported by Pérez-Obiol and Julià (1994). The core was resampled for sedimentary facies analysis and isotopic studies on charophyte remains; there are no data for the upper part of the Late Glacial sequence. Zeroual (1995) provided a comprehensive and multidisciplinary study of Lake Isli environmental history based on five sediment cores and 13 ^{14}C dates on terrestrial macrofossil and gastropod shells (Table 6.1); in this chapter his analyses of sedimentary facies and stable isotopes are summarized.

Total organic and inorganic carbon was measured by colorimetric techniques. Oxygen and carbon isotopic composition on charophyte samples (Banyoles) and on ostracod shells and bulk sediments (Isli) are reported in the conventional delta notation relative to a PDB standard.

GEOGRAPHIC LOCATION

Site description

Lake Banyoles (42°08′N, 2°45′E, 175 m altitude) is located in northeastern Spain some 25 km from the Mediterranean Sea (Figure 6.1). Climate can be characterized as a

Table 6.1   Radiocarbon dates from the Banyoles and Isli cores

Depth	Sample	[14]C age (yr BP)	Lab. number
Lake Banyoles			
0.77–0.84 m*	bulk organic matter	5794 $\pm$ 60	Lv- 1920
1.17–1.24 m*	bulk organic matter	6417 $\pm$ 90	Lv- 1921
19.30 m	pollen concentrate	22 890 $\pm$ 310	AA- 21 054
Lake Isli			
33 cm (IS 5)	bulk organic matter	295 $\pm$ 60	ETH- 7888
493 cm (IS 5)	bulk organic matter	1705 $\pm$ 55	ETH- 7890
667.9 cm (IS 5)	bulk organic matter	2860 $\pm$ 60	ETH- 7889
789.4 cm (IS 5)	bulk organic matter	3620 $\pm$ 70	ETH- 7891
222.6 cm (IS 6)	bulk organic matter	3050 $\pm$ 75	ETH- 13 018
710.7 cm (IS 6)	gastropod shell	34 850 $\pm$ 410	ETH- 7892
558.7 cm (IS 6)	bulk organic matter	20 000	ETH- 11 990
587.8 cm (IS 6)	bulk organic matter	28 000	ETH- 11 992
12 cm (TA)	gastropod shell	10 460 $\pm$ 75	ETH- 1194
92 cm (TA)	gastropod shell	7225 $\pm$ 65	ETH- 1195
16 cm (TB)	gastropod shell	10 400 $\pm$ 95	ETH- 13 021
80 cm (TB)	gastropod shell	9415 $\pm$ 75	ETH- 11 997
170 cm (TB)	gastropod shell	7890 $\pm$ 70	ETH- 11 998

* From Perez-Obiol and Julià (1994)

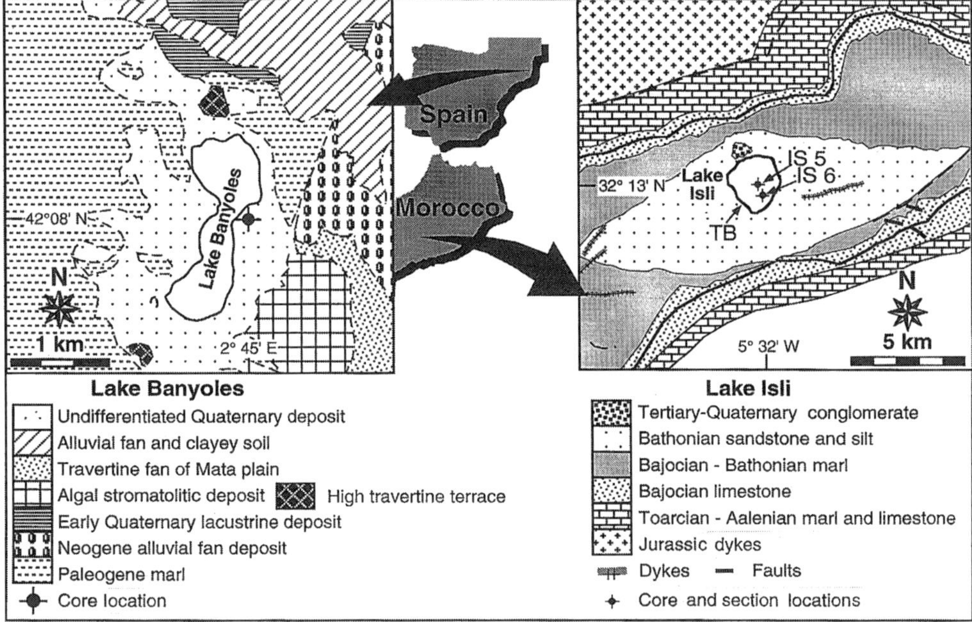

Figure 6.1 Geographical and geological location of the two studied sites: Lake Banyoles and Lake Isli

humidMediterranean type, with an annual rainfall of 750 mm with minimal precipitation during the summer and December. Vegetation is dominated by evergreen oak forest. Lake Banyoles is a hydrologically open lake, connected to a large karst system developed in Paleogene limestone and gypsum rocks.

Lake Isli (32°13′N, 5°32′W, 2270 m altitude), in the High Atlas Mountains, Morocco (Figure 6.1), is currently a mesosaline, cold monomictic lake, with a maximum depth of 95 m. Climate is semi-arid with an average precipitation of 244 mm. The rains are more frequent during winter and spring, with minimal values during the summer. Vegetation is very scarce. The lake basin originated during the Quaternary as a result of strike-slip tectonic activity controlled by an ENE–WSW compression. The lake is topographically closed, without surface inlet or outlet. However, there are evidences of two paleo-outlets at 45 and 40 m above current lake level.

THE BANYOLES SEQUENCE

Pollen

The pollen stratigraphy (Figure 6.2) shows a clear interstadial event between 30 000 and 27 000 yr BP, a Full Glacial period, with minor oscillations, that ends abruptly around 14 420 yr BP, and a classic Late Glacial sequence with the re-expansion of open vegetation, corresponding to the Younger Dryas event, at about 12 000 yr BP (U-series age). Subsequently, a strong increase in arboreal taxa (a short warming phase) is followed by a decrease in *Pinus* which is interpreted as a cold event around 11 000 yr BP (Pérez-Obiol and Julià, 1994). Arboreal taxa spread progressively during the Holocene in response to ameliorating climate and tree cover maximum was reached around 6000 yr BP.

Sedimentology

The Banyoles sequence was composed of carbonate-rich littoral calcarenites and muds capped by a peat layer (Figure 6.2). Carbonate deposits were mostly composed of low-magnesium calcite (up to 98%), with minor amounts of clays (illite and chlorite), quartz and feldspars. Biogenic components were dominant: charophytes and encrusting aquatic vegetation, gastropods, and ostracods. Organic carbon content was low (commonly < 1%) and inorganic carbon values were about 30%. Most of the core was composed of alternating coarse (gravel or sand size) calcarenites and fine (silt or mud) biogenic facies that record changes in lake level, productivity and other parameters in the littoral zone. Seven facies were identified based on lamination, color, grain size and other sedimentological properties (Figure 6.2). Coarser facies are interpreted as clastic deposition in more littoral (shallower and more influenced by wave action) areas; fine-grained facies are interpreted as deposited in relatively deeper areas. The Full Glacial was

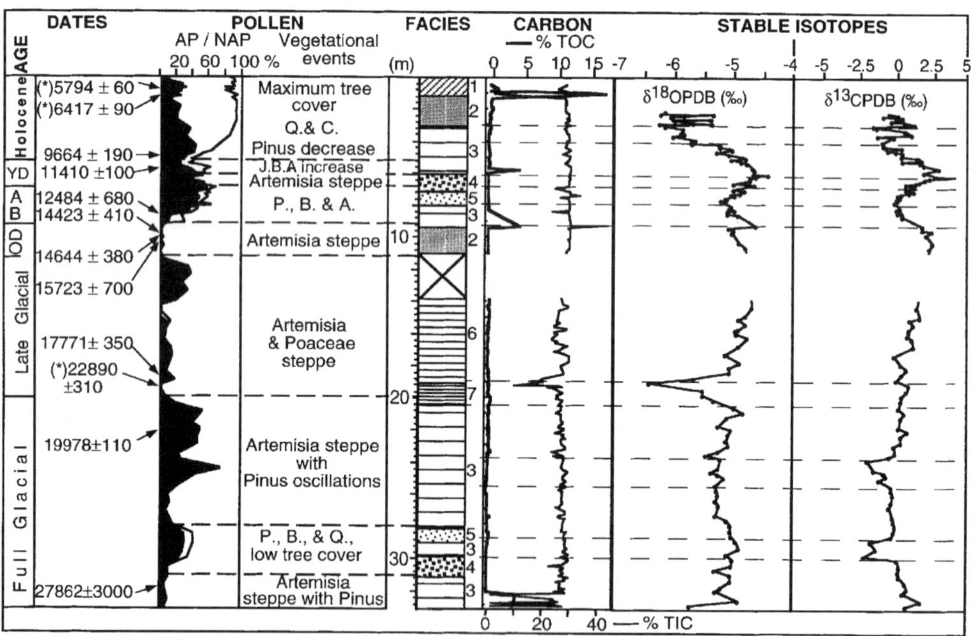

Figure 6.2   Pollen, sediment, and stable isotope stratigraphies of the Banyoles core. The upper part of the Late Glacial sequence was not sampled for sedimentological or isotope analyses. Chronology is constrained by three ^{14}C AMS dates (*) and nine U/Th dates. Pollen curves show the *Pinus* percentage (in black) and the arboreal/non-arboreal pollen ratio (AP/NAP). Vegetational events show only main changes: Q, *Quercus*; C, *Corylus*; J, *Juniperus*; B, *Betula*; A, *Acer*; P, *Pinus*. Sedimentological facies as follows: 1, peat; 2, fine-grained, massive muds; 3, alternating of muds and calcarenites; 4, coarse (gravel-size) calcarenites; 5, fine (sand-size) calcarenites; 6, banded muds and silts (with some coarse layers); 7, fine-grained, laminated, black muds

characterized by deposition of banded muds, silts and calcarenites arranged in fining-upward sequences (Facies 3). A distinctive period of coarser deposition (gravels and sands; Facies 4 and 5) between 28.17 and 30.70 m is correlated with a pollen-defined interstadial. Other *Pinus* oscillations, however, do not show such clear correlations with changes in lacustrine deposition.

An abrupt change in sedimentary facies occurred during the transition from Full Glacial to Late Glacial. Fine lamination, darker colors, and reduced total inorganic carbon values favor an interpretation as deposition in deeper water during this transition. The fine-grained, laminated, black muds (Facies 7) deposited around 22 890 ^{14}C yr BP are interpreted as the deepest facies at the coring site. The dominance of finer facies (muds and silts, Facies 6) and the more restricted occurrence of coarse calcarenitic facies suggest deposition in generally deeper water during Late Glacial than during Full Glacial times. A conspicuous 20 cm thick layer composed of angular carbonate clasts, up to 2 cm long (Facies 4), is correlated with the Younger Dryas pollen chronozone. The clastic and reworked nature of this layer suggests that the Younger Dryas period in Banyoles was characterized by low lake levels.

Banded muds and silts (Facies 3) deposited during the early Holocene suggest a lake level recovery. The occurrence of massive muds (Facies 2) with abundant gastropods during the middle Holocene is interpreted as shallower deposition. After 6417 $\pm$ 90 ^{14}C yr BP, water table was low enough to allow development of a peat bog (Facies 1).

Stable isotopes

The $\delta^{18}O$ record

Isotope techniques are proving most successful for deciphering paleoenvironmental information and contributing to paleoclimatic reconstructions (Talbot, 1990; Talbot and Kelts, 1990; Swart et al., 1993; Valero-Garcés et al., 1995). The main parameters controlling the $\delta^{18}O$ compositions of precipitating carbonates in lakes are the evaporation–precipitation ratio, air and water temperature, and changes in the source of the waters (Talbot, 1990; Chivas et al., 1993; Valero-Garcés et al., 1995).

The small range (about 2‰) and relatively low $\delta^{18}O$ values in the Banyoles charophyte samples were expected in a hydrologically open system dominated by regional groundwater input with a relatively short residence time. Relatively high $\delta^{18}O$ values during the Full Glacial suggest relatively low effective moisture (precipitation minus evaporation) during that period. In addition, ocean waters were isotopically heavier due to the large ice volume during this time, and meteoric water would have carried the same signal. In the absence of major changes in water sources, decreasing $\delta^{18}O$ trends are interpreted as an increase in the water balance (more groundwater input or increased effective moisture), a colder mean air temperature, a warmer lake water temperature, or a combination of any of these; increasing $\delta^{18}O$ trends indicate a more negative water budget in the lake, decreasing lakewater temperature or increasing air temperature. Increasing trends occurred during the 30–27 ka interstadial, at the end of the Full Glacial, at the beginning of the Late Glacial, and around 12 ka, reaching the highest values of the whole period at the top of the Younger Dryas sequence. Decreasing trends are identified at the end of the 30–27 ka interstadial, at the Full Glacial–Late Glacial transition, during the Older Dryas, and the Bølling-Allerød. The Early Holocene sequence showed decreasing $\delta^{18}O$ values until about 8 ka, and reflects the climate amelioration and higher variability prevalent until the end of the carbonate lacustrine deposition at about 6.5 ka.

The $\delta^{13}C$ record

According to the $\delta^{13}C$ curve, six different isotopic units were differentiated in the Banyoles record, with vegetational changes in the watershed potentially explaining some of the observed $\delta^{13}C$ variability. A higher proportion of arboreal cover (C3 plants) versus steppe grasses (C4 plants) could contribute to isotopically lighter dissolved inorganic carbon in groundwaters, and consequently cause some of the decreasing trends (Hakänsson, 1985; Valero-Garcés et al., 1995). The negative event at 29–27 ka correlates with increasing arboreal pollen, as do the most negative values between 25.5 and 24 ka. Increasing $\delta^{13}C$ values after 24 ka could reflect either the dominance of steppe plants in the watershed or a relative increase in littoral organic productivity (Mckenzie, 1985). The transition between Full Glacial and Late Glacial is not clearly marked in the $\delta^{13}C$ record indicating that the main change in the lacustrine carbon budget occurs after the largest *Pinus* oscillation. Decreasing $\delta^{13}C$ trends correlate with periods of rapid increase in the arboreal cover in the two Full Glacial interstadials, the Bølling-Allerød, and the early Holocene. Increasing $\delta^{13}C$ trends and the heaviest $\delta^{13}C$ values correspond to periods of steppe development during the Late Glacial, especially during the Older and Younger Dryas.

LAKE ISLI SEQUENCE

Pollen

Pollen analyses on core IS6 (Lamb et al., 1994) and correlation with other sites in the Atlas mountains (Lamb et al., 1989; Zeroual, 1995) indicate a steppe vegetation dominated by *Artemisia*, Gramineae and Chenopodiaceae in the region during the Full Glacial (45–28 ka). Between 20 000 and 8500 yr BP the steppe vegetation was dominated by Gramineae instead of *Artemisia*. A period of climatic amelioration is identified between 14 000 and 11 800 yr BP. Pollen assemblages dominated by *Pinus, Quercus* and *Juniperus* from 4000 yr BP reflect the development of a Mediterranean forest north of the Lake Isli area. The colonization by the Mediterranean forest could have occurred earlier, but the presence of a sedimentary hiatus between 8500 and 4000 yr BP prevents a more accurate dating.

Sedimentology

A synthetic sedimentary profile including information from three cores is shown in Figure 6.3. The basal coarse clastic and encrusted charophyte-rich sediments were interpreted as littoral deposition during the first stages of lake evolution at about 30–35 ka. The overlying reddish and greyish muds with carbonate layers and encrusted calcite indicate that littoral deposition lasted until 28 ka. A fall in lake level and consequent erosion of the deposited sediment causes a long hiatus between 28 000 and 20 000 yr BP. The fine-grained nature of the sediments, the absence of carbonate concretions and the higher content of detrital minerals (clay minerals, quartz, plagioclases) suggest deeper deposition and, consequently, higher lake levels between 20 000 and about 10 400 yr BP (Unit C). At

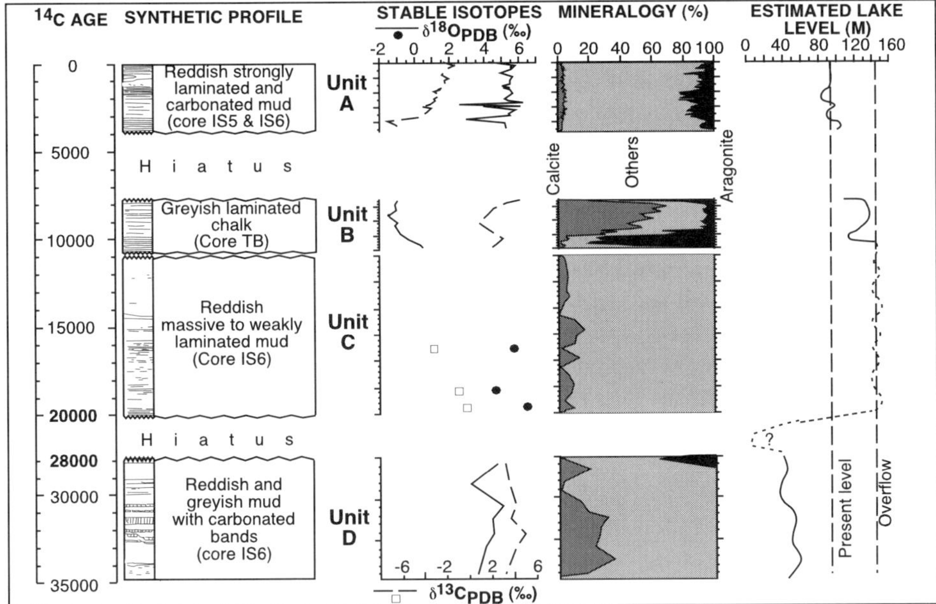

Figure 6.3   Sediment, mineralogy and stable isotope stratigraphies of the Isli core. Chronology is constrained by 15 ^{14}C AMS dates

this time Lake Isli had a surface outlet, and a maximum depth of 160 m. A rapid drop in lake level occurs between 10 400 and 9900 yr BP, causing lake water chemical concentration and, consequently, dominance of authigenic aragonite deposition. The increase in calcite content between 9900 and 9700 yr BP indicates a progressive refreshening of the lake waters as lake levels increase. Lake level remains high between 9700 and 7900 yr BP as indicated by the dominance of calcite in the greyish laminated sediments. The interval between 7900 and 7200 yr BP is characterized by aragonite deposition which indicates chemical concentration of lake water and another low lake level episode. None of the cores recorded sedimentation between 7225 and 4000 yr BP. Reddish laminated carbonate muds with very low calcite content, likely of detrital origin, were deposited during the last 4000 years (Unit A). Two intervals of higher aragonite deposition marked periods of increased water concentration at 3500–2500 and 2100–1500 yr BP.

Stable isotope

The $\delta^{18}O$ record

Stable isotope analyses were performed on ostracod shells of *Eucypris crassoïdes* (cores IS6 and IS5) and bulk sediments (core TB). During deposition of the basal Unit D (34 800–28 000 yr BP), $\delta^{18}O$ values increased from 3.5 to about 5‰. This trend suggests isotopic concentration of the lake water; however, mineralogy and diatoms (Lamb et al., 1994) indicated that water remained fresh without reaching chemical saturation for aragonite precipitation. Low lake water temperatures, low effective moisture, and isotopically heavier rainfall during Full Glacial could have contributed to high $\delta^{18}O$ values during deposition of this unit. The few samples from Unit C show high oxygen isotope values suggesting that arid and cold conditions remained until 16 ka. However, sedimentological evidence indicates that lake levels were the highest at this period and the lake had a surface outlet. Some possible explanations of this apparent contradiction between high lake levels and high isotope values could involve limnological changes (the large increase in lake surface would have allowed evaporative $\delta^{18}O$ enrichment of the lake waters), or changes in moisture sources and/or seasonality. There are no isotopic data for the rest of the Late Glacial. The dominance of aragonite favors the hypothesis that lower effective moisture rather than colder lake water temperatures was responsible for the relatively high $\delta^{18}O$ values during the Younger Dryas, between 7900–7200 and 4000–2500 yr BP. The transition to the Holocene witnesses an increase in effective moisture as indicated by decreasing $\delta^{18}O$ values and the dominance of calcite. Low $\delta^{18}O$ variability suggested that the lake had reached an isotopic equilibrium since 2500 yr BP.

The $\delta^{13}C$ record

The $\delta^{13}C$ values from samples of the basal Unit D show little variability and the highest values of the whole section (average 3.7‰). High biological productivity indicated by the abundance of encrusted carbonates, relatively longer residence times and higher evaporation conditions would have favored water depletion in $^{12}CO_2$ (Mckenzie, 1985; Hakänsson, 1985; Valero-Garcés et al., 1995). Lower $\delta^{13}C$ values (about 3‰) during deposition of Unit C could reflect lower organic productivity and increased groundwater input. Low organic carbon content, low ostracod diversity and abundance, and sedimentological evidence for a high lake level favor an oligotrophic, hydrologically open, freshwater lake as a depositional environment for Unit C. Carbon isotopic compositions were higher again during deposition of Unit B as a result of high organic productivity in the littoral zone. Decreasing values during the transition to the Holocene follow similar trends in $\delta^{18}O$ values suggesting increased groundwater influx to the lake. The large positive shift in $\delta^{13}C$ values during the early Holocene would be a reflection of increasing organic productivity in a lake with progressively less groundwater contribution. The $\delta^{13}C$ values during the last 4000 years were lower compared with both ostracod (Units D and C) and bulk sediment (core TB) samples. Lower values between 4000 and 3500 yr BP indicate higher groundwater influx, corresponding to more dilute waters. The abrupt positive shift after 3500 yr BP was a reflection of increased residence time in the lake and likely increased productivity during a more concentrated water episode. Values remain relatively high during the last 2000 years.

DISCUSSION

Our study of two lacustrine time series in the western Mediterranean provides a record of limnological changes that can be related to variations in precipitation, temperature, seasonality and atmospheric circulation patterns. The main abrupt arid/humid transitions in Banyoles occur during the 27–30 and 24–26 ka interstadials, the Full Glacial–Late Glacial transition, the Older Dryas and Younger Dryas, and at about 8 ka and 6 ka. In Lake Isli, humid phases occurred during 34.8–28 ka, 20–11 ka, and the mid-Holocene; a long arid period occurred during 28–20 ka, and shorter arid phases during the Younger Dryas, about 10 ka, 8–7 ka, 3–4 ka, and about 2 ka. Aridity in the western

Mediterranean region is related to the intensity and location of the winter rain belt, which is strongly sensitive to atmospheric circulation patterns, especially the subtropical anticyclone belt (the Azores High) (Flohn and Fantechi, 1984; Kutzbach et al., 1993; Huntley and Prentice, 1993).

Vegetation changes during the 30–27 ka and 26–24 ka periods in Banyoles were large enough to cause a negative excursion of $\delta^{13}C$ associated with increased C3 plants (trees) in the landscape. However, the slightly positive $\delta^{18}O$ excursion does not support a significant increase in temperature or effective moisture during the 30–27 ka interstadial. This period corresponds with deposition of the Heinrich layer 3 (27 ka) in the central North Atlantic (Bond et al., 1992). The Heinrich Event 3 rarely influenced sedimentation on the Portuguese margin (Baas et al., 1997) and, consequently, a large meltwater signal is not expected. During this period, relatively lower lake levels deduced by the occurrence of more littoral deposits suggest a decrease in effective moisture and favor a scenario of warmer temperatures, more conducive to tree growth but also increasing evaporative conditions. A change in moisture source, with a higher contribution of summer precipitation, would also explain heavier lake water isotope composition. The 27–30 ka period in Lake Isli corresponded to increasingly arid conditions.

The long hiatus between 28 and 20 ka reflects high aridity during the Full Glacial in the Atlas Mountains. In Banyoles, the pollen (*Pinus* and *Artemisia*) and the isotope curves show complex oscillations. Higher *Pinus* content in Banyoles during the late Full Glacial (23–19 ka, U/Th age) also correlates with increasing $\delta^{18}O$ values, although there was no associated negative $\delta^{13}C$ excursion. The $\delta^{18}O$ isotope curve for Banyoles suggests increased effective moisture (lower oxygen isotopic compositions) or changes in moisture composition during Full Glacial and Late Glacial periods of decreasing arboreal cover: 27–26 ka, Late Glacial onset (19–18 ka, U/Th age; around 22 ka ^{14}C age) and Older Dryas (15–14 ka). The largest $\delta^{18}O$ negative excursion at the Late Glacial onset correlates with an abrupt decrease in arboreal pollen and the deepest sedimentary facies in the Banyoles core. This large negative shift would reflect the decrease in $\delta^{18}O$ oceanic water composition as a result of the beginning of the northern ice sheets melting process (Heinrich Event 2, 21–22 ka).

Both Lake Isli and Lake Banyoles records support the interpretation of changes in precipitation seasonality as an explanation for the apparent contradiction in Late Glacial reconstructions for the Mediterranean between aridity (treeless vegetation) and increased effective moisture (high lake levels) (Huntley and Prentice, 1993; Roberts and Wright, 1993; Harrison et al., 1993). Summer conditions of slightly lower temperatures, some precipitation and increased cloudiness would have been critical for lake water balance, and extremely cold winters would have prevented tree growth (Roberts and Wright, 1993). On the other hand, a shift of the summer precipitation toward winter would result in a water stress increase for the vegetation and a runoff increase for the lakes (Prentice et al., 1992).

Vegetational change during the Older Dryas was more significant in northern (Lake Banyoles (Pérez-Obiol and Juliá, 1994) and Pyrenean sites (Jalut et al., 1992)) than in southern Spain (Padul site, Granada (Pons and Reille, 1988)). The absence of significant vegetational or sedimentological changes in Lake Isli associated with this arid phase, suggests a decreasing N–S gradient. A generally cooler and drier episode about 11 ka, ascribed to the Younger Dryas, occurs in many Iberian pollen records, but the signal is characteristically weak in comparison with the Atlantic and northern European sites

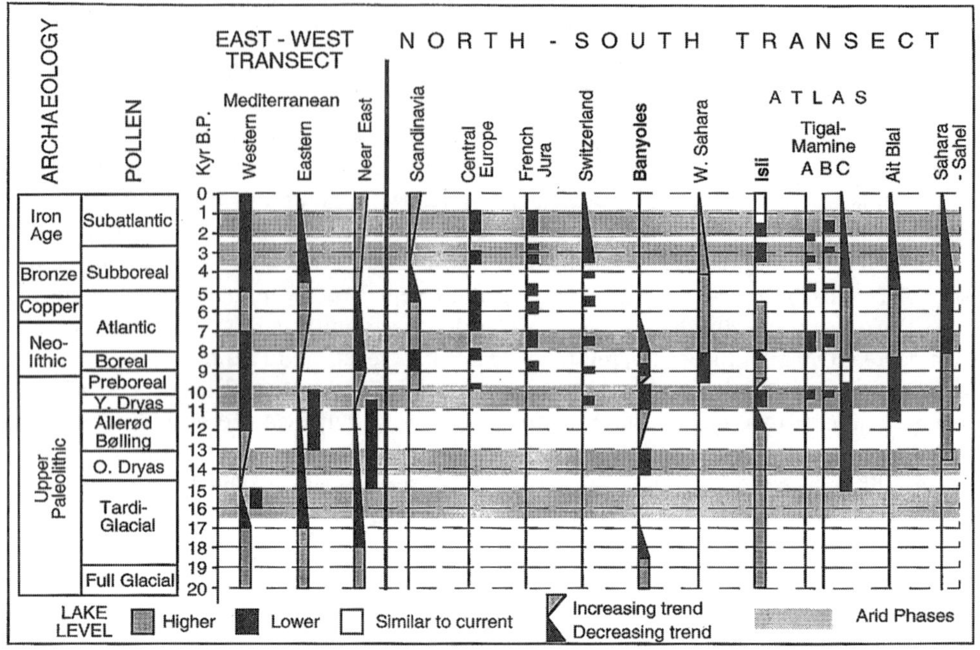

Figure 6.4 A compilation of lake level changes in western Europe, Mediterranean regions and northern Africa during the last 20 ka along north–south and east–west transects

(Peñalba et al., 1997). Geochemical studies on ostracod shells in the Banyoles core (Wansard, 1996) indicate that the mean summer temperature during the Younger Dryas was similar to that of the Last Glacial Maximum (about 8 °C cooler than today). The intensity of the Younger Dryas signal seems to decrease southward, and its timing and occurrence are less defined on the African continent. Lower lake levels and large positive $\delta^{18}O$ excursions single out the Younger Dryas as an arid period of low effective moisture in both regions. The Isli record confirms the presence of a low effective moisture period in northern Africa associated with the Younger Dryas (Gasse et al., 1990; Zeroual, 1995).

Drier conditions in the Mediterranean western Africa, the Atlas Mountains, southern Spain (Padul) and eastern Spain (Banyoles) during the early Holocene sharply contrast with the moister Sahara and Sahel (Zeroual, 1995). A major change at about 4 ka towards drier and cooler conditions is evident in all North African sites (Lamb et al., 1989; Zeroual, 1995), in southern Europe (Huntley and Birks, 1983), and in southern Spain (Padul site; Pons and Reille, 1988), and has been interpreted as the onset of the modern Mediterranean dry summer regime (Street-Perrott and Perrott, 1993). A N–S transect of lake records (Figure 6.4) including the monsoon-dominated regions (Sahara) and the westerlies-dominated regions (the Atlas mountains and the Iberian Peninsula) suggests that during the Full Glacial and until the Younger Dryas, arid and humid intervals in the western Mediterranean regions were out of phase with those of the Sahara (Zeroual, 1995). On the other hand, the Holocene is characterized by the appearance of similar climatic variations in all three regions (Zeroual, 1995). The different geographic distribution of Holocene aridity patterns compared with the Older and Younger Dryas and the

Late Glacial arid period (stronger in the northern records) suggests a different causal mechanism for Holocene aridity in the western Mediterranean. An intensification of the subtropical anticyclone (the Azores High), and correspondent northern shift of the Mediterranean winter belt could be the causal mechanism for Holocene droughts (Huntley and Prentice, 1993). Roberts et al. (1994) suggest that arid episodes in northern Africa have occurred approximately every 2–3 ka during the Holocene. Such abrupt and rapid changes imply a periodicity that cannot be explained by orbital forcing. Broecker et al. (1990) postulated that changes in North Atlantic Deep Water formation could be responsible for these short-lived arid periods.

The Banyoles core provides further evidence of the impact of iceberg discharge on the western Mediterranean climate. Large environmental changes in Banyoles correlate with the three most recent Heinrich layers deposited during the last 30 ka on the Portuguese continental margin: H1 (13.6–15.9 ka), H2 (21.0–22.0 ka), and H3 (27 ka) (Bond et al., 1992; Baas et al., 1997). The two most recent ones correspond with decreased arboreal percentages and δ^{18}O negative shifts (Older Dryas and Full Glacial–Late Glacial transition) reflecting large meltwater influx into the oceans. Paleoclimatic reconstruction for the arid Older Dryas (H1) and Younger Dryas (H0) periods shows the Atlantic oceanic polar front displaced southwards to the latitude of Spain and the main position of the Intertropical Convergence Zone (ITCZ) moved south of the geographical equator (Gasse et al., 1990). These dry periods have been explained by the displacement of the Mediterranean winter rain belt south over northern Africa, promoting more frequent incursions of cold, northerly air. A colder North Atlantic during periods of more intense ice-calving may have stabilized the lower atmosphere and helped block the moisture over Iberia and North Africa (Overpeck, 1989). However, the Heinrich Event 2 correlates with the deeper facies in Banyoles, which indicates an abrupt increase in effective moisture in the western Mediterranean associated with the main ice-calving phase during the Full Glacial–Late Glacial transition. The Heinrich Event 3 (27 ka) had a weak influence on the Portuguese slope (Baas et al., 1997) and that could explain the absence of a meltwater isotopic signal in the Banyoles site during the 27–30 ka interstadial. Lower lake levels, as deduced by the presence of shallower sedimentary facies, and an increase in the arboreal cover during this interstadial could be attained by changes in seasonal distribution of precipitation.

CONCLUDING REMARKS

The hydrological balance of lakes is sensitive to effective moisture fluctuations (precipitation minus evaporation), and moisture source changes, and therefore it reflects past environmental and climate changes. The main arid/humid transitions in Banyoles occurred during the 27–30 and 24–26 ka interstadials, the Full Glacial–Late Glacial transition, the Older Dryas and Younger Dryas, and at about 8 ka and 6 ka. In Lake Isli a long arid period occurred during 28–20 ka, and shorter arid phases during the Younger Dryas, about 10 ka, 8–7 ka, 3–4 ka, and about 2 ka. The Banyoles and Isli records showed clear teleconnections between the main climate events in the Iberian Peninsula and northern Africa and differences in the Deglaciation and Holocene dynamics. The location and intensity of the winter rain belt are the main factors controlling precipitation in the Mediterranean regions. Both depend upon the atmospheric circulation features, mainly the subtropical Azores High. Large environmental changes in Banyoles correlate with the most recent Heinrich events recorded on the Portuguese margin (H0, 10.5 ka; H1,

13.6–15.9 ka; H2, 21.0–22.0 ka; H3, 27 ka) and give further evidence of the impact of iceberg discharge on the western Mediterranean climate during Deglaciation. The occurrence of a number of abrupt and rapid arid/humid transitions during the Holocene clearly indicates that other mechanisms besides orbital forcing have controlled the location and intensity of the Azores High.

ACKNOWLEDGEMENTS

The research in Lake Banyoles was supported by the USA National Science Foundation (EAR-941 8657) and in Lake Isli by the Swiss National Science Foundation. Ramón Julià (Institut Jaume Almera-CSIC, Barcelona) graciously provided the Banyoles core, and Santi Giralt and Loren Hoppe helped with the sampling. We are grateful to Professor B. Kubler (Neuchatel University, Switzerland). We also thank an anonymous reviewer for constructive comments.

REFERENCES

Baas, J.H., Mienert, J., Abrantes, F. and Prins, M.A., 1997. Late Quaternary sedimentation on the Portuguese continental margin: climate-related processes and products. *Palaeogeography, Palaeoclimatology, Palaeoecology*, **130**, 1–23.

Bond, G., Heinrich, H., Broecker, W.S., Labeyre, L.D., McManus, J., Andrews, J., Huon, S., Jantschik, R., Clasen, S., Simet, C., Tedesco, K., Klas, M., Bonani, G. and Ivy, S., 1992. Evidence from massive discharge of iceberg into the North Atlantic ocean during the last glacial period. *Nature*, **360**, 245–249.

Broecker, W.S., Bond, G. and Klas, M., 1990. A salt oscillator in the glacial Atlantic? 1. The concept. *Paleoceanography*, **5**, 469–477.

Chivas, A.R., De Deckker, P., Cali, J., Chapman, A., Kiss, E. and Shelley, J., 1993. Coupled stable isotope and trace element measurements of lacustrine carbonates as paleoclimatic indicators. In P. Swart, K. Lohmann, J. Mckenzie and S. Savin (eds), *Climate Change in Continental Isotopic Records*, American Geophysical Union, Geophysical Monograph 78, 113–122.

Flohn, H. and Fantechi, R., 1984. *The Climate of Europe: Past, Present and Future*, Reidel, Dordrecht.

Gasse, F., Téhet, R., Durand, A., Gibert, E. and Fontes, J.C., 1990. The arid–humid transition in the Sahara and the Sahel during the last deglaciation. *Nature*, **346**, 141–146.

Guiot, J., Harrison, S. and Prentice, C.I., 1993. Reconstruction of Holocene precipitation patterns in Europe using pollen and lake-level data. *Quaternary Research*, **40**, 139–149.

Hakänsson, S., 1985. A review of various factors influencing the stable carbon isotope ratio of organic lake sediments by the change from glacial to post-glacial environmental conditions. *Quaternary Science Reviews*, **4**, 135–146.

Harrison, S., Prentice, C. and Bartlein, P., 1992. Influence of insolation and glaciation on atmospheric circulation in the North Atlantic sector: implications of general circulation model experiments for the Late Quaternary Climatology of Europe. *Quaternary Science Reviews*, **11**, 283–299.

Harrison, S.P., Prentice, I.C. and Guiot, J., 1993. Climatic controls on Holocene lake-level changes in Europe. *Climate Dynamics*, **8**, 189–200.

Huntley, B. and Birks, H.J.B., 1983. *An Atlas of Past and Present Pollen Maps for Europe: 0–13,000 years ago*, Cambridge University Press, Cambridge.

Huntley, B. and Prentice, I.C., 1993. Vegetation and climates of Europe. In H.E. Wright Jr, J.E. Kutzbach, T. Webb III, W.F. Ruddiman, F.A. Street-Perrot and P.J. Bartlein (eds), *Global Climates since the Last Glacial Maximum*, University of Minnesota Press, Minneapolis, 136–167.

Jalut, G., Monserrat Marti, J., Fontugne, M., Delibrias, G., Vilaplana, J.M. and Juliá, R., 1992.

Glacial to interglacial vegetation changes in the northern and southern Pyrenées: deglaciation, vegetation cover and chronology. *Quaternary Science Reviews*, **11**, 449–480.

Kutzbach, J.E., Guetter, P.J., Behling, P.J. and Selin, R., 1993. Simulated climatic changes: results of the COHMAP climate-model experiments. In H.E. Wright Jr, J.E. Kutzbach, T. Webb III, W.F. Ruddiman, F.A. Street-Perrot and P.J. Bartlein (eds), *Global Climates since the Last Glacial Maximum*, University of Minnesota Press, Minneapolis, 24–93.

Lamb, H.F., Eicher, U. and Switsur, V.R., 1989. An 18,000-year record of vegetational, lake level and climatic change from the Middle Atlas, Morocco. *Journal of Biogeography*, **16**, 65–74.

Lamb, H.F., Duigan, C.A., Gee, J.H.R., Kelts, K., Lister, G., Maxted, R.W., Merzouk, A., Niessen, F., Tahri, M., Whittington, R.J. and Zeroual, E., 1994. Lacustrine sedimentation in a high altitude, semi-arid environment: the paleolimnological record of Lake Isli, High Atlas, Morocco. In A.C. Millington and K. Pye (eds), *Environmental Change in Drylands: Biogeographical and Geomorphological Perspectives*, Wiley, Chichester, 147–161.

Mckenzie, J.A., 1985. Carbon isotopes and productivity in the lacustrine and marine environment. In W. Stumm (ed.), *Chemical Processes in Lakes*, Wiley, New York, 99–118.

Overpeck, J.T., 1989. Century to millennium-scale climatic variability during the Late Quaternary. In R.S. Bradley (ed.), *Global Change of the Past*, Vol. 2, UCAR, Office for Interdisciplinary Earth Studies, Global Change Institute, 139–173.

Overpeck, J.T., Peterson, L.C., Kipp, N., Imbrie, J. and Rind, D., 1989. Climate change in the circum-North Atlantic region during the last deglaciation. *Nature*, **338**, 553–557.

Peñalba, M.C., Arnold, M., Guiot, N., Duplessy, J.C. and de Beaulieu, J.L., 1997. Termination of the Last Glaciation in the Iberian Peninsula inferred from the pollen sequence of Quintanar de la Sierra. *Quaternary Research*, **48**, 205–214.

Pérez-Obiol, R. and Julià, R., 1994. Climatic change on the Iberian Peninsula recorded in a 30,000-yr pollen record from Lake Banyoles. *Quaternary Research*, **41**, 91–98.

Pons, A. and Reille, M., 1988. The Holocene and Upper Pleistocene pollen record from Padul (Granada, Spain): a new study. *Palaeogeography, Palaeoclimatology, Paleoecology*, **66**(13), 243–263.

Prentice, I.C., Guiot, J. and Harrison, S.P., 1992. Mediterranean vegetation, lake levels and palaeoclimate at the Last Glacial Maximum. *Nature*, **360**, 658–660.

Roberts, N. and Wright, H.E. Jr, 1993. Vegetation, lake-level and climatic history of the Near East and Southwest Asia. In H.E. Wright Jr, J.E. Kutzbach, T. Webb III, W.F. Ruddiman, F.A. Street-Perrott and P.J. Bartlein (eds), *Global Climates since the Last Glacial Maximum*, University of Minnesota Press, Minneapolis, 194–220.

Roberts, N., Lamb, H.F., El Hamouti, N. and Barker, P., 1994. Abrupt Holocene hydro-climatic events: palaeolimnological evidence from North-West Africa. In A.C. Millington and K. Pye (eds), *Environmental Change in Drylands: Biogeographical and Geomorphological Perspectives*, Wiley, Chichester, 163–175.

Seret, G., Guiot, J., Wansard, G., de Beaulieu, J.L. and Reille, M., 1992. Tentative paleoclimatic reconstruction linking pollen and sedimentology in La Grande Pile (Vosges, France). *Quaternary Science Reviews*, **11**, 425–430.

Street-Perrott, F.A. and Perrott, R.A. 1993. Holocene vegetation, lake levels, and climate of Africa. In H.E. Wright Jr, J.E. Kutzbach, T. Webb III, W.F. Ruddiman, F.A. Street-Perrot and P.J. Bartlein (eds), *Global Climates since the Last Glacial Maximum*, University of Minnesota Press, Minneapolis, 318–356.

Swart, P., Lohmann, K., Mckenzie, J. and Savin, S. (eds) 1993. *Climate Change in Continental Isotopic Records*, American Geophysical Union, Geophysical Monograph 78.

Talbot, M.R., 1990. A review of the palaeohydrological interpretation of carbon and oxygen isotopic ratios in primary lacustrine carbonates. *Chemical Geology (Isotope Geoscience Section)*, **80**, 261–279.

Talbot, M.R. and Kelts, K., 1990. Paleolimnological signatures from carbon and oxygen isotopic ratios in carbonates from organic carbon-rich lacustrine sediments. In B.J. Kyrtz (ed.), *Lacustrine Basin Exploration – Case Studies and Modern Analogs*, American Association of Petroleum Geologists, Memoir 50, 99–112.

Valero-Garcés, B.L., Kelts, K. and Ito, E., 1995. Oxygen and carbon isotope trends and sedimentological evolution of a meromictic and saline lacustrine system: the Holocene Medicine Lake

Basin, North American Great Plains, USA. *Palaeogeography, Palaeoclimatology, Palaeoecology*, **117**, 253–278.

Wansard, G., 1996. Quantification of paleotemperature changes during isotopic stage 2 in the La Draga continental sequence (NE Spain) based on the Mg/Ca ratio on freshwater ostracods. *Quatenary Science Reviews's*, **15**, 237–245.

Yu, G. and Harrison, S.P., 1995. *Lake Status Records from Europe: Data Base Documentation*, Paleoclimatology Publications Series, Report No. 3, World Data Center for Paleoclimatology, Boulder.

Zeroual, E., 1995. *Enregistrements climatiques dans les sediments du Lac Isli (Haut Altas du Maroc) Variations des influences climatiques sahariennes et mediterranéennes (de 34000 ans BP a nos jours)*, PhD thesis, University of Neuchatel, Switzerland.

PART II

Palaeohydrological Interpretations from Depositional Sequences and Geomorphological Changes

CHAPTER 7

Depositional Model for Cold-climate Tundra Rivers

C. KASSE

The Netherlands Centre for Geo-Ecological Research (ICG), Vrije Universiteit, Amsterdam, The Netherlands

INTRODUCTION

During the last decade Weichselian Middle Pleniglacial (isotope stage 3; *c.* 60–26 ka) fluvial sequences have been investigated intensively in The Netherlands and Germany. The sedimentary environments, provenance of the sediment, climate and vegetation have been studied in both exposures and bore holes (Van Huissteden and Vandenberghe, 1988; Van Huissteden, 1990; Ran, 1990; Kasse et al., 1995). Relations between channel pattern and climate have been established and changes in river pattern have been attributed to climate change (Ran et al., 1990; Vandenberghe, 1992; Mol, 1997). In these studies the fluvial sedimentary environment has been interpreted as sandy anastomosing because of the supposed multichannel nature with generally sandy interchannel deposits. According to Nanson and Knighton (1996), however, the term 'anastomosing' should be applied for low-energy river systems with mostly fine-grained or organic deposition (cf. Smith and Smith, 1980; Smith, 1983, 1986). Therefore, in accordance with Nanson and Knighton, the more general term 'anabranching', defined as a system of multiple channels characterized by vegetated or otherwise stable alluvial islands, is used in this study. As opposed to the braided system, the islands of the anabranching system usually persist for decades or centuries, support well-established vegetation and have relatively stable banks.

The aim of this chapter is: (a) to describe the sedimentary facies and depositional environment of the Middle Pleniglacial fluvial system; (b) to compare this fluvial system with existing floodplain classifications; and (c) to establish the factors that favoured floodplain formation during the Middle Pleniglacial.

The investigated sections have been formed during the Weichselian Middle Pleniglacial which is correlated with oxygen isotope stage 3 (*c.* 60 to 26 ka; Vandenberghe, 1992). The study areas Grouw and Dinkel in The Netherlands (Figure 7.1) are part of the southern North Sea Basin, a generally subsiding area of low relief and unconsolidated sediments. In Grouw the Holocene and Weichselian sequence was temporarily exposed in two building pits for a tunnel below the Prinses Margriet Canal (Kasse et al., 1995). The Dinkel valley was investigated mainly by drillings (Van Huissteden, 1990). The Weichselian Middle Pleniglacial fluvial sequences in these sites have been deposited in depressions of Saalian glacial origin by relatively small local rivers (Boorne, Dinkel).

Palaeohydrology and Environmental Change. Edited by G. Benito, V. R. Baker and K. J. Gregory
© 1998 John Wiley & Sons Ltd.

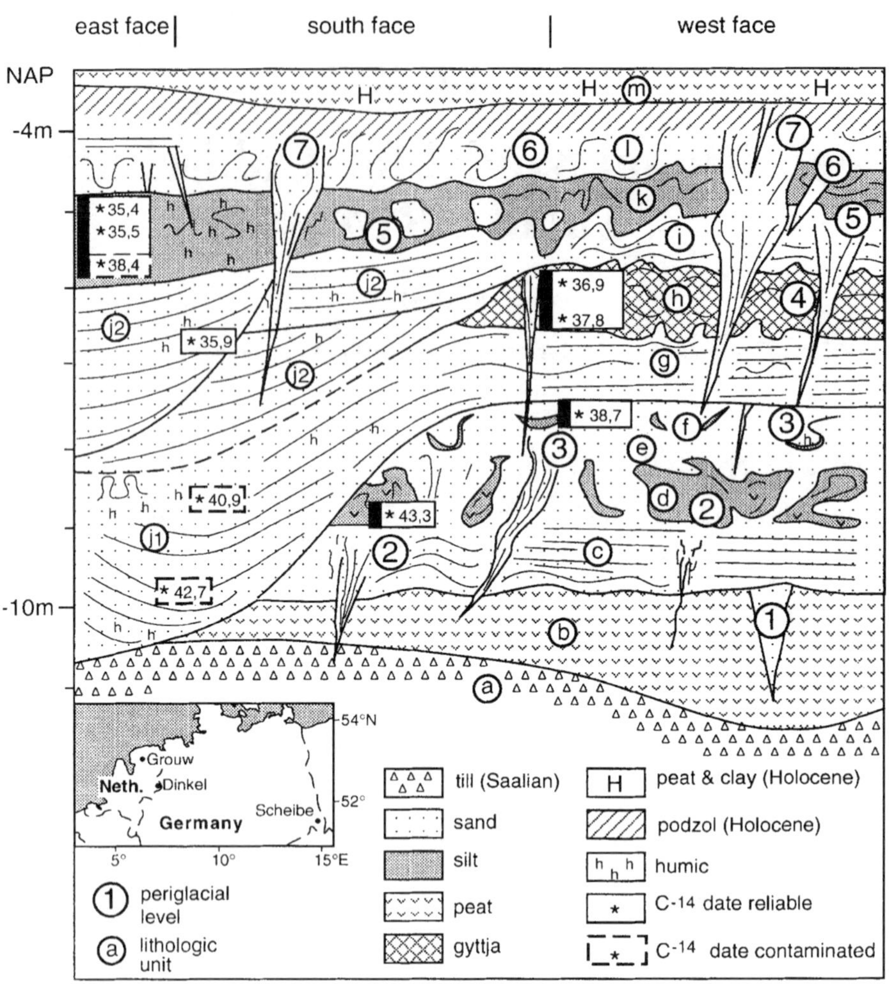

Figure 7.1   Location map of the study areas in The Netherlands and Germany and composite section of exposure at Grouw (strong vertical exaggeration). The eastern and part of the southern pit face are dominated by stacked channel sequences separated by large-scale unconformities. The western and part of the southern pit face contain an 8000 year record of overbank sedimentation. The south face is *c.* 50 m long

In eastern Germany, the Weichselian fluvial sequence has been investigated in the large brown coal pit at Scheibe in the Niederlausitz (Mol, 1997). The sediments are part of a Tertiary and Quaternary sequence deposited at the transition from the Bohemian Central German landmass in the south to the North German Basin in the north. The Weichselian sequence has been deposited in a former proglacial outwash valley (Lausitzer Urstromtal) of Saalian age. During the Weichselian local rivers such as the Spree and Neisse flowed through this valley towards the west.

CLIMATE

The climate during the Middle Pleniglacial was cold and the vegetation was tundra (Ran, 1990; Kasse et al., 1995). Frost cracks are abundant in the fluvial sequences and ice-wedge casts occur occasionally. By comparison with recent periglacial environments a mean annual temperature lower than -1 to $-5°C$ has been inferred (compared with the recent mean annual temperature of $c.$ $10°C$) (Vandenberghe, 1992). The Middle Pleniglacial was a period with intense seasonal frost and periodic phases of permafrost (Ran et al., 1990; Kasse et al., 1995). Erosion of the generally sandy ice-pushed ridges (up to 100 m above present-day sea level) and proglacial sands of the previous Saalian glacial led to rapid fluvial aggradation in the former glacial basins and valleys.

SEDIMENTARY FACIES

The Middle Pleniglacial sequences in Grouw, Dinkel and Scheibe, respectively, are $c.$ 6 m, 10 m and 9 m thick and span at least 8 ka (43 to 35 ka), 22 ka ($c.$ 50 to 28 ka) (Van Huissteden, 1990) and $c.$ 18 ka (46 to 28 ka) (Mol, 1997). The standard errors of the ^{14}C dates increase from $c.$ 200 years for the younger dates to $c.$ 2000 years for the oldest ones.

Two large-scale sedimentary facies have been distinguished in these Middle Pleniglacial fluvial sequences: a channel and an overbank facies (Van Huissteden, 1990; Kasse et al., 1995; Mol, 1997). The two facies types are illustrated in Figure 7.1. In the east and south face of the pit, the channel sediments are fine- to medium-grained (100–500 μm) and they reveal several generations of infilling (stacked channels) generally with low-angle dipping beds. In the west face the overbank facies is characterized by predominantly fine sand (100–250 μm) alternating with more silty or organic units. The overbank facies is often strongly cryoturbated and contains numerous frost cracks and ice-wedge casts formed by thermal contraction during subaerial exposure.

The channel facies is characterized by large-scale cross-bedding at the base of the channel formed by the migration of megaripples or dunes (Figure 7.2 left). Locally, the tabular cross-bedded sets (up to 50 cm high) consist of high-angle foreset bundles, separated by low-angle erosional surfaces (Kasse et al., 1995). These structures have been interpreted as falling stage features or reactivation surfaces separating different flood events (Collinson, 1970). It suggests that the Pleniglacial river system experienced large discharge fluctuations (Jones, 1977). This idea is supported firstly by the presence of silt drapes within the foreset cross-bedding which may be associated with deposition from suspension during the falling water stage in the channel. Secondly, frost cracks have been observed in the channel sequence. Since these can only be formed during extreme cooling of the surface, their presence indicates subaerial exposure and periodic very low or no discharge, probably during the winter. It is proposed that in spring the accumulated winter snow melted and in a short phase meltwater was transported through the system. Flow velocity first increased and the dunes became larger and migrated faster over the channel floor. At the end of the flood event flow velocity decreased, dune migration came to an end and silt was deposited from suspension. The presence of small-scale current ripples in the silts indicates fluvial deposition, although it has been argued that the silt originally is of aeolian origin (loess) (Van Huissteden, 1990).

The major part of the channel infill consists of alternating bedding of sand and humic, sandy silts (Figure 7.2 right and Figure 7.3). The sands were probably deposited during

Figure 7.2 Lacquer peels (110 cm high) from the channel facies at Grouw. Left: large-scale cross-bedding formed by dune migration at the base of the channel that erosively overlies Eemian (*c.* 130–120 ka) peat. Right: channel infilling with alternating sand–silt bedding reflecting the episodic and flashy character of the discharge. The high-water stage sands are horizontally bedded or trough cross-bedded; the low-water stage sandy silts are small-scale cross-laminated or massive

Figure 7.3 Detail of the alternating bedding with climbing ripple cross-lamination in silty fine sand associated with rapid sedimentation in the channel during the falling water stage (exposure Grouw). Trowel is 25 cm long

the snowmelt peak discharge and the silt beds during the waning flow stage. The well-developed climbing ripple cross-lamination points to rapid deposition of fine-grained previously suspended sediment in a decreasing flow (Figure 7.3). The lower boundary of the silty bed is normal graded while the upper boundary is erosive due to the next flood event. These sand–silt couplets reflect, like the reactivation surfaces in the dune foresets, the strong variation in flow velocities and flashy hydrographs of the discharges in the Middle Pleniglacial tundra rivers. The sand–silt couplets, locally with moss mats in the silty beds, may reflect annual cycles and therefore high sedimentation rates. However, the long time span of the total sequence, as shown by the ^{14}C dates, indicates that major hiatuses must be present. Indeed, several generations of channel infilling, separated by large-scale unconformities or erosional boundaries, have been found (Figure 7.1, units j1 and j2).

The overbank facies consist of an alternation of fine sand units (*c.* 0.5–1.0 m thick) and more silty or organic units (*c.* 0.1–0.5 m thick) (Figure 7.4). The sandy units are dominant (*c.* 75% of the section in Grouw). In comparison with the channel facies the overbank facies is often strongly cryoturbated and contains frost cracks and ice-wedge casts. These periglacial features were formed by seasonal frost (cracks) or permafrost (ice wedges) penetrating the soil, later followed by permafrost degradation and periglacial loading. They therefore point to repeated subaerial exposure, which is an argument in favour of deposition in an overbank or alluvial island environment.

The sandy overbank units are predominantly horizontally laminated and small-scale current rippled in those sections that are not disturbed by cryoturbation (Figure 7.5). The horizontal lamination occurs at the base of short (10 cm) fining-upward sequences while small-scale current ripple cross-lamination is found in the upper part of the sequences

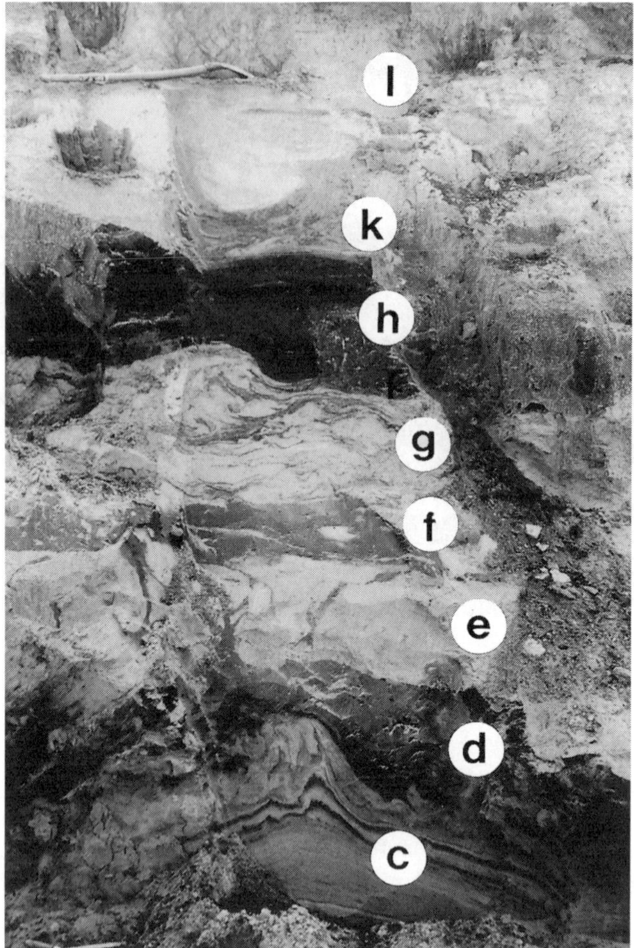

Figure 7.4 Overbank facies with sedimentary cycles of fine sand units (c,e,g,l) and silty or organic units (d,f,h,k) (exposure Grouw). The overbank units are affected by several phases of cryoturbation. Spade for scale is 120 cm long. The letters refer to units in Figure 7.1

(Figure 7.5). Since both bedding types occur within one fining-up sequence, it is concluded that the horizontal lamination is of fluvial origin as well. The fining-up sequences probably represent a change in flow velocity from upper flow regime horizontal lamination into lower flow regime current ripple lamination. The current ripples are locally visible by the draping of organic detritus in the ripple troughs. These sedimentary structures point to strong fluctuations in flow velocity and rather shallow water. The dominance of horizontal bedding and the absence of major channels in the overbank environment may be attributed to sheet flood or crevasse splay deposition during the inundation of the overbank. The upper part of Figure 7.5 shows the upper part of sandy overbank unit c (see Figure 7.1). The increasing disturbance by cryogenic processes (frost cracks and disturbed lamination) may be related to a decreasing sedimentation rate and increasing subaerial exposure at the end of the overbank depositional cycle.

Figure 7.5   Photograph of lacquer peel (110 cm high) from overbank unit c (exposure Grouw). In the lower part of the peel, thin (*c.* 10 cm) fining-upward sequences with horizontal bedding grading upward into small-scale cross-lamination point to sheet flood deposition on the crevasse splay surface. The upper part of unit c is more disturbed by cryogenic processes

A characteristic feature, often encountered in the overbank sequence, is the gradual upward change from peat into lacustrine organic sediment (gyttja) or silt and fluvial sand. This coarsening-upward trend can be explained by channel avulsion when crevasse splays extend into low-lying marshy or lacustrine interchannel areas. This agrees with observations on avulsion deposits in present-day environments (Smith et al., 1989; Smith and Pérez-Arlucea, 1994) where coarsening-upward sequences were formed by crevasse splays extending into interchannel areas. Van Huissteden (1990) noted that locally the horizontally bedded sediments merged laterally into delta cross-bedding. He interpreted this feature as the distal facies of crevasse splays, terminating in shallow lakes (c. 0.5–1.0 m deep) on the floodplain.

ALLUVIAL ARCHITECTURE

The geometry of the sedimentary facies has often been used and misused to reconstruct the river channel pattern (Bridge, 1985, and references cited there). The channel/overbank ratio demonstrates the dynamics of the fluvial system (Allen, 1965). Highly dynamic braided systems are characterized by rapid channel migration and therefore have a high channel/overbank ratio. Meandering and anastomosing systems are relatively stable with lateral migration within well-defined channel belts and channel displacement is by the process of avulsion. Consequently, the channel/overbank ratio is low. Bridge and Leeder (1979) and Mackay and Bridge (1995), however, stressed that, in addition to the nature and frequency of lateral channel migration, the coarse to fine ratio in alluvial sequences is also determined by the vertical floodplain aggradation rate, the ratio of floodplain width to channel-belt sand body width, compaction and vertical tectonic movements.

The alluvial architecture of the Middle Pleniglacial tundra rivers is characterized by the following features.

(i) *The low ratio of channel and overbank sediments.* The alluvial architecture has been studied in extensive exposures (Grouw, Scheibe) or by extrapolation between series of bore holes (Dinkel; Figure 7.6). In the Grouw exposure (Figure 7.1) *c.* 50% of the sequence consists of channel sediments. Furthermore, the exposed channel seems to have been rather stable without much lateral migration or erosion, which is illustrated by the stacking of overbank units beside the channel from at least 38.7 to 35.5 ka (Figure 7.1). This stability may be explained by the presence of tundra vegetation and rather consistent loamy, peaty or ground-ice layers in the overbank sequence. Both the low channel to overbank ratio and the bank stability have been associated with an anastomosing or anabranching river pattern (Kasse et al., 1995). Figure 7.6 is one of the many cross-sections over the Dinkel valley (Van Huissteden, 1990). It illustrates the low ratio of channel to overbank sediments. The channel deposits consist of medium to coarse sand; the overbank deposits contain fine sand, silt, clay and organic materials. In general 40 to 55% of the total thickness of the Middle Pleniglacial sequence consists of channel sands (Van Huissteden, 1990). Because of the scarcity of lateral accretion deposits and the dominance of crevasse splay sand sheets and abundance of marshy to lacustrine deposits (low channel to overbank ratio) it was concluded that these Middle Pleniglacial sequences in the Dinkel valley were deposited by an anastomosing river. The Middle Pleniglacial sequence in eastern Germany (exposure Scheibe) is very similar to the Dinkel valley

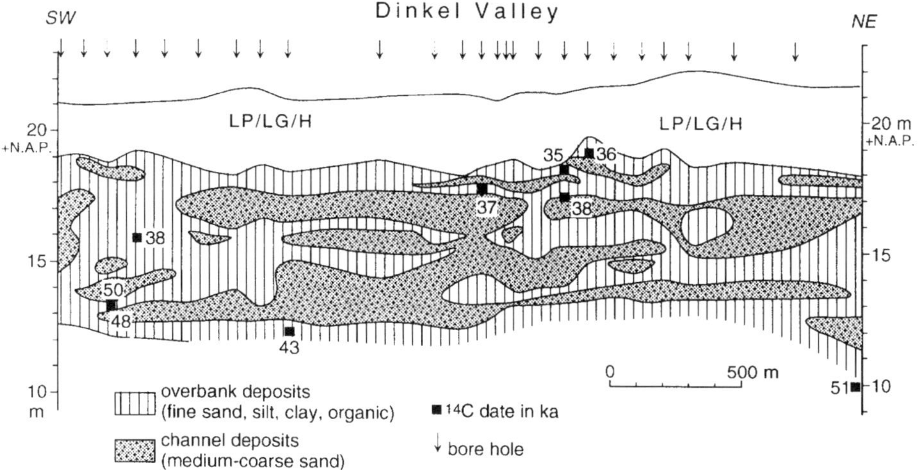

Figure 7.6   Cross-section over the Dinkel valley, eastern Netherlands, illustrating the low channel to overbank ratio (after Van Huissteden, 1990). LP/LG/H, Late Pleniglacial/Late Glacial/Holocene deposits; N.A.P., Dutch Ordnance Datum

(Mol, 1997). The channel to overbank ratio is low and the sequences are interpreted also as being deposited by an anastomosing fluvial system of the Spree or Neisse.

(ii) *The relatively high ratio of overbank sand to overbank fines (silt, clay, organic material).* In the investigated sites of Scheibe, Grouw and the Dinkel valley, 80%, 75% and 70%, respectively, of the overbank sediments consist of fine sand. Low-energy anastomosing river systems, described by Smith and Smith (1980) and Smith (1983, 1986), are characterized by standing water pools in which clay settles from suspension and plants accumulate as peat. Consequently, the overbank sand/fines ratio in such environments is low. The Middle Pleniglacial tundra rivers, however, have a high overbank sand to fines ratio, which means a more dynamic overbank environment. The large volume of sand in the overbank deposits suggests that the floodbasins played an important role in transporting water and sediment during major floods.

DISCUSSION

The Middle Pleniglacial fluvial systems in The Netherlands and Germany have been interpreted previously as sandy anastomosing (Van Huissteden, 1990; Kasse et al., 1995; Mol, 1997). However, in accordance with Bridge (1993), it is difficult to prove the coevality and connection of anabranching river segments in the Middle Pleniglacial alluvial sequences. It therefore cannot be excluded that the alluvial architecture described above was formed by a single channel system characterized by frequent channel avulsions. In the following, the causes for the formation of the Middle Pleniglacial alluvial architecture are discussed and the river pattern will be compared with existing river pattern classifications (e.g. Nanson and Croke, 1992; Nanson and Knighton, 1996).

A reconstruction of the depositional processes in the Middle Pleniglacial river valleys is presented in Figure 7.7. The reconstruction is based on the study of lithology, sedimen-

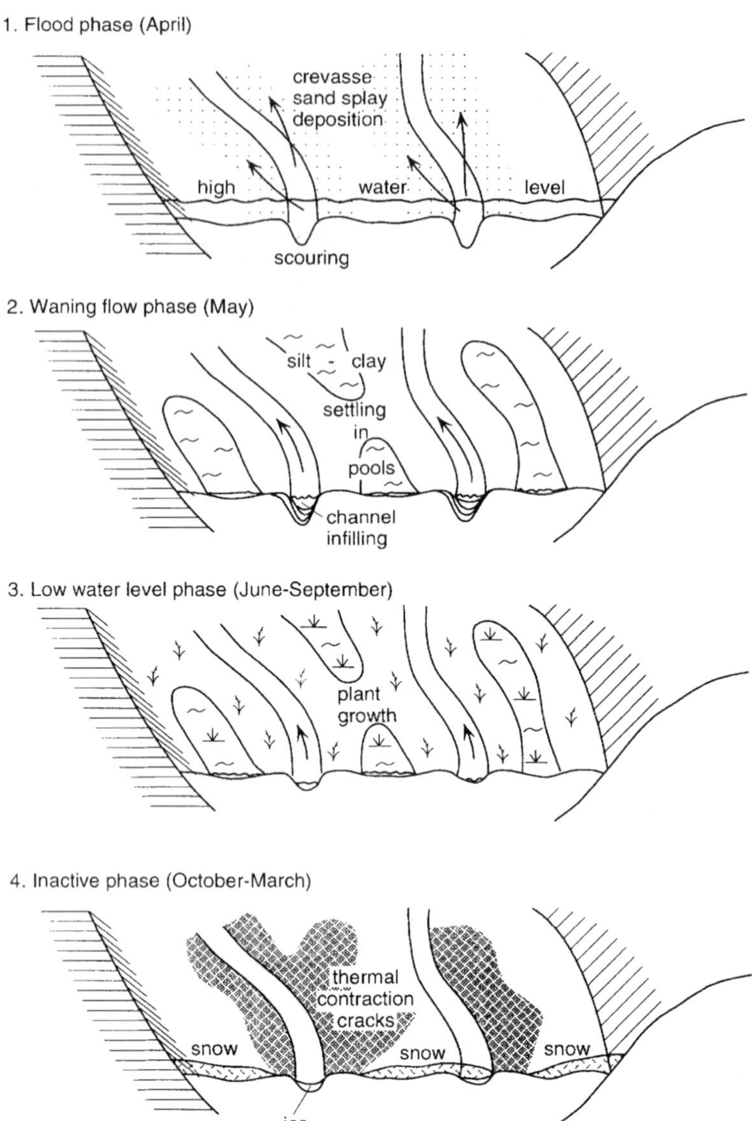

Figure 7.7   Reconstruction of the seasonal depositional processes in the Middle Pleniglacial tundra river valleys

tary sequences and vegetation. The vegetation during the Middle Pleniglacial in The Netherlands and Germany was arctic tundra to shrub tundra (Ran, 1990; Kasse et al., 1995). The months are hypothetical and based on present-day analogues.

During the flood phase (1) the floodplain was inundated by snow meltwater. The channels were scoured and sand was deposited on the overbanks as crevasse splays (Grouw, Scheibe) or crevasse deltas if lakes existed on the floodplain (Dinkel; Van Huissteden, 1990). The flow velocities were high enough to transport medium to coarse sand in the channels and fine sand on the overbank, probably because of the moderate floodplain gradient of 0.0002–0.0004 (20–40 cm km^{-1}).

During the waning flow phase (2) water level dropped, the flow velocity decreased and the water retreated into the channels. The channels were filled first with sand, later with silt forming a couplet of alternating bedding, and locally with climbing ripple bedding. Lateral migration reflected by lateral accretion bedding, a characteristic of meandering rivers, is of minor importance in the exposures, probably because the flow phase was of short duration. In pools on the floodplain some silt and clay settled from suspension.

In the summer period (3) plant growth was abundant on the floodplain as indicated by rooting of the overbank sediment. Locally, moss mats that floated on stagnant water were found in the channel infillings (Grouw; Kasse et al., 1995). This indicates that the fluvial system was rather inactive in this phase and river flow was limited.

In the long winter period (phase 4) the river was most likely inactive, because the floodplain was deeply frozen. Thermal contraction cracks (frost cracks) were formed in snow-free patches on the floodplain, predominantly in the overbank sediment, but locally also in the inactive channels.

The studies in The Netherlands and Germany show that the following factors determine the specific alluvial architecture of the Middle Pleniglacial cold-climate tundra rivers.

(1) *The large width and more or less unconfined character of the valleys.* The Weichselian rivers flowed through large Saalian age subglacial basins (5–10 km wide or more) or proglacial outwash valleys (up to 20 km wide). In comparison with the Saalian meltwater systems the Weichselian rivers are underfit. The unconfined character promoted the overbank deposition. The importance of valley width on the distribution of channel belt bodies within overbank fines has been demonstrated also in fluvial model experiments (Bridge and Leeder, 1979). Indeed, more confined rivers like the Thames reveal a braided channel pattern during the Middle Pleniglacial (Gibbard, 1994).

(2) *The rapid vertical aggradation of the floodplain.* Vertical aggradation in the investigated sites is *c.* 0.5 m ka^{-1}. Van Huissteden (1990) reported somewhat lower mean sedimentation rates for the whole Middle Pleniglacial unit (0.16–0.35 m ka^{-1}). These sedimentation rates are lower than those reported for Holocene anastomosing systems (e.g. > 1.5 mm yr^{-1} (Törnqvist, 1993) and 6 mm yr^{-1} (Smith, 1983)), probably because the Weichselian values are based on a longer period and hiatuses may be present in the sequence. The high sedimentation rates in the basins are probably closely related to periglacial erosion (solifluction, sheet flow erosion, wind erosion) of surrounding higher topography such as till plateaus, sandur plains and ice-pushed ridges of Saalian age. The vertical aggradation enabled the channel to maintain its course. Channel abandonment by means of avulsion occurred at the end of a depositional cycle, when the channel belt had aggraded above the level of the surrounding floodplain. This promoted breaching of the levees and crevasse splay development as indicated by the drowning of overbank peats and the general coarsening-upward trend in the individual overbank units. Ultimately, the crevasse channels evolved into a new course in the lower-lying former interchannel area (Ran and Van Huissteden, 1990).

(3) *The limited lateral migration of the channels.* Limited lateral migration is probably caused by two elements. First, these periglacial tundra rivers were probably charac-

terized by strong seasonal discharge variations related to the spring snowmelt (Woo and Winter, 1993). Because of the shortness of the active flow stage only limited lateral migration could occur and the channels did not develop into meandering channels. Secondly, lateral migration will have been limited by river bank cohesion. The well-developed tundra vegetation on the overbank environment may have stabilized the sediment. In addition the loamy and organic beds in the overbank sequence, despite their limited thickness, may have restricted bank erosion. Finally, especially important in cold-climate rivers, the seasonal or permanently frozen state of the overbank sequence and the presence of ground ice will have limited lateral erosion. Even sandy overbank sediments, normally easily eroded, will be cohesive under such frozen conditions.

(4) *The sandy nature of the source material.* According to Van Huissteden (1990) the generally sandy nature of the overbank sequence may be explained by: (i) later compaction especially reducing the thickness of the loamy and organic units; (ii) restricted organic matter production in the cold Middle Pleniglacial climate. The low organic production will have reduced the overbank sedimentation rate. Channel belt aggradation may have been faster than overbank aggradation and consequently sandy crevasse splays developed and avulsion occurred; (iii) the nival hydrologic regime and blockage of channels by ice dams. The Middle Pleniglacial hydrologic regime was probably characterized by a relatively short snow meltwater phase. In combination with snow or ice dams meltwater floods may have been diverted out of the channels promoting extensive overbank sedimentation of sand. In addition to the foregoing, the sandy character of the Middle Pleniglacial fluvial systems may be determined by the sandy nature of the source material. The sediment sources in the river catchments consist of unconsolidated, mostly sandy (fluvio)glacial and ice-pushed preglacial sediments. Nanson et al. (1986) described a similar relationship for the anastomosing Cooper Creek river in Australia, where the transported sediment consists mostly of sand-sized clay aggregates and the floodplain therefore consists of clay.

The comparison of the Middle Pleniglacial river with existing floodplain classifications (Nanson and Croke, 1992; Nanson and Knighton, 1996) is shown in Table 7.1. Important elements are the channel/overbank ratio and the coarse/fine ratio of the floodplain sediments.

The Middle Pleniglacial systems have a low channel/overbank ratio which is comparable with the more organic anastomosing systems (Smith and Smith, 1980; Smith, 1983, 1986), and with the inorganic anastomosing system (Nanson et al., 1986). However, the coarse/fine ratio of the anastomosing types C2a and C2b is low, which means that the floodplains are dominantly muddy or organic, while the Middle Pleniglacial systems have a high coarse/fine overbank ratio. In this respect they have more in common with the braided system. The peaked discharge character of the Middle Pleniglacial systems resembles the semi-arid inorganic anastomosing type (C2b) as described from Australia (Nanson et al., 1986); however, instead of sand, type C2b is dominated by clay.

Recently, the anastomosing or anabranching floodplain types have been summarized by Nanson and Knighton (1996). In addition to the well-established cohesive (clay/organic) anastomosing floodplains (Nanson and Croke, 1992), sand-dominated anabranching floodplain types are described (types 2 and 4). They occur in tropical

Table 7.1 Comparison of the Middle Pleniglacial tundra rivers with floodplain classifications

Floodplain type (Nanson and Croke, 1992; Nanson and Knighton, 1996)	Ratio channel/ overbank	Ratio gravel–sand/silt–clay	Specific stream power	Dominant sediment
Braided (B1)	high	high	(high) medium	gravel/sand
Meandering (B3)	medium	medium	medium	sand
Anastomosing organic-rich (C2a/1b)	low	low	low	clay/organic
Anastomosing inorganic (C2b/1c)	low	low	low	clay
Anabranching, sand-dominated, island-forming (2)	low	high	low	sand
Middle Pleniglacial (this study)	low	medium/high	medium	sand

monsoonal or semi-arid climates and are characterized by a monsoonal or episodic flow regime. Because of the generally non-cohesive nature of sand, these types require the combination of stabilizing bank vegetation and a low stream power to prevent the channels from braiding or meandering (Nanson and Knighton, 1996). Despite the difference in climatic setting, the cold-climate Middle Pleniglacial rivers of northwestern Europe reveal comparable environmental conditions. The discharge was strongly episodic as it was related to the snowmelt period and floodplain sedimentation was predominantly sandy. The stabilizing role of the tundra vegetation and the presence of deep seasonal frost or permafrost enhanced bank stability and the tendency towards lateral migration and meandering was suppressed. It is concluded, therefore, that the Middle Pleniglacial tundra rivers of northwestern Europe are comparable with modern sand-dominated, island-forming anabranching rivers.

CONCLUSIONS

(1) The cold-climate (mean annual temperature -1 to $-4\,^{\circ}$C), low-relief tundra rivers of the Middle Pleniglacial in northwestern Europe are characterized by channel and overbank environments. Alternating bedding of sand and silt and reactivation phenomena (erosional surfaces in the dune foresets, silt drapes) in the channel sediments reflect the peaked discharge of these rivers probably related to the short snowmelt period.

(2) The alluvial architecture of the floodplain is characterized by the low ratio of channel (*c.* 50% of the total sequence) to overbank sediments and by the absence of lateral accretion deposits in the channel belts which suggests an anastomosing or anabranching river style. The overbank sequence is dominated by sand (70–80%) which was deposited in crevasse splays or crevasse deltas during avulsion phases. Despite the sandy character of the system, lateral channel migration associated with river meandering was very limited, because the flood phase was short and the river banks were stabilized by tundra vegetation and seasonal or permanent ground ice.

(3) The development of the specific floodplain architecture is favoured by: the unconfined character of the rivers flowing in former proglacial and subglacial basins; the relative-

ly high floodplain aggradation rate of 0.5 m ka^{-1}; the sandy source material derived from preglacial and Saalian glacial sandy deposits in the catchments; the flashy discharge related to the snow-melt period; the high bank stability because of the tundra vegetation and presence of ground ice in the overbank sequence.

(4) The Middle Pleniglacial systems differ from the cohesive (organic to mud-dominated) anastomosing river types by their predominantly sandy character. These tundra rivers possibly resemble monsoonal or semi-arid, sand-dominated, island-forming anabranching rivers with an episodic flow regime.

ACKNOWLEDGMENTS

The manuscript benefited from the critical review by Dr P.J. Ashworth, University of Leeds. Figures 7.1, 7.2, 7.3, 7.4 and 7.5 have been reproduced from the *Mededelingen Rijks Geologische Dienst* with permission of the Nederlands Instituut voor Toegepaste Geowetenschappen TNO.

REFERENCES

Allen, J.R.L., 1965. A review of the origin and characteristics of recent alluvial sediments. *Sedimentology*, **5**, 89–191.

Bridge, J.S., 1985. Paleochannel patterns inferred from alluvial deposits: a critical evaluation. *Journal of Sedimentary Petrology*, **55**, 579–589.

Bridge, J.S., 1993. Description and interpretation of fluvial deposits: a critical perspective. *Sedimentology*, **40**, 801–810.

Bridge, J.S. and Leeder, M.R., 1979. A simulation model of alluvial stratigraphy. *Sedimentology*, **26**, 617–644.

Collinson, J.D., 1970. Bedforms of the Tana River, Norway. *Geografiska Annaler*, **52A**, 31–56.

Gibbard, P.L., 1994. *Pleistocene History of the Lower Thames Valley*, University Press, Cambridge.

Jones, C.M., 1977. Effects of varying discharge regimes on bed-form sedimentary structures in modern rivers. *Geology*, **5**, 567–570.

Kasse, C., Bohncke, S.J.P. and Vandenberghe, J., 1995. Fluvial periglacial environments, climate and vegetation during the Middle Weichselian in the northern Netherlands with special reference to the Hengelo Interstadial. *Mededelingen Rijks Geologische Dienst*, **52**, 387–414.

Mackay, S.D. and Bridge, J.S., 1995. Three-dimensional model of alluvial stratigraphy: theory and application. *Journal of Sedimentary Research*, **B65**, 7–31.

Mol, J., 1997. *Fluvial response to climate variations. The Last Glaciation in eastern Germany*, Thesis, Vrije Universiteit, Amsterdam.

Nanson, G.C. and Croke, J.C., 1992. A genetic classification of floodplains. *Geomorphology*, **4**, 459–486.

Nanson, G.C. and Knighton, A.D., 1996. Anabranching rivers: their cause, character and classification. *Earth Surface Processes and Landforms*, **21**, 217–239.

Nanson, G.C., Rust, B.R. and Taylor, G., 1986. Coexistent mud braids and anastomosing channels in an arid-zone river: Cooper Creek, central Australia. *Geology*, **14**, 175–178.

Ran, E.T.H., 1990. Dynamics of vegetation and environment during the Middle Pleniglacial in the Dinkel valley (The Netherlands). *Mededelingen Rijks Geologische Dienst*, **44**, 139–205.

Ran, E.T.H. and Van Huissteden, J., 1990. The Dinkel Valley in the Middle Pleniglacial: dynamics of a tundra river system. *Mededelingen Rijks Geologische Dienst*, **44**, 209–220.

Ran, E.T.H., Bohncke, S.J.P., Van Huissteden, J. and Vandenberghe, J., 1990. Evidence of episodic permafrost conditions during the Weichselian Middle Pleniglacial in the Hengelo Basin (The Netherlands). *Geologie en Mijnbouw*, **69**, 207–218.

Smith, D.G., 1983. Anastomosed fluvial deposits: modern examples from Western Canada. In J.D.

Collinson and J. Lewin (eds), *Modern and Ancient Fluvial Systems*, International Association of Sedimentologists, Special Publication 6, Blackwell, Oxford, 155–168.

Smith, D.G., 1986. Anastomosing river deposits, sedimentation rates and basin subsidence, Magdalena River, northwestern Columbia, South America. *Sedimentary Geology*, **46**, 177–196.

Smith, D.G. and Smith, N.D., 1980. Sedimentation in anastomosed river systems: examples from alluvial valleys near Banff, Alberta. *Journal of Sedimentary Petrology*, **50**, 157–164.

Smith, N.D. and Pérez-Arlucea, M., 1994. Fine-grained splay deposition in the avulsion belt of the lower Saskatchewan River, Canada. *Journal of Sedimentary Research*, **B64**, 159–168.

Smith, N.D., Cross, T.A., Dufficy, J.P. and Clough, S.R., 1989. Anatomy of an avulsion. *Sedimentology*, **36**, 1–23.

Törnqvist, T.E., 1993. Holocene alternation of meandering and anastomosing fluvial systems in the Rhine-Meuse delta (Central Netherlands) controlled by sea-level rise and subsoil erodibility. *Journal of Sedimentary Petrology*, **63**, 683–693.

Vandenberghe, J., 1992. Geomorphology and climate of the cool oxygen isotope stage 3 in comparison with the cold stages 2 and 4 in The Netherlands. *Zeitschrift für Geomorphologie Neue Folge, Supplement-Band*, **86**, 65–75.

Van Huissteden, J., 1990. Tundra rivers of the last glacial: sedimentation and geomorphological processes during the Middle Pleniglacial in Twente, eastern Netherlands. *Mededelingen Rijks Geologische Dienst*, **44**, 1–138.

Van Huissteden, J. and Vandenberghe, J., 1988. Changing fluvial style of periglacial lowland rivers during the Weichselian Pleniglacial in the eastern Netherlands. *Zeitschrift für Geomorphologie Neue Folge, Supplement-Band*, **71**, 131–146.

Woo, M.-K. and Winter, T.C., 1993. The role of permafrost and seasonal frost in the hydrology of northern wetlands in North America. *Journal of Hydrology*, **141**, 5–31.

CHAPTER 8

Tectonic versus Climatic Controls on the River Maas Dynamics During the Late Glacial

MARGRIET HUISINK

The Netherlands Centre for Geo-Ecological Research (ICG), Vrije Universiteit, Amsterdam, The Netherlands

INTRODUCTION

The Maas (Meuse) is a rain-fed river which drains an area of 33 000 km^2. It flows from France through Belgium, and enters The Netherlands from the south (Figure 8.1). There it flows through the southern part of the North Sea Basin which is filled with Quaternary sediments. An important part of the headwaters of the Maas is formed by the relatively high Ardennes, which are the result of long-term tectonic uplift. The Maas crosses part of the Roer Valley rift system, namely the South Limburg Block, the Roer Valley Graben, the Peel Block (Peelhorst), and finally flows through part of the Venlo Graben, where the study area is situated south of Nijmegen (Figure 8.1). Since the hinterland has been subjected to uplift the Maas has had a tendency to erode and has formed a series of terraces. During the Late Glacial (13–10 ka) the Maas incised several times and deposited sediments afterwards in an increasingly narrowing floodplain. Five terraces are recognized from the Weichselian late Pleniglacial (oxygen isotope stage 2) to the Holocene.

The alternation of erosion and deposition resulted in a complex morphology, which is enhanced by the fact that incision is often equally balanced by deposition. During the Late Glacial the sea level was some 100 m lower than at present (Jelgersma, 1966), but because the sea level rose considerably during the Holocene, sedimentation of Holocene sediments became dominant until roughly this area. Upstream erosion prevailed, while downstream younger sediments are deposited on top of older terraces. Reconstructions of the Maas fluvial activity were made by Pons (1957), Pons and Schelling (1951), Berendsen et al. (1995), Vandenberghe et al. (1994), Kasse et al. (1995), Bohncke et al. (1993) and Van den Berg (1994a,b, 1996). Recently, a morphological and sedimentological study was undertaken to refine the reconstruction of the river evolution (Huisink, 1997).

However, the relative importance of either climate-related factors or tectonic factors on the river system is still not completely understood. The focus of this chapter is therefore to study the impact of both climatic and tectonic controls on changing fluvial patterns and phases of incision and accumulation. More insight into the tectonic control on river patterns is achieved by the construction of longitudinal profiles of each terrace by

---

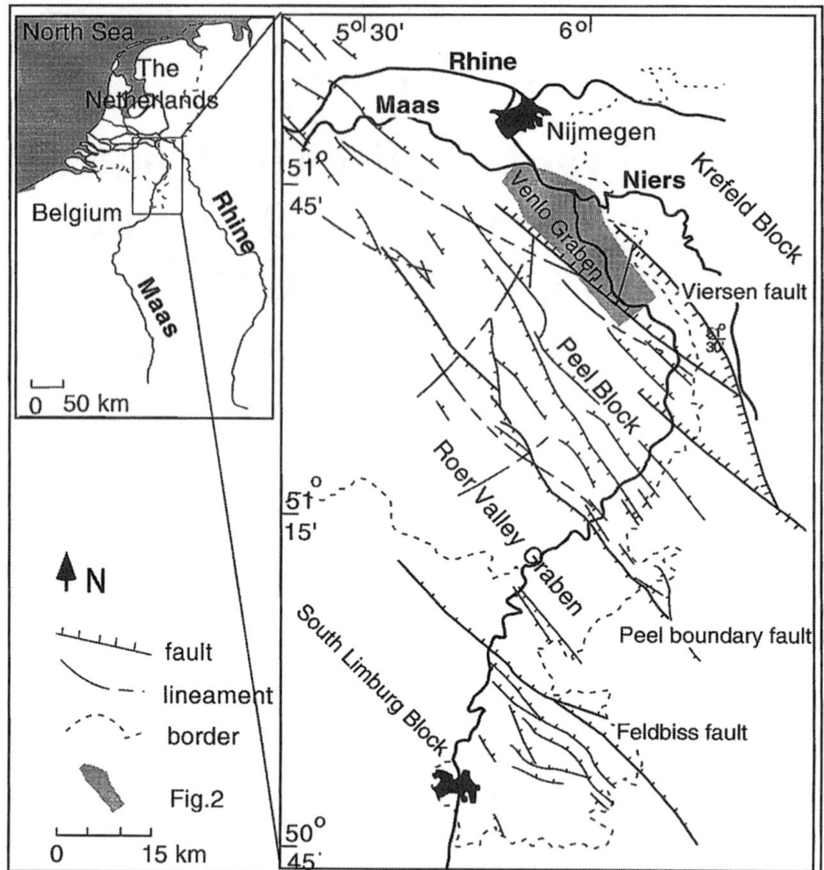

Figure 8.1   Location of the study area and major faults (faults according to Van Montfrans (1975); lineaments with signs of horizontal motion according to Van den Berg (1994b))

multiple linear regression analysis. Locally steep terrace slopes are recognized which are explained by local tectonic movements. The effect of these steeper terrace slopes on river morphology and sedimentology will be discussed.

LATE WEICHSELIAN TERRACE STRATIGRAPHY

The terrace stratigraphy of the Maas valley, south of Nijmegen, has been established by Huisink (1997) and is shown in Figure 8.2. Detailed sediment descriptions and ages of sediments are provided by Huisink (1997). The oldest investigated Weichselian terrace level is the Rijkevoort terrace of Pleniglacial age. The terrace sediments were deposited by a braided river. Part of the valley was abandoned during the Late Pleniglacial and became covered by 0–4 m of aeolian sands (so-called coversands).

The Pleniglacial river system changed at the beginning of the Late Glacial (13 ka), when minor channels of the braided system became abandoned and larger ones started to incise up to 3 m. The morphology shows the abandoned channel scars of the former braided river and low-sinuosity channels with immature levees of the new river system

(Vierlingsbeek terrace). Upstream of Beugen the channel morphology of the Vierlings-beek terrace changes (Figure 8.2). At some locations several small, low-sinuosity channels merged into one large, more curved channel (Vandenberghe et al., 1994). This river system represents a transition between the former braided river system of the Rijkevoort terrace and the meandering river of the Broekhuizen terrace. The incision of the major channels continued until finally one highly sinuous channel remained active in the Allerød (12–11 ka, Broekhuizen terrace).

At the onset of the Younger Dryas (11 ka) the river system changed rapidly to a braided form (Wanssum terrace). A distinct incision took place whereafter the floodplain was partly filled in by sediment. Aeolian processes became important in the later part of the Younger Dryas when large parabolic dunes were formed on the east banks of the Maas. The sediment

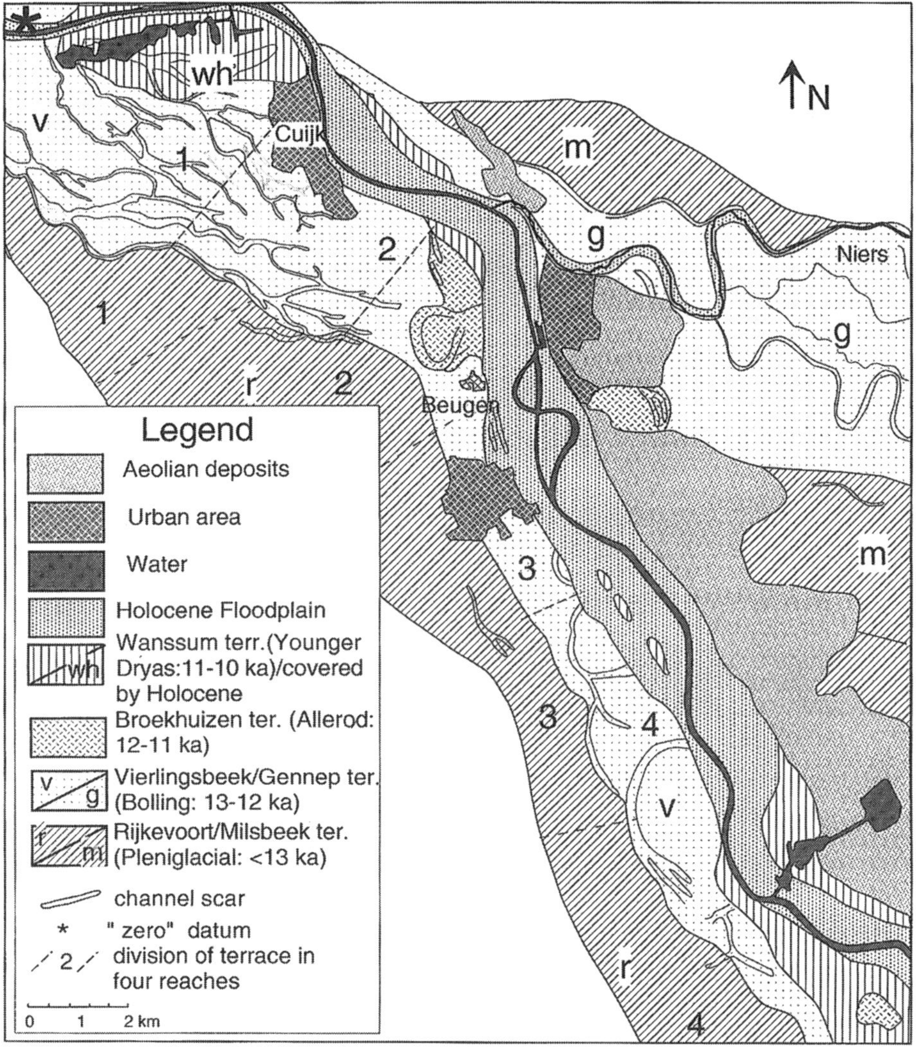

Figure 8.2   Late Weichselian terraces and Holocene floodplain of the Maas

for these dunes originated from the Younger Dryas floodplain, as shown by heavy mineral analysis (Huisink, 1997). Finally the Maas incised the Younger Dryas floodplain at the onset of the Holocene and changed again into a meandering river. The dimensions of the Holocene floodplain are largely comparable to those of the Younger Dryas floodplain. The Holocene Maas was a low-sinuosity river confined to a relatively small floodplain in which the channel shifted. The Holocene floodplain widens downstream of Nijmegen where five Holocene river systems were recognized by Berendsen et al. (1995).

CALCULATING LONGITUDINAL PROFILES OF RIVER TERRACES

Longitudinal profiles of the Late Glacial terraces are constructed to improve the terrace stratigraphy and to establish the effects of tectonism on river pattern changes. A reliable reconstruction requires a large number of altitudes and an objective way of plotting. A multiple linear regression (m.l.r.) analysis is therefore used instead of the more traditional perpendicular projection of ground point data to a hand-drawn river axis transect (e.g. Merritts et al., 1994). An m.l.r. analysis determines the relation between the independent variables (x and y coordinates) and a dependent variable (Z or altitude) by fitting a planar surface through the data. The assumption that a regression surface resembles a terrace can be made in this area because the valley is more or less straight. When a large bend occurs in a valley a curved surface would be needed, resulting in a more complicated regression equation. Since the x and y coordinates are equally important, a stepwise m.l.r. analysis has not been used.

The data points are collected from altitude maps with a scale of 1 to 10 000. Only the highest points on the terrace surfaces are used, mostly the points on top of levees or pointbar sediments. Detailed sedimentary and morphological analysis is necessary to exclude aeolian deposits. A minimum of 54 and a maximum of 133 points for each terrace are used in the m.l.r. analyses. The output of the m.l.r. analysis for each terrace is an equation representing the best fit of a surface through the observed points (Figure 8.3). This equation consists of a constant and two regression coefficients which are calculated by the method of least squares. These regression coefficients explain the effect of the coordinates on the altitude. The standard error of estimate is a measure of the scatter of the data around the regression line and the R^2 goodness of fit represents the degree of association between the measured and estimated altitudes. The slope of the surface, calculated by the m.l.r. analysis, is the steepest longitudinal terrace profile and is determined by taking the root of the combined squared regression coefficients (x coeff.^{2} + y coeff.^{2} = surface slope^2).

The surface, calculated by the m.l.r. analysis, is a three-dimensional feature. In order to visualize the surface slopes in two dimensions, the altitude has to be plotted against one axis. This can be accomplished by rotation of the x–y coordinate configuration in the direction of the steepest surface slope. To obtain this rotation, the most distal point in the area is taken as the 'zero' point (Figure 8.2). A line is calculated through this point, the so-called zero line, which has a direction perpendicular to the surface slope direction. The coordinates of all points are recalculated in distances from this zero line. The distances of points can now be plotted against altitude and a linear graph is constructed which represents the longitudinal terrace profile. The longitudinal profiles of all terraces are plotted in Figure 8.4 for comparison. The Holocene floodplain was divided into two reaches, up- and downstream of Cuijk (Figure 8.2) because the direction change in the

Holocene floodplain appears to be too large to consider the whole floodplain as one flat surface.

RESULTS OF THE MULTIPLE LINEAR REGRESSION ANALYSIS

The reliability of these reconstructions (Figure 8.3) is high: the R^2 value varies between 0.92 (Allerød terrace) and 0.99 (late Pleniglacial terrace). The Holocene floodplain downstream of Cuijk (Figure 8.3f) is poorly constructed ($R^2 = 0.6$), therefore only the more reliable reconstruction of the floodplain upstream of Cuijk (Figure 8.3e) is discussed further. The scatter of points around the regression line is normal for all terraces (68% of the points are within plus or minus the standard error of estimate). Points with a large deviation from the regression line were checked again and eventually corrected if in error.

The oldest terraces are relatively steep, 41.8 cm km^{-1} for the Rijkevoort terrace (late Pleniglacial) and 35.1 cm km^{-1} for the Vierlingsbeek terrace (Bølling) when compared with the slopes of the Broekhuizen (Allerød, 23.5 cm km^{-1}) and Wanssum (Younger Dryas, 25.1 cm km^{-1}) terraces. The longitudinal profiles of the Broekhuizen and Wanssum terraces correspond well with downstream values of 22 cm km^{-1} (Verbraeck, 1990) and 25.4 cm km^{-1} (Berendsen et al., 1995), respectively. The Rijkevoort and Vierlingsbeek terraces appear to be too steep, as downstream 27.5 cm km^{-1} is observed by Berendsen et al. (1995) for their 'generation 2' river system which comprises both the Rijkevoort and Vierlingsbeek terraces (Huisink, 1997). Upstream also, a gentler slope value of 22 cm km^{-1} for the Rijkevoort terrace (Figure 8.4) can be derived from Van den Broek and Maarleveld (1963).

In order to see where exactly on the Vierlingsbeek and Rijkevoort terraces the break in slope occurs, a subdivision of the data into four separate data sets (reaches) is made. From each data set, consisting of 34 (Vierlingsbeek terrace) and 17 (Rijkevoort terrace) altitude observations, a separate m.l.r. analysis has been made (Table 8.1). The location of the four reaches is shown on Figure 8.2. The accuracy of the analyses decreases and the R^2 value varies between 0.6 and 0.9, since the amount of data used in the analysis diminishes. The most downstream reach of the Vierlingsbeek terrace (reach 1, Table 8.1) has a slope of 27.0 cm km^{-1}, which matches the value observed downstream by Berendsen et al. (1995). The slope increases drastically to 39.6 cm km^{-1} from the first to the second reach (reach 2, Table 8.1), and is maintained in the upstream reaches (36.2 cm km^{-1} and 34.8 cm km^{-1}). The Rijkevoort terrace also shows a steeper slope value in the second reach of 46 cm km^{-1}, which decreases upstream into 38 cm km^{-1} for the third and fourth reaches (Table 8.1). It is interesting to see the change in direction of the terrace slope as well: the most upstream reaches show a south–north direction, whereas the more downstream reaches indicate a change in direction towards the northwest.

TECTONIC ACTIVITY INFERRED FROM GRADIENT LINES

One way to explain the breaks in slope on the oldest terraces (Figure 8.4) is to consider the confluence of the Niers. The Niers was comparable in size with the Maas, considering the palaeovalley dimensions, until the Allerød, when it stopped serving as part of the Rhine river system (Huisink, 1997). The extra input of sediment might explain the lower terrace slopes downstream of the confluence, but does not explain the break in slope upstream (Figure 8.4), where a gentler slope is seen. The position of this break in slope of the

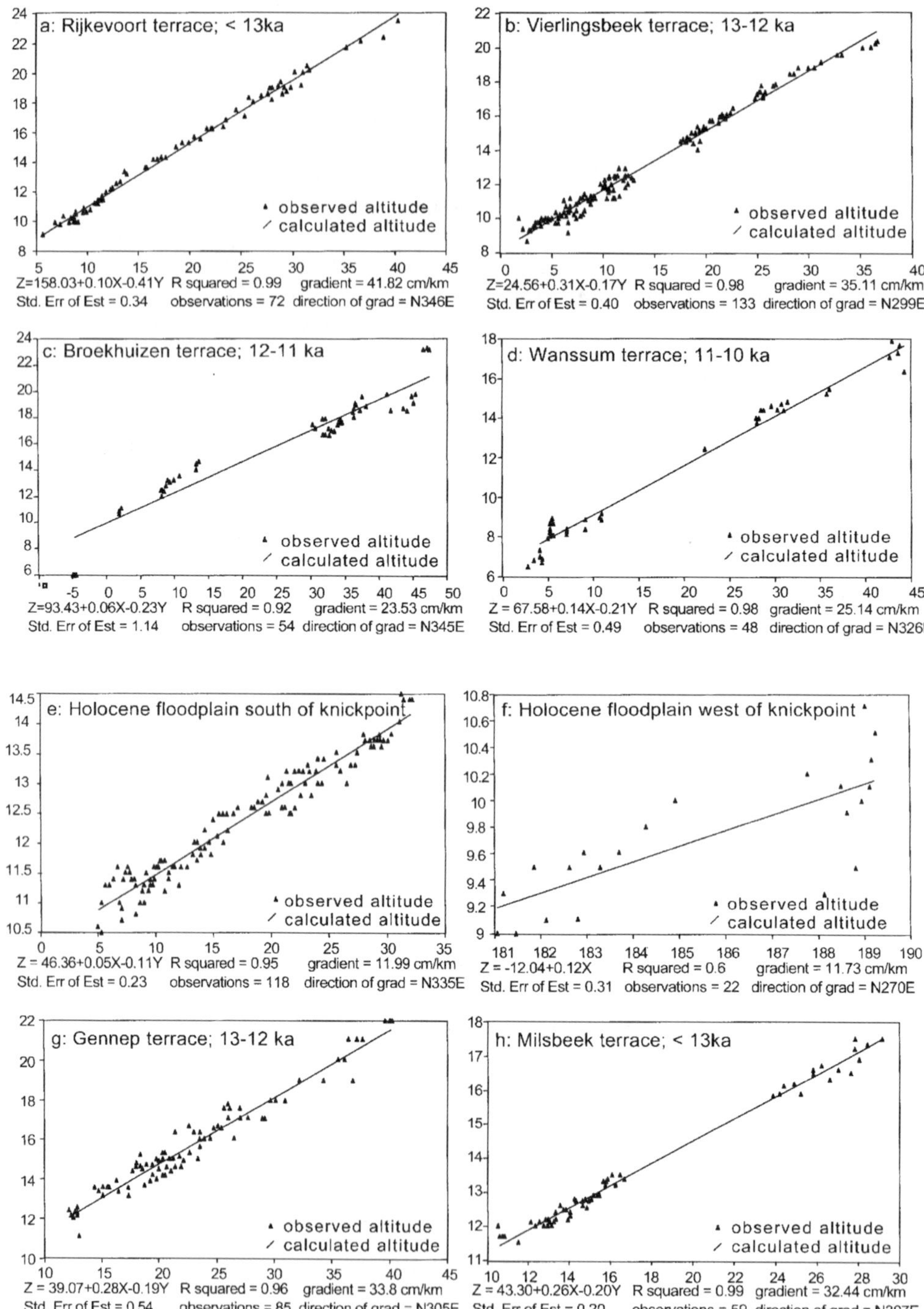

Figure 8.3   Multiple linear regression curves of Late Weichselian Maas and Niers terraces and Holocene Maas floodplain. Vertical axis is the observed altitudes (in m) above Dutch ordinal altitude (NAP), horizontal axis is horizontal distance upstream from arbitrary datum (in km)

Table 8.1 Regression analyses for the Vierlingsbeek (13–12 ka) and Rijkevoort (< 13 ka) terraces, where each terrace is divided into four reaches; the location of each reach is shown on Figure 8.2

	Reach 1	Reach 2	Reach 3	Reach 4
Vierlingsbeek terrace				
Regression equation	$Z = -69.45 + 0.26X + 0.08Y$	$Z = -39.47 + 0.39X - 0.06Y$	$Z = 38.05 + 0.30X - 0.20Y$	$Z = 155.45 + 0.002X - 0.35Y$
R^2	0.61	0.90	0.90	0.99
slope (cm km^{-1})	27.64	39.64	36.20	34.8
Std error of est.	0.34	0.28	0.45	0.18
No. of observations	34	33	33	33
Direction of slope	N286E	N278E	N304E	N360E
Rijkevoort terrace				
Regression equation	$Z = 96.25 + 0.15X - 0.28Y$	$Z = 152.91 + 0.17X - 0.42Y$	$Z = 138.42 + 0.11X - 0.36Y$	$Z = 162.53 + 0.02X - 0.37Y$
R^2	0.79	0.96	0.96	0.97
slope (cm km^{-1})	31.59	45.75	37.56	37.45
Std. error of est.	0.27	0.21	0.28	0.26
No. of observations	18	18	18	18
Direction of slope	N331E	N338E	N343E	N356E

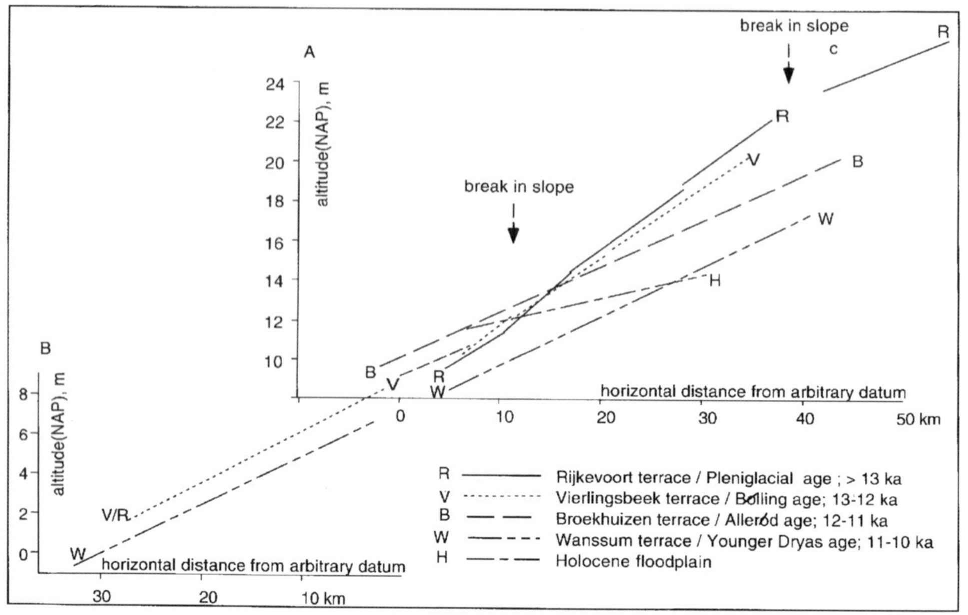

Figure 8.4 Compilation of longitudinal profile data for the late Weichselian and Holocene terraces along the River Maas (data from this study, A); data for the upper reach (c) of the Rijkevoort terrace from Van den Broek and Maarleveld (1963), and data from the lower reach (B) from Berendsen et al. (1995)

Rijkevoort and Vierlingsbeek terraces is located where the Maas crosses a fault which marks the boundary between the Peel Block and the Venlo Graben (Figure 8.1).

The 'steeper' slope segments and breaks in slope of the two oldest terraces are probably best explained by considering tectonic movements in this area. The area is part of the Roer Valley rift system, which has been subjected to tectonic movements for a long time. The Roer Valley Graben, for instance, has been subjected to differential subsidence from the Oligocene onwards. Displacements along the Peel Boundary Fault (Figure 8.1), which were 0.8 mm yr^{-1} during the Quaternary, still occur, which is illustrated by the earthquake of 13 April 1992 (Geluk et al., 1994). The faults related to the Peel Block (Figure 8.1) are locally seen in the landscape as scarps, for instance in the northern part of this area where 7 m difference in altitude marks the boundary between the Rijkevoort terrace in the Venlo Graben and the Peel Block (Huisink, 1997).

Tectonic movements of either the rising Peel Block or the subsiding Venlo Graben may possibly have led to tilting of blocks in the Venlo Graben, causing small scarps more or less perpendicular to the main faults. This is in accordance with a lineament analysis by Van den Berg (1994b), who showed that the main faults are orientated in a NW–SE direction, but a second important direction, namely NE–SW, is also seen in lineaments occasionally with signs of horizontal movements (Figure 8.1). Tilting is not unknown in this area as tilting of the Peel Block is demonstrated by Van den Berg (1994b). The river possibly reacted by incising upstream of such a scarp and depositing sediment downstream which resulted in a relatively steeper gradient in the area near the scarp in comparison with gradients in up and downstream, tectonically unaffected parts of the valley. The preservation of such scarps is poor in this energetic fluvial environment. The tectonic activity responsible for the local steepening of the Rijkevoort and Vierlingsbeek terraces was apparently active before the formation of the Broekhuizen terrace (Allerød), since the longitudinal profile of this terrace can be correlated downstream without breaks in slopes.

IMPACT OF TECTONIC MOVEMENTS VERSUS CLIMATIC CHANGES ON FLUVIAL STYLES

The slopes of the Broekhuizen and Wanssum terraces (23.5 and 25.4 cm km^{-1}, respectively) are quite similar. The palaeoriver styles, however, are distinctly different: a highly sinuous meandering river in contrast to a braided river. This is probably the most convincing evidence that the Late Glacial Maas did not change its pattern due to tectonic influences. The braided river system of the late Pleniglacial is recognized in large parts of the Maas valley, both up- and downstream (Huisink, 1997; Berendsen et al., 1995; Van den Berg and Schwan, 1996). Changes in channel dimensions or sediment characteristics have not been observed, regardless of differences in slope values. This also shows the minor influence of longitudinal profiles on the fluvial pattern.

The channel morphology on the Vierlingsbeek terrace (Bølling) changes downstream of Beugen, between reaches 2 and 3 (Figure 8.2). The location of this change, however, does not coincide with the break in slope of the longitudinal profiles, which was found between the first and second reach (Table 8.1). The channel patterns on these parts of the terrace are identical, regardless of the large difference in terrace slope (39.6 versus 27.0 cm km^{-1}). Therefore, locally steeper longitudinal terrace profiles, related to tectonic movements, probably did not account for the change in morphology.

A more plausible explanation for the change in morphology on the Vierlingsbeek terrace was the effect of the Niers confluence, a tributary of the Rhine river system at that time. Large amounts of water and especially sediment were added to the Maas system which is shown in the heavy mineral composition of the sediment (Huisink, 1997). Extra input of sediment, especially in a transitional river, might be the reason for the more braided-like pattern downstream of Beugen. So the influence of sediment yield on river morphology seems to be far larger than the influence of longitudinal terrace profiles on river morphology in the Late Glacial Maas valley. The effect of the locally steeper terrace profile resulted only in deeper incised channels on the Vierlingsbeek terrace. Detailed sections across channel scars on the steeper parts of this terrace show deeper incised channels (up to 6 m; Huisink, 1997) as compared to channels on more gently sloping parts of the terrace (up to 3 m).

The importance of tectonic movements, inferred from longitudinal terrace profile analyses, was most probably not large on the river evolution of the Maas during the Late Glacial. Although the uplift of the hinterland of the Maas on a Quaternary time scale resulted in the tendency to erode, and the formation of terraces, it could not account for the development of four terraces in such a short period during the late part of the Pleniglacial and Late Glacial. The phases of incision and deposition are better explained by the distinct climatic changes during the Late Glacial which affected sediment and water supply significantly.

The change in pattern from a braided river in the Pleniglacial via a transitional system into finally one highly sinuous meandering channel in the Allerød is explained by reduced sediment load and more regular discharge. The warming of the Late Glacial resulted in an increasingly dense cover of vegetation, which in turn diminished sediment load. The combination of permafrost thaw and vegetation development increased soil water storage capacity, which, combined with decreasing importance of snowmelt as a major supplier of discharge, resulted in a more regular discharge regime. The Maas reacted to the reduced sediment yield and more regular discharge regime compared to the previous period by incision, whereby the Rijkevoort terrace was formed. Minor channels became abandoned, while larger ones incised and became more curved. This trend proceeded until finally one highly sinuous channel remained active in the Allerød, when a closed wooded vegetation was present. During this process the Vierlingsbeek terrace (transitional system) was thus formed.

At the onset of the cold Younger Dryas an incision occurred due to a rapid change in discharge and a delay in the disappearance of vegetation, whereby bank stability was maintained. Later on the vegetation changed and a more open landscape (yielding higher sediment load) together with irregular discharges resulted in a braided river pattern (Wanssum terrace). Another incision took place at the onset of the Holocene when vegetation changed fast into a closed forest in response to higher temperatures. Discharge became much more regular and a meandering river evolved which incised in the former floodplain. The meandering river remained confined to a narrow floodplain in which the river shifted and deposited merely fine sediments.

CONCLUSIONS

The changes in steepness of terrace slopes did not influence the Late Glacial Maas river styles. For instance, the longitudinal terrace profiles of the highly sinuous meandering river of the Allerød and the braided river of the Younger Dryas are more or less identical, while the morphology is distinctly different. The four phases of incision and accumulation that occurred since the onset of the Late Glacial were most probably not triggered by tectonic activity, but can be better explained by climatically induced factors.

Local tectonic activity, expressed by changes in steepness of the longitudinal terrace profiles on the Rijkevoort and Vierlingsbeek terraces, most likely did not lead to changes in channel morphology. Instead, changes in channel pattern observed on the Vierlingsbeek terrace were probably caused by an extra input of sediment by a tributary. River styles could be correlated through large parts of the valley, regardless of local steeper terrace slopes. On the other hand, the local tectonic activity did possibly lead to locally more deeply incised channels in areas with steeper terrace profiles. Faults also influenced the flow direction of the river, which is more or less perpendicular to the faults.

Longitudinal terrace profiles, calculated by using multiple regression analysis, prove to be useful in the morphological mapping of terraces. They can be used for correlation purposes by connecting isolated terrace fragments.

ACKNOWLEDGEMENTS

This research was funded by the NWO, Dutch Organization for Scientific Research, and is incorporated in The Netherlands Center for Geo-Ecological Research (ICG). I thank K.R. Vincent, J. Vandenberghe, K. Kasse, K. van Huissteden, J. Mol and W. Hoek for their critical comments on this chapter or for their useful discussions.

REFERENCES

Berendsen, H., Hoek, W. and Schorn, E., 1995. Late Weichselian and Holocene channel changes of the rivers Rhine and Meuse in the Netherlands (Land van Maas en Waal). *Paläoklimaforschung*, **14**, 151–172.

Bohncke, S., Vandenberghe, J. and Huijzer, A.S., 1993. Periglacial palaeoenvironments during the Late Glacial in the Maas valley, The Netherlands. *Geologie en Mijnbouw*, **72**, 193–210.

Geluk, M.C., Duin, E.J.Th., Dusar, M., Rijkers, R.H.B., Van den Berg, M.W. and Van Rooijen, P., 1994. Stratigraphy and tectonics of the Roer Valley Graben. *Geologie en Mijnbouw*, **73**, 129–141.

Huisink, M., 1997. Sedimentological and morphological changes of a lowland river as a response to climatic change: the Maas; The Netherlands. *Journal of Quaternary Science*, **12**, 209–223.

Jelgersma, S., 1966. Sea level changes in the last 10 000 years. In *International Symposium on World Climate from 8000 – 0 BC*, Royal Meteorological Society, 54–69.

Kasse, K., Vandenberghe, J. and Bohncke, S., 1995. Climatic change and fluvial dynamics of the Maas during the Late Weichselian and early Holocene. *Paläoklimaforschung*, **14**, 123–150.

Merritts, D.J., Vincent, K.J. and Wohl, E.E., 1994. Long river profiles, tectonism, and eustasy: A guide to interpreting fluvial terraces. *Journal of Geophysical Research*, **99**(B7), 14031–14050.

Pons, L.J., 1957. De geologie, de bodemvorming en de waterstaatkundige ontwikkeling van het Land van Maas en Waal en een gedeelte van het Rijk van Nijmegen. *Mededelingen Stichting Bodemkartering, Bodemkundige Studies*, Vol. 3.

Pons, L.J. and Schelling, J., 1951. De laatglaciale afzettingen van de Rijn en de Maas. *Geologie en Mijnbouw*, **13**, 293–297.

Van den Berg, M.W., 1994a. Neotectonics of the Roer Valley rift system. Style and rate of crustal deformation inferred from syn-tectonic sedimentation. *Geologie en Mijnbouw*, **73**, 143–156.

Van den Berg, M.W., 1994b. Patterns and velocities of recent crustal movements of the Dutch part of the Roer Valley rift system. *Geologie en Mijnbouw*, **73**, 157–168.

Van den Berg, M.W., 1996. *Fluvial sequences of the Maas. A 10 Ma record of neotectonics and climate change at various time-scales*, Thesis, University of Wageningen.

Van den Berg, M.W. and Schwan, J.C.G., 1996. Millennial climatic cyclicity in Weichselian Late Pleniglacial to Early Holocene fluvial deposits of the river Maas in the Southern Netherlands. In M.W. Van den Berg, *Fluvial Sequences of the Maas*, Thesis, University of Wageningen, 99–121.

Vandenberghe, J., Kasse, C., Bohncke, S. and Kozarski, S., 1994. Climate-related river activity at the Weichselian–Holocene transition: a comparative study of the Warta and Maas rivers. *Terra Nova*, **6**, 476–485.

Van den Broek, J.M.M. and Maarleveld, G.C., 1963. The Late-Pleistocene deposits of the Meuse. *Mededelingen Geologische Stichting*, **16**, 13–24.

Van Montfrans, H.M., 1975. Toelichting bij de ondiepe breukenkaart met de diepteligging van de Formatie van Maassluis, 1:600,000. In W.H. Zagwijn and C.J. van Staalduinen (eds) *Toelichtingen bij geologische overzichtskaarten van Nederland*, Rijks Geologische Dienst, Haarlem.

Verbraeck, A., 1990. De Rijn aan het einde van de laatste ijstijd: de vorming van de jongste afzettingen van de Formatie van Kreftenheye. *K.N.A.G. Geografisch Tijdschrift*, **23**(4), 328–339.

CHAPTER 9

Late Pleistocene and Holocene Siberian River Valley Geomorphogenesis as a Result of Palaeogeographical Cyclic Changes

ANATOLY F. YAMSKIKH

Department of Geoecology, Krasnoyarsk State University, Krasnoyarsk, Russia

INTRODUCTION

Reconstruction of valley formation and stratigraphical subdivision of river valley deposits is one of the most difficult problems in Siberian geomorphology. The stratigraphy of river valley deposits is important for palaeogeographical, geomorphological, and palaeoecological reconstructions, and for investigations of neotectonic movements in the Siberian palaeoperiglacial belt. The relief and deposits of river valleys provide good indicators of landscape changes. Most archaeological sites are concentrated in the river valleys. Despite many Siberian river valley investigations, there is little agreement among scientists about terrace ages, or about the dynamics of their formation.

Most studies of Siberian terrace sequences interpret the terrace strata in terms of cyclic theories of terrace formation: every terrace corresponds to a specific palaeoclimatic and hydrological cycle. Application of this theory to the specific deposits of Siberian river valleys has led to discussions of various geological and geomorphological problems.

Estimates of terrace ages vary by a factor of 10 or more. For example, the first terrace of the Yenisei river valley is variously dated Late Pleistocene Glacial (Q_{III}^{2}–Q_{III}^{4}) (Gromov, 1948), beginning of the Holocene (Nagorsky, 1941), Middle Würm Interstadial–Late Würm Glacial (Feniksova, 1977), Late Würm Glacial (Arkhipov, 1973; Borisov, 1984), Middle Würm Interstadial–beginning of Late Würm Glacial. Detailed investigation of the first terrace deposits and horizon dating show progressive development from the Middle Würm Megainterstadial to the end of the Holocene. The lowest terrace strata are channel gravel and floodplain–channel alluvium (dated 25.8–20.4 to 14.4–12.9 ka). The upper part is dated from 10.4 to 0.4–0.2 ka.

The polycyclic theory of terrace formation in the Siberian palaeoperiglacial belt solved many of these problems. In contrast to cyclic terraces, polycyclic terraces form during several palaeoclimatic and hydrological cycles. For example, formation of the middle terrace (35–55 m high) lasted, with several interruptions (indicated by buried soils), from Middle Riss and the Riss–Würm Interglacial up to the Late Würm. Surfaces of polycyclic terraces have several levels which form terrace level sequences.

Polycyclic terrace formation is well demonstrated for the Late Pleistocene–Holocene terrace formation in the central and southern Siberian palaeoperiglacial belt (Yamskikh, 1993). Data for northern (Kind, 1974), western (Arkhipov, 1973), and northeastern (e.g. Shofman, 1980; Tomirdiaro, 1980) Siberian terrace deposits indicate that polycyclic valley geomorphogenesis is typical for the whole Siberian palaeoperiglacial belt.

Polycyclic river valley geomorphogenesis is caused by the palaeohydrological river regime of the Siberian palaeoperiglacial belt (Yamskikh, 1993). Some peculiarities of the Ob river valley (upper part) hydrological conditions for sediment accumulation and valley relief formation are described by Voikov (1976).

Most Siberian researchers recognize the existence of Pleistocene dammed lake basins in the river valleys. Large glacial dammed lakes existed in Ob and Yenisei river basins (middle part) during the Last Glaciation in western Siberia. Glacial dammed basins also formed in the lower part of the Yenisei valley during the Last Glaciation (Goncharov, 1989). A large lacustrine basin existed in the Tuva depression (Borisov and Minina, 1980). Dammed lakes also formed in the valley of the river Lena (Osadchi, 1982). The Darkhat glacial dammed lake existed in the upper part of the Yenisei river basin (Grosswald, 1987).

REGIONAL SETTING AND METHODS OF INVESTIGATION

The Siberian region of northern Asia stretches from the Ural mountains (in the west) to the Chukotka peninsula and the Baikal region (in the east) (Figure 9.1). The western Siberian depression, central Siberian plateau, mountains raised by recent tectonism (Altay, Sayan and ridges of northeastern Siberia) and the Taimyr peninsula form the main features of Siberian relief. Siberian tectonics, orography and palaeogeography determine the complex river systems.

Palaeohydrological river regime reconstructions employ several methods: morphostructural, lithological, geomorphological, palaeoclimatic, palaeontological, palaeohydrological and geochronological. Detailed investigations were done for key Late Pleistocene and Holocene sections in central and southern Siberia. Special attention has been devoted to the multilayer Palaeolithic, Neolithic, Bronze and Iron Age archaeological sites. These investigations provided important evidence for environmental fluctuations and palaeohydrological river regime changes.

PECULIARITIES OF THE PALAEOHYDROLOGICAL RIVER REGIME CHANGES IN THE SIBERIAN PALAEOPERIGLACIAL BELT

The extreme continental climate, large variations of snowfall intensity, permafrost distribution, landscape changes and the resulting river regime are the main polycyclic factors of terrace formation in the Siberian palaeoperiglacial belt. Extreme continentality, variable barometric maxima and irregular precipitation are the main features of the modern Siberian climate. The 'cold pole' of the northern hemisphere is located in northeastern Siberia (min. $-71\,°C$ at Oymyakon). The climate continentality of this region is 95–97%.

This extreme continental cold climate was typical for the Siberian region during the Pleistocene and Holocene. A trend of winter temperature decreasing from west (-30 to

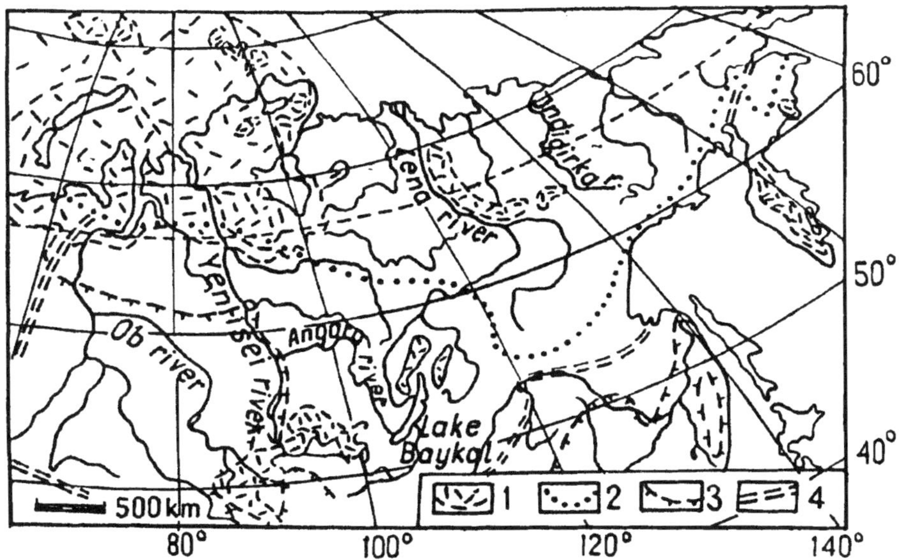

Figure 9.1   General map of Siberia. Legend: 1, boundaries of Late Würm Glaciation; 2, boundaries of continuous permafrost expansion; 3, boundaries of 'island-like' permafrost expansion; 4, boundaries of Siberian region

$-32\,°C$; in the north -42 to $-50\,°C$) to east (-54 to $-60\,°C$ in northeastern Siberia) was well expressed during the Late Würm (Frenzel et al., 1992). A similar temperature trend occurs in southern Siberia (from -28 to $-32\,°C$ in the west to $-45\,°C$ in the east). Differences in the modern precipitation distribution (central parts of western Siberia, 500–600 mm; 100–200 mm in winter) were sharper during the Last Glaciation. Asymmetry of the permafrost and glacier distribution resulted from these differences. Generally, precipitation in northern Siberia (compared to modern values) decreased by 100–250 mm, while it increased in southern Siberia by 1.5 times during glaciations.

The dynamics of eastern European and Siberian glaciations is very important for palaeohydrological reconstructions. Siberian glaciations began earlier than European ones. This was caused by earlier cooling and higher snowfall intensity in Siberia. This was typical, for example, for the Early Würm Glaciation. Cooling of the Late Pleistocene climate led to glacier expansion in Europe, but anticyclonic air masses limited glacier expansion in Siberia (Velichko, 1987, 1993). The Siberian climate was affected by seasonal central Asian and Arctic barometric maxima, which varied greatly for different Siberian regions during Pleistocene glacial–interglacial periods. In regions covered by glaciers these maxima were barriers to the western Atlantic cyclones. The asymmetry of western and eastern glaciation was a consequence of this process. The influence of the central Asian barometric maxima was weaker during the cold season. The cyclonic tracks moved to the south from the glaciers (in central and southern Siberia). This caused increasing snowfall intensity, especially in south Siberia. River runoff in the south of the palaeoperiglacial belt increased during glaciations (caused by snowmelt and glacial meltwater). The role of central Asian barometric maxima increased during the cold seasons of interstadials. Cyclones moved to the north. Interglacial epochs are characterized by decreasing river runoff. Flow was more uniform than during glacial periods. Lower levels of precipitation were typical for northeastern Siberia (caused by relief and by barometric maxima influenced by glaciers located in the north of western Siberia).

Spore–pollen spectra of the terrace deposits reflect cyclic changes of vegetation and climate. Cold-steppe and tundra–steppe spectra with forest vegetation along river valleys (*Larix sibirica, Larix daurica, Pinus sibirica, Pinus sylvestris, Betuia sect. albae*, bush forms of *Betula sect., nanae* and *Alnaster* sp.) were typical of this period. Chenopodiaceae, *Artemisia, Ephedra,* Poaceae dominated among herbaceous plants. Ligneous and herbaceous vegetation was more diverse during interglaciations and interstadials. Also widespread was *Abies sibirica, Picea obovata* (*Abies sibirica* indicates climatic warming and average seasonal river runoff in Siberia).

Two climatic and vegetation zones can be distinguished in the palaeoperiglacial belt of central and southern Siberia during the Late Pleistocene: (1) a cryosemiarid zone located to the south from the Western Sayan mountains to the Obruchev ridge; and (2) a cryosemihumid zone located in the north, occupying the largest Siberian palaeoperiglacial belt. Dry steppe landscapes (modern semidesert) prevailed in the southern zone. Spore–pollen diagrams of periglacial deposits of some terraces in the cryosemihumid zone indicate 10–12 cycles of vegetation change. These cycles reflect sequences of stadial–interstadial climatic changes (Yamskikh and Yamskikh, 1992). Horizons of basal alluvium (normal alluvial suites) were formed during interglacials and interstadials. Spore–pollen spectra of boreal forests, similar to modern Siberian taiga, are typical of these deposits. Palynological diagrams of bedded 20–30 m thick strata of Late Pleistocene terraces reflect polycyclic vegetation changes. Alluvial loess and other deposits in the lower parts of the cyclic units contain microfossils of cold-steppe and tundra–steppe vegetation. Spore–pollen spectra indicate forest vegetation. The upper parts of cyclic units usually contain soils and are characterized by forest and forest–steppe spore–pollen spectra. Pleistocene spore–pollen diagrams of northeastern Siberia show no forest vegetation. Cold-steppe and tundra–steppe spore–pollen spectra are typical of these deposits, corresponding to the Pleistocene palaeoclimates of this region.

Holocene deposits contain more microfossils. Spore–pollen spectra vary in space. Section Merzlyi Yar is the key section of Holocene deposits. It is located in the upper part of Yenisei river valley in the Todza depression (southern Siberia). Holocene strata are deposited on Middle and Late Würm alluvium. Periglacial pollen spectra are typical of these deposits. The central part of these strata has a radiocarbon age of 20 400 ± 200 yr BP (GIN 3667). Holocene alluvial and lacustrine–alluvial deposits are intercalated with peat and soil horizons, layers of fossil forest, and prominent ice wedges. The thickness of the Holocene strata is about 15–18 m.

An increase of *Abies sibirica* indicates Holocene warming of the climate. A decrease of *Abies sibirica* and increase of *Picea obovata, Pinus sibirica, Pinus svivestris, Larix* sp., *Betula sect. albae* and the bush form of *Betula sect. nanae* indicate Holocene cooling of the climate. There are maxima of *Abies sibirica* abundance in the Allerød, at 10.4, 8.7–8.5, 6.7, about 5.2, and 2.7–2.5 ka. Several hundred years later (after maxima of *Abies sibirica*) there were maxima of *Picea obovata* and *Pinus sibirica* abundance at 11.2–11.0 ka (the largest maximum), about 9.5 ka (secondary maximum), 8.2–8.0 ka, about 6.0, 2.3 ka BP (secondary maximum), and during the Little Ice Age. Cryogenesis, indicated by a polygonal net of ice-wedge formation, developed after the *Picea obovata* and *Pinus sibirica* maxima. Lacustrine–alluvial deposits accumulated during these periods. Their spore–pollen diagrams contain both ligneous and herbaceous species. The spore–pollen spectra reflect metachronous changes of temperature and precipitation. Deposits of the Merzlyi Yar section reflect the change of mean river regime, including high floods and the Todza dammed lake formation. The temperature depression of the southern Siberian mountains results from global climatic fluctuations, corresponding to stages of Alpine mountain glaciations. Stadial moraines and some features of river morphology indicate mountain glaciation in Altay, the Sayan mountains and in eastern Siberia.

Permafrost is one of the most important factors in the irregular river runoff and dam formation. The largest part of the Siberian region is covered by permafrost (except the southern part of western Siberia and the plains of southern Yenisei Siberia) (Figure 9.1). Central and northeastern Siberia are occupied by continuous permafrost. Nearly all river runoff occurs during the warm season. High river ice dams (up to 35 m high) lead to the formation of dammed lakes in the basins of these rivers.

Permafrost occupied the whole territory of Siberia during Pleistocene cold climates. Horizons of cryogenic features, polygonal nets of fossil ice wedges, and layers of solifluction in Pleistocene and Holocene deposits indicate different ranges of rhythmical permafrost expansion. This can be distinguished in the loess soil, alluvial soil, and diluvial soil cyclic units. Permafrost was also widespread during the Middle Würm megainterstadial.

The greatest permafrost thickness (800–1500 m) is typical of northeastern Siberia. It occurs in the so-called 'edoma' deposits containing thick ice wedges. Some researchers hypothesize a cryogenic–aeolian origin for these deposits (Tomirdiaro, 1980). We support the lacustrine–alluvial theory of their origin (Popov, 1967). In the continental part of northeastern Siberia edoma deposits form plains that penetrate into mountains along the river valleys in the form of 'gulfs'. It is interesting that radiocarbon dates of different layers of edoma deposits generally coincide with dates for polycyclic terraces (located in the Yenisei valley and in the river valleys of the Baikal region).

Buried pedocomplexes indicate climate warming, well-developed vegetation cover, less variable seasonal river runoff, and river valley relief stabilization. Polycyclic terrace and valley slope deposits of the cryosemihumid zone of the palaeoperiglacial belt usually contain several buried soil horizons. Grey forest, meadow–steppe (meadow–chernozem-like) soils are widespread among these buried soils. Alluvial soils, meadow–swamp soils, and peat–gleyic soils dominate the Holocene buried soils of low elevation terraces and the floodplain. Early and Middle Pleistocene soils are usually absent. Detritus-rich horizons of edoma deposits have been dated by Tomirdiaro (1980).

It is remarkable that the main soil formation epochs of different regions are approximately synchronous. Thus, the phases of valley relief stabilization were synchronous (Figure 9.2). Full profile soils formed during the warmest periods of the Pleistocene and Holocene. Weakly developed soils formed on the frequently flooded floodplain and low terraces. Sedimentation interruption occurred simultaneously on the river valley slopes and terraces. Soils indicate these interruptions.

PALAEOHYDROLOGICAL RIVER REGIME OF THE SIBERIAN PALAEOPERIGLACIAL BELT

Snow meltwater, with lesser influences of rain and ground water, determines the modern river regime (Table 9.1). High spring floods are typical for all Siberian rivers except those located in the east of the Baikal region and in the centre of northeastern Siberia (where

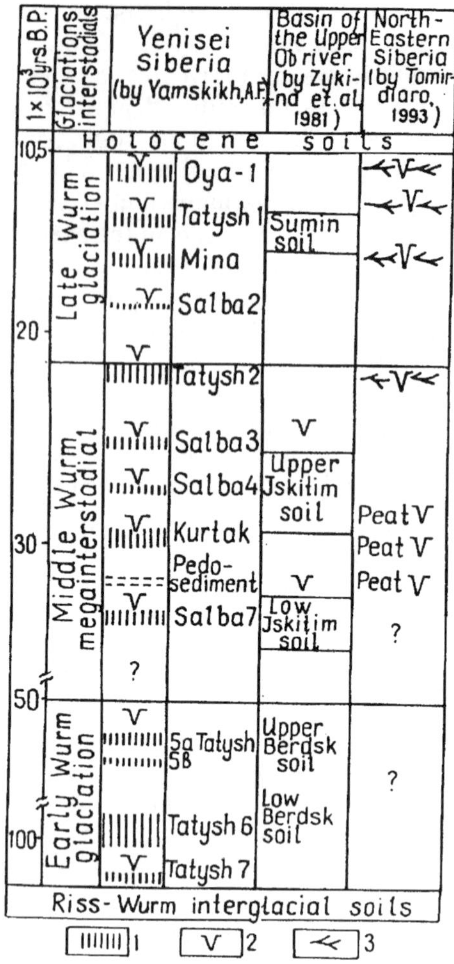

Figure 9.2   Correlation of Late Pleistocene soil formation and cryogenic processes in Siberian palaeoperiglacial belt. Legend: 1, buried soil; 2, cryogenic horizons; 3, root-rich horizons

floods during warm seasons are typical). Unit discharges in the plains increase from south $(0.5\,l\,s^{-1}\,km^{-2})$ to north $(20\,l\,s^{-1}\,km^{-2})$, in northeastern Siberia $(0.5-10\,l\,s^{-1}\,km^{-2})$, and in southern Siberia $(6-25\,l\,s^{-1}\,km^{-2})$.

Modern flood discharges in the middle part of the Yenisei river valley reach 3.7×10^4 $m^3 s^{-1}$. Decreasing evaporation and infiltration, and increasing suspended load cause high rates of river discharge in southern Siberia, of about 1.5×10^5 to 2×10^5 $m^3 s^{-1}$, without outflows from dammed lakes. Modern floods in the same part of the Yenisei river valley reach about 6–15 m in height (also without floods caused by outflows from dammed lakes). The influence of relief barriers on Yenisei river level fluctuations is important, with flood levels on the plain parts of the river valleys lower by 5–8 m than those in the valleys located in mountainous regions. The northern direction of river flow leads to an important role of river ice dams on river level fluctuations.

Table 9.1 Morphometric characteristics and runoff of Yenisei, Ob and Lena rivers (Siberia)

River	River length (km)	Basin area (km^2)	Annual runoff (km^3)	Mean annual runoff $(m^3 s^{-1})$
Yenisei (with river Selenga)	5540	226×10^4	623	19.8×10^3
Ob (from river Irtysh)	5570	289×10^4	390	12.4×10^4
Lena	4440	249×10^4	508	16.1×10^4

Archaeological investigations of the Palaeolithic and Neolithic cultural layers show evidence of regular changes of elevation along the river valleys. Multilayer Late Palaeolithic archaeological sites (up to 24 cultural layers) are widespread in the palaeoperiglacial belt. Archaeological sites are located in the terrace complexes from 9–10 to 70–80 m high. All multilayer archaeological sites occur in the cryosemihumid zone, where large amplitudes of the river level fluctuation were well expressed (Yamskikh, 1992). Low elevation archaeological sites correspond to the buried soils. Cultural horizons of higher archaeological sites usually occur at the boundaries between different genetic types of deposits (alluvial, lacustrine–alluvial, and diluvial deposits). Changes of elevation during interstadials (low elevation terraces) and stadial cold climates (high terraces and river valley slopes) were, respectively, 10–15 and 30–40 (up to 60) m.

Results of geological and geomorphological investigations, palaeogeographical data, evidence of sudden abandonment of archaeological sites, and palaeoreconstructions reveal that ancient people settled on different geomorphological levels. This can be attributed to high floods, and the formation and outflow of dammed lakes. It is remarkable that in southern Siberia there are no similar-aged archaeological sites located in a succession of different heights. In the cryosemiarid zone of the palaeoperiglacial belt, Early, Middle, and Late Pleistocene stadial–interstadial archaeological sites are located on the modern surface as relict archaeological sites.

Phases of high flood and dammed lake formation alternated with stages of average river regime. This occurred during Late Pleistocene stadial–interstadial rhythms (Table 9.2). Holocene river regime changes reflect climatic fluctuations during the last 10 500 years (Table 9.3). The results derive from studies of key southern and central Siberian section complexes.

Siberian river valley morphology, with alternating broad and narrow parts of the river valley, promotes dammed lake formation. River ice and glacial dams formed at the entrance of the river channels to mountains. Lacustrine–river systems formed during river ice movements. Pleistocene and Holocene terraces and ancient cliffs, located on the slopes of the modern lake depressions, are consequences of this process. Heights of lacustrine terraces in the middle part of the Yenisei river basin (from 1–3 to 50–70 m high) are similar to the river terrace heights. Northern glacial dammed lakes affected river ice dam formation in the upper parts of the river valleys. The horizons of lacustrine–alluvial deposits contain layers of buried forest which indicate the beginning of dammed lake formation. The duration of the river ice dams varied from several days up to 1.5–2 weeks (during river ice movements). Atamonovo, Syr, Bateny Abakan, Maina and Todza lakes were the largest river ice-dammed lakes in Yenisei river basin (Figure 9.3). Outflows from these lakes were catastrophic.

Large dammed glacial lakes also appeared during the Late Pleistocene in the Ob and

Table 9.2 Palaeohydrological river regime changes and palaeoecological fluctuations in central and southern Siberia during stadial–interstadial phases of the Late Pleistocene

^{14}C, age (ka)	Palaeoclimatic, hydrological, and ecological events
10.5–11.1	Glaciation – third thermal Late Würm minimum; increased precipitation; several phases of high floods indicated by horizons of fossil forest and layers of rhythmically bedded periglacial sediments; cryogenesis. 10 990 $\pm$ 180 ^{14}C yr BP (LE 4676).
11.1–11.8	Lowered variability of seasonal runoff; river relief stabilization; soil formation; first maximum of coniferous forests (*Picea obovata, Abies sibirica*) during Late Würm and Holocene. 11 600 $\pm$ 500 ^{14}C yr BP (GIN 403).
11.8–12.0	Stadial cooling; high floods; periglacial vegetation. 11 900 $\pm$ 150 ^{14}C yr BP (SOAN 1444).
12.0–13.0	Interstadial warming; lowered variability of seasonal river runoff; soil formation; boreal forests; archaeological sites located on low geomorphological surfaces. 12 980 $\pm$ 130, 12 910 $\pm$ 100 ^{14}C yr BP (LE 2133).
13.0–15.5	Glaciation; second thermal Late Würm minimum; high floods and dammed lake formation; archaeological sites located on high geomorphological surfaces; periglacial vegetation; cryogenesis. 13 650 ^{14}C yr BP (GIN 3666), 14 300 $\pm$ 100 ^{14}C yr BP (LE 1457), 14 600 $\pm$ 200 ^{14}C yr BP (GIN 2101), 15 200 $\pm$ 150 ^{14}C yr BP (LE 2383), 15 500 ^{14}C yr BP (LE 2299).
15.5–16.5	Interstadial warming; low levels of floods; buried soils; archaeological sites located on low geomorphological surfaces. 16 176 $\pm$ 180 ^{14}C yr BP (LE 2135A), 16 540 $\pm$ 170 ^{14}C yr BP (LE 2135B), 16 600 $\pm$ 200 ^{14}C yr BP (GIN 2102).
16.5–22	Glaciation; periglacial tundra–steppe vegetation; high levels of floods and dammed lakes, cryogenesis. 17 660 $\pm$ 700 ^{14}C yr BP (GIN 2863).
	Short-term interstadial, weakly developed soil formation, low levels of floods. 18 470 $\pm$ 730 ^{14}C yr BP (LE 4629).
	First thermal Late Würm minimum; high levels of floods and dammed lakes; mass extinction of some species of Mamuthus fauna; disappearance of *Coeiodonta antiquitatis (Blum)*; slope processes; periglacial vegetation; cryogenesis; archaeological sites located on high geomorphological surfaces. 19 500 $\pm$ 200 ^{14}C yr BP (GIN 2859), 19 700 $\pm$ 200 ^{14}C yr BP (GIN 2861), 20 100 $\pm$ 100 ^{14}C yr BP (GIN 2863), 20 400 $\pm$ 200 ^{14}C yr BP (GIN 3667).
22–45	Megainterstadial.
	Thermal maximum of the megainterstadial; boreal forests; decreasing amplitude of river level fluctuations; formation of full-profile soils. 22 100 $\pm$ 80 ^{14}C yr BP (GIN 2465).
	Cooling, high levels of rivers and dammed lakes; cryogenesis; archaeological sites located on high geomorphological surfaces. 22 930 $\pm$ 350 ^{14}C yr BP (LE 3611), 23 470 $\pm$ 200 ^{14}C yr BP (LE 2833a), 24 170 $\pm$ 230 ^{14}C yr BP.
	Warming of climate; weakly developed soil formation (pedosediments); river valley relief stabilization. 25 110 $\pm$ 800 ^{14}C yr BP (LE 4628).
	Cooling of climate, high levels of alluvial and lacustrine sediment accumulation; activation of slope processes. 27 470 $\pm$ 200 ^{14}C yr BP (LE 2833).
	River valley relief stabilization (c. 29–30 ka); soil formation.
	Cooling of climate (c. 30–33 ka), high levels of floods and dammed lakes, archaeological sites located on high geomorphological surfaces; activation of slope processes; cryogenesis. 31 650 $\pm$ 520 ^{14}C yr BP (LE 3352).
	Lowered variability of seasonal river runoff during warming of climate (c. 33–45 ka), fluctuations of precipitation, soil formation. 38 640 $\pm$ 1000 ^{14}C yr BP (LE 4625).
50–110	Early Würm Glaciation.
	Glaciation; high levels of floods and dammed lakes; two weak interstadials with relatively low levels of floods and weakly developed soil formation (soils Tatysh 5a and 5b, soils separated by 'cold' alluvial loess) (about 60–70 ka).
	Interstadial; low levels of floods; long-term soil formation (about 100 ka).
	Glaciation; high levels of floods.
	Soil complex formation (consisting of 3–4 weakly developed soils) intercalated in loess-like sandy floodplain deposits.
>110	Interglacial Riss–Würm Period; lowered variability of the seasonal river runoff with trend of increasing annual discharges; boreal forests; formation of normal alluvial suites.

Table 9.3   Palaeohydrological river regime changes and palaeoecological fluctuations in central and southern Siberia during the Holocene

^{14}C, age (ka)	Palaeohydrological river regime phases; periods of dammed lake formation and valley relief stabilization
0–0.4	Activation of erosion; high levels of floods; extinction of forests on floodplain and low terraces about 400 yr BP. 200 ± 50 ^{14}C yr BP (GIN 4282), 380 ± 30 ^{14}C yr BP (GIN 2267).
0.4–0.5	River valley relief stabilization; soil formation on floodplains; lowered variability of river runoff. 460 ± 30 ^{14}C yr BP (GIN 2265), 480 ± 60 ^{14}C yr BP (GIN 4256).
0.5–1.5	Cooling of the climate; high levels of floods and dammed lakes in the river valleys. 720 ± 40 ^{14}C yr BP (GIN 4307), 930 ± 40 ^{14}C yr BP (GIN 4285), 1070 ± 100 ^{14}C yr BP (GIN 4279).
1.5–1.8	Increasing precipitation; high levels of floods; alluvium accumulation on low terraces; activation of slope processes; degradation of forests on low terraces. 1425 ± 50 ^{14}C yr BP (KRIL 399).
1.8–2.5	River valley relief stabilization; soil formation on the terraces; decreased variability of seasonal river runoff. 1960 ± 80 ^{14}C yr BP (LE 437). Increasing precipitation; swamp formation; high levels of floods; cryogenesis. 2200 ± 40 ^{14}C yr BP (GIN 2251), 2260 ± 55 ^{14}C yr BP (LE 4375), 2380 ± 50 ^{14}C yr BP (GIN 3657).
2.5–3.0	Decreased variability of seasonal river runoff; soil formation on low terraces. 2790 ± 65 ^{14}C yr BP (KRIL 402), 2830 ± 80 ^{14}C yr BP (LE 4377).
3.0–4.1	High levels of floods and dammed lakes; horizon of fossil forest; cryogenesis. 3000 ± 30 ^{14}C yr BP (GIN 2253), 3170 ± 50 ^{14}C yr BP (GIN 2255). Increasing precipitation; increasing river discharges; prevalence of *Pinus sibirica*, *Picea obovata* in mountainous taiga belt. 3520 ± 60 ^{14}C yr BP (LE 792–49).
4.1–5.0	Third Holocene climatic optimum; mean seasonal runoff, river valley relief stabilization; soil formation. 4260 ± 17 ^{14}C yr BP (GIN 4260), 4380 ± 120 ^{14}C yr BP (GIN 4257), 4580 ± 60 ^{14}C yr BP (LE1230), 4780 ± 80 ^{14}C yr BP (GIN 4283), 4820 ± 280 ^{14}C yr BP (GIN 4265).
5.0–6.0	Increasing precipitation and river discharges; activation of erosion; high levels of floods and dammed lakes with short period of mean seasonal runoff. 5930 ± 50 ^{14}C yr BP (GIN 2264).
6.0–6.3	Dammed lakes; swamps; horizon of buried forest; increasing precipitation. 6140 ± 40 ^{14}C yr BP (GIN 2252), 6270 ± 110 ^{14}C yr BP (GIN 2268–2), 6280 ± 40 ^{14}C yr BP (GIN 2256).
6.3–7.5	Second Holocene optimum; river valley relief stabilization; mean seasonal river regime; soil formation. 6530 ± 150 ^{14}C yr BP (GIN 2259), 6580 ± 150 ^{14}C yr BP (GIN 2259a), 6670 ± 100 ^{14}C yr BP (GIN 2257a), 6690 ± 80 ^{14}C yr BP (GIN 2257), 6830 ± 210 ^{14}C yr BP (LE 1227).
7.5–8.2	Increasing precipitation; swamps; high levels of floods and dammed lakes; degradation of forests. 8190 ± 60 ^{14}C yr BP (GIN 2258).
8.2–9.9	First climatic Holocene optimum; mean seasonal river runoff; stabilization of river valley relief; soil formation on low terraces. 8300 ± 230 ^{14}C yr BP (LE 1226), 8370 ± 40 ^{14}C yr BP (GIN 2261). Two phases of high levels of floods and dammed lakes; horizons of fossil forest; increasing river discharges. (1) 8590 ± 80 ^{14}C yr BP (GIN 2261), 8790 ± 40 ^{14}C yr BP (GIN 2260); (2) 9160 ± 9480 ^{14}C yr BP.
9.9–10.4	River relief stabilization; mean seasonal river runoff; soil formation, on terrace surfaces also. 10 360 ± 110 ^{14}C yr BP (LE 491-489), 10 450 ± 50 ^{14}C yr BP (GIN 2263).

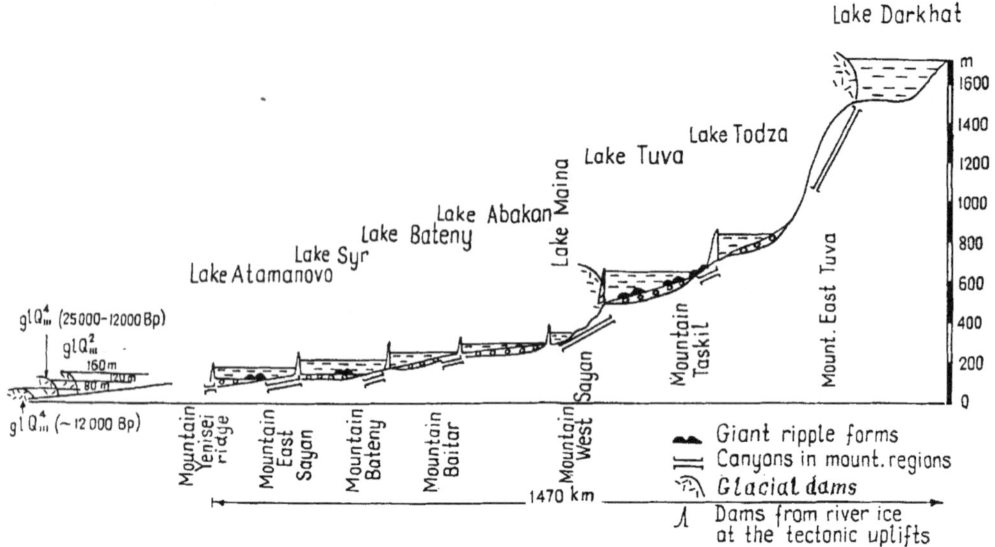

Figure 9.3   Series of repeatedly dammed lakes in the upper and middle part of Yenisei river valley during the Late Glaciation (longitudinal profile)

Yenisei river basins. The Mansa lake stretched to the south from the northern Siberian glacier. It occupied more than half of this part of western Siberia.

Three spillways connected the dammed lakes of the Ob and Yenisei river basins. Large Yenisei river discharges caused dammed lake outflows from the Yenisei river basin into the Ob dammed basin. Water then moved to the central Asian closed drainage basin. A sequence of dammed lakes also appeared in the Lena river basin (Osadchi, 1982). Heights of dams in the Yenisei river basin during the Last Glaciation were about 120–160 m (Goncharov, 1989). Lakes Darkhat and Todza were the largest ones in the upper part of the Yenisei river valley. The volume of the Darkhat dammed lake was about 250 km^3 (Grosswald, 1987). Velocities of water movement during outflows reached 15–20 m s^{-1}; discharges were 3.1×10^5 to 4×10^5 m^{3}s^{-1}. Outflows from the Darkhat lake were repeated every 100–300 years and more frequently during deglaciations. Todza lake cliffs and levels of lacustrine sediment accumulation show that the depth of the Todza lake was about 35–40 m. Discharges during outflows from the Todza and Tuva dammed lakes were about 5.1×10^4 to 1.2×10^5 m^{3}s^{-1}. Large dammed glacial lakes also existed in the Altay mountains (Baker et al., 1993). The Kura'i lake volume was 1000 km^3, maximum discharges during outflows were 1.6×10^7 to 1.8×10^7 m^{3}s^{-1}. The velocity of water movement was 20 m s^{-1} (maximum 45 m s^{-1}).

SIBERIAN-TYPE VALLEY GEOMORPHOGENESIS

Pliocene, Early, Middle and Late Pleistocene erosional and depositional macrocycles correspond to the formation of strath terraces for the high, middle and low height terrace complexes. Geological and geomorphological consequences of erosion–deposition processes were studied in detail for the Late Pleistocene–Holocene macrocycle because of the excellent preservation of its sediments and landforms. Early and Middle Pleistocene sediments are destroyed or reworked by the processes of complex denudation.

Polycyclic sediments are intercalated with diluvial–proluvial layers of rock debris and sand. Their accumulation occurred during cryoarid phases of glacial cycles, which caused erosional and depositional processes. These types of sediments are widespread in the southern cryosemiarid zone of the Siberian palaeoperiglacial belt. Loess formation was weakly developed in this zone; terrace strata consist of coarse clastic alluvium.

Multiheight terrace complexes exist in the valleys of the great Siberian rivers; multilevel floodplains and terraces are 4–7 to 240–250 m high. Multiple levels of one terrace and similar heights of neighbouring terraces in the palaeoperiglacial belt can be explained by simultaneous alluvial deposition on different elevation terrace complexes (caused by high levels of floods and dammed lakes).

Giant ripple forms occur on the river terraces of the Altay mountains and Tuva region as a result of outflows from dammed lakes. These are composed of boulder–gravel sediments. Outflows from glacial-dammed and river ice-dammed lakes formed sandy landforms on terraces (up to 20 m high) in the Minusa–Krasnoyarsk part of the Yenisei river valley, in river valleys of the Baikal region, and other areas. These sediments were subsequently reworked by wind and became aeolian landforms. Well-developed spill-ways connect the Abakan and Yenisei river valleys in the Minusa depression. They connect large modern lacustrine depressions in the foothills of the Kuznetski Alatau mountains and in other regions.

Correlation of general cross-sections for the great Siberian rivers of the palaeoperiglacial belt shows: (1) the young ages of the terraces; and (2) similar ages of completion for some terrace complex formations (Figure 9.4). Polycyclic valley geomorphogenesis is well expressed for the Yenisei river valley (Central Siberia) (Yamskikh, 1996). In western Siberia (Ob river basin) the Pleistocene climate was less continental, and the region was less tectonically active. That is why polycyclic sedimentation and terrace formation are well expressed only in the southeastern region, at the upper part of the Ob river basin. In northeastern Siberia 'edoma' deposits occurred in conditions of extreme continental climate and low precipitation. Nevertheless, terrace profiles reveal the evidence of polycyclic sedimentation.

CONCLUSION

Polycyclic terrace formation is the main consequence of the simultaneous accumulation of alluvium and dammed basin sediments on different terrace surfaces. This specific type of valley geomorphogenesis is named 'Siberian'. This term reflects the geographical location and specificity of natural process dynamics. Hydrological factors are the most important processes of Siberian valley geomorphogenesis. Anomalous catastrophic floods and dammed basins were important features of the palaeohydrological river regime in the Siberian palaeoperiglacial belt. Powerful streams formed as a result of outflows from glacial and river ice-dammed basins. These streams formed deep river valleys in their narrow (mountainous) parts and deposited 'giant ripple forms' and ridge sandy landforms in broad (plain) areas.

Tectonic and hydroclimatic factors caused the formation of polycyclic terrace complexes. Each polycyclic terrace complex (Early, Middle and Late Pleistocene) has a normal alluvial basal sequence. Above this, extremely high floods and dammed lakes,

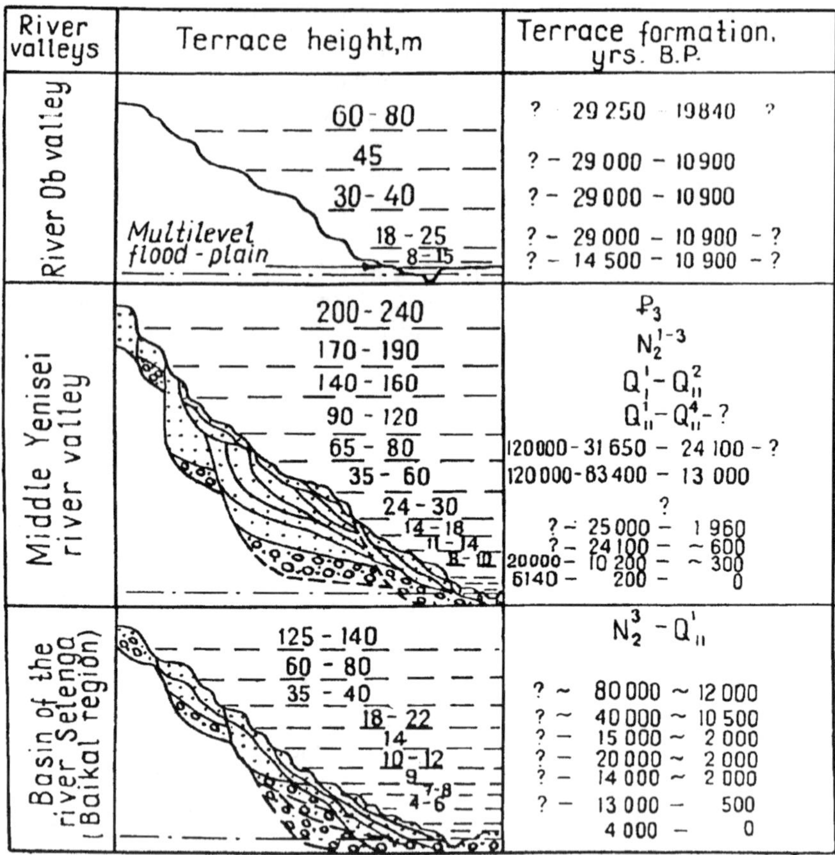

Figure 9.4   Timing of polycyclic terrace formation in the great Siberian river valleys of the palaeoperiglacial belt (Ob river valley data from Arkhipov (1973), and data for valleys of the Selenga river basin from Bazarov et al. (1982))

formed during the warming–cooling rhythms (inside macrocycles), caused polycyclic strata formation. The termination of alluvial deposition occurred simultaneously in some parts of the river valleys on the low and middle terraces.

Climatic rhythms do not always perfectly correspond to the palaeohydrological rhythms. This results from different intensities of warm and cold phases, and from metachronous fluctuations of temperature and precipitation.

Climatic–landscape conditions of stadial–interstadial periods variously influenced erosional–depositional processes. On one hand, increasing river discharges in warm climatic conditions and river regime led to erosion. On the other hand, flood discharges with high contents of suspended load were powerful agents of erosion (in the narrow parts of the river valleys) and of deposition (in broad parts of river valleys). Coarse gravel sediments accumulated on the bottoms of river valleys. At the same time, sandy deposits accumulated at high geomorphological levels. This model of erosional–depositional process changes can be ascribed to metachronous temperature and precipitation changes.

Powerful Quaternary floods and outflows from dammed lakes were not catastrophic from the palaeogeographical point of view. The regular repetition of extreme palaeohydrological events during the Pleistocene and Holocene makes these events typical of this region.

REFERENCES

Arkhipov, S.A., 1973. Stratigraphy, terrace and buried valleys geochronology of upper part of Ob River Basin. In Saks et al. (eds), *Pleystocen Sibiri i Prilegauschikh Regionov*, Nauka, Moscow, 7–21 (in Russian).

Baker, V.R., Benito, G. and Rudoy, A.N., 1993. Palaeohydrology and Late Pleistocene superflooding, Altay Mountains, Siberia. *Science*, **259**, 348–350.

Bazarov, D.B., Konstantinov, M.V. and Imektenov, L.B., 1982. *Geolgiya i Kultura Drevnikh Poseleniy Zapadnogo Zabaykaliya*, Nauka, Novosibirsk (in Russian).

Borisov, B.A., 1984. Altay-Sayan mountainous region. In I.I. Krasnov (ed.), *Stratigraphiya SSSR. Chetvertichnaya Systema, Polytom 2*. Nedra, Moscow, 331–351 (in Russian).

Borisov, B.A. and Minina, E.A., 1980. Edge moraines in south Siberian mountains and their maintenance for stratigraphy and palaeogeography of Pleistocene. In I.I. Krasnov et al. (eds), *Chetvertichnaya Geoiogiya i Geomorphologiya. Distancionnye Metody*, Nauka, Moscow, 21–24 (in Russian).

Feniksova, V.V., 1977. *Verkhniy Pleystocen Ugo-Vostoka Zapadno-Sibirskoy Nizmennosti*, lzd-vo Moskovskogo Universiteta, Moscow (in Russian).

Frenzel, B., Pesci, M. and Velichko, A.A. (eds), 1992. *Atlas of Palaeoclimates and Palaeoenvironments of the Northern Hemisphere. Late Pleistocene – Holocene*, Geographical Research Institute, Hungarian Academy of Sciences. Stuttgart Fisher, Budapest.

Goncharov, S.V., 1989. *Poslednee Oledenenie Zapadnoy Sibiri i Lednikovo-Podprudnye Ozera v Basseyne Srednego Eniseya*, Institut Geographii, Moscow (in Russian).

Gromov, V.I., 1948. *Palaeontologicheskoe i Arkheologicheskoe Obosnovanie Stratigraphii Kontinentainykh Otiozeniy Chetvertichnogo Perioda na Territorii SSSR*, Trudy Geoiogicheskogo Instityta, Seriya Geoiogicheskaya, No. 17 (in Russian).

Grosswald, M.G., 1987. Last glaciation of Sayan-Tuva mountainous region: morphology, intensity of income, dammed lakes. In V.M. Kotlyakov and M.G. Grosswald (eds), *Vzaimodei'stvie Oledeneniya s Atmospheroi i Okeanom*, Nauka, Moscow, 152–171 (in Russian).

Kind, N.V., 1974. *Geokhronologiya Pozdnego Antropogena po Isotopnym Dannym*, Nauka, Moscow (in Russian).

Nagorsky, M.P., 1941. Osnovnye Etapy Chetvertichnoy lstorii Zapadno-Sibirskoy Nizmennosti. *Vestnik Zapadno-Sibirskogo Geologicheskogo Upravieniya*, **3**, 36–56 (in Russian).

Osadchi, S.S., 1982. Problem of pluvial and glacial epochs correlation on the territory of the northern part of Zabaykalie region. In *Pozdnekainozoyskaya lstoriya Ozer v SSSR, Novosibirsk*, 62–71 (in Russian).

Popov, A.I., 1967. *Merziotnye Yavieniya v Zemnoy Kore Kryoiytologiya*. lzd-vo MGU, Moscow (in Russian).

Shofman, L.L., 1980. Late anthropogene geology and palaeogeography of non-glacial region of north eastern part of Siberian platform. In N.K. Ivanova and N.V. Kind (eds), *Geokhronologiya Chatvertichnogo Perioda*, Nauka, Moscow, 223–230 (in Russian).

Tomirdiaro, S.V., 1980. *Lessovo-Ledovaya Formaciya Vostochnoi Sibiri v Pozdnem Pleistcene i Golocene*, Nauka, Moscow (in Russian).

Tomirdiaro, S.V., 1993 Lednikovyi geomorphogenez v gornykh i gornykh territoriyah. Magadan Book Publisher, Magadan. [In Russian].

Velichko, A.A., 1987. Sovremennoe Sostoyanie Konceptcii Pokrovnykh Oledeneni' Zemli. *Izvestie AN SSSR. Ser. Geographicheskaya*, **3**, 21–34 (in Russian).

Velichko, A.A., 1993. *Evoluciya Landshaftov i Klimatov Severnoy Evrasii. Pozdnepleystocenovye-Holocenovye Elementy Prognoza. 1. Regionainaya Palaeo-geographiya*. Nauka, Moscow (in Russian).

Voikov, I.A., 1976. Influence of climate and erosion basement fluctuations on river valley development (Example of Ob River Basin). In N.A. Florensov et al. (eds), *Problemy Ekzogennogo Rel'efoobrazovaniya*, Khiga 11, Nauka, Moscow, 191–240 (in Russian).

Yamskikh, A.F., 1992. The reconstructions of the sedimentological cyclicity in the river valleys of the southern Siberia in palaeolithic studies. In A.P. Derevyanko (ed.), *Khronostratigraphiya Palaeolita v Severnoi, Centrainoy, Vostochnoy Azii i Amerike*, Nauka, Novosibirsk, 83–91 (in Russian).

Yamskikh, A.F., 1993. *Osadkonakopienie i Terrasoobrazovanie v Rechnykh Dolinakh Uznoi Sibiri*, Krasnoyarsk State University Press, Krasnoyarsk (in Russian).

Yamskikh, A.F., 1996. Late Quaternary intra-continental river palaeohydrology and polycyclic terrace formation: the example of south Siberian river valleys. In J. Branson, A.J. Brown and K.J. Gregory (eds), *Global Continental Changes: the Context of Palaeohydrology*, Geological Survey Special Publication, No. 115, 181–190.

Yamskikh, A.F. and Yamskikh, G.U., 1992. Palaeolithic archaeological sites on high geomorphological levels: Geology and palaeogeography (Example of Kurtak-4 and Shienka archaeological sites). In A.P. Derevyanko (ed.), *Khronostratigraphiya Palaeolita v Severnoi, Centrainoy, Vastochnoy Azii i Amerike*, Nauka, Novosibirsk, 92–105 (in Russian).

Zykina, V.S., Volkov, I.A. and Dergacjeva, M.I., 1981. Verkhnechetvertichnye Otlozeniya i Paleopochvy Reki Ob. Moscow, Nauka, 1–209, [In Russian].

CHAPTER 10

Palaeohydrological Changes in the Upper Dneper Valley, Belarus, During the Last 20 000 Years

TOMASZ KALICKI

Institute of Geography, Polish Academy of Sciences, Krakow, Poland

AND

ALEXANDER F. SANKO

Institute of Geological Sciences, National Academy of Sciences, Minsk, Belarus

INTRODUCTION

The Dneper is the biggest river in Belarus (Figure 10.1). Its springs are situated on the Valdai Plateau and it has its estuary in the Black Sea near Kherson. The study area included the *c.* 75 km reach of the valley between Orsha and Shklov. The Dneper valley upstream of Orsha has a slightly sinuous course; its shape resembles valleys of 'underfit' rivers with confined meanders (cf. Lewin and Brindle, 1977). The main direction of the valley is parallel to the Orsha Plateau which marks the boundary of the maximum advance of the Vistulian ice sheet (Figure 10.1). Downstream of Orsha the valley dissects the fluvioglacial moraine Orsha–Mogilev Plain from the older glaciation. It has a meridional and straight course. There are two 'dead branches' of the valley in this section (Figure 10.2).

The Dneper in Orsha has a mean annual discharge of $125\,\mathrm{m^3s^{-1}}$ (mean maximum $937\,\mathrm{m^3s^{-1}}$ and absolute maximum $2000\,\mathrm{m^3s^{-1}}$) while the surface area of the river basin is about $20\,000\,\mathrm{km^2}$. The river has a snow regime with a distinct, long-lasting high water level period (from March to May) and sometimes summer rainfall floods. The longitudinal slope is small (0.1–0.3‰), and the relation of the width of the channel to the width of the valley is about 0.1. The Dneper flows imprisoned in its narrow valley (meander indicator is 1.1–1.3).

Research in the Dneper valley was a continuation of the studies on Belarus rivers of the Black Sea catchment area, which, during the maximum extent of the ice sheet of the last glaciation carried away proglacial waters to the south. This type of river valley is typical of the whole Russian Plain, therefore the results from the Dneper valley are probably representative for all this territory.

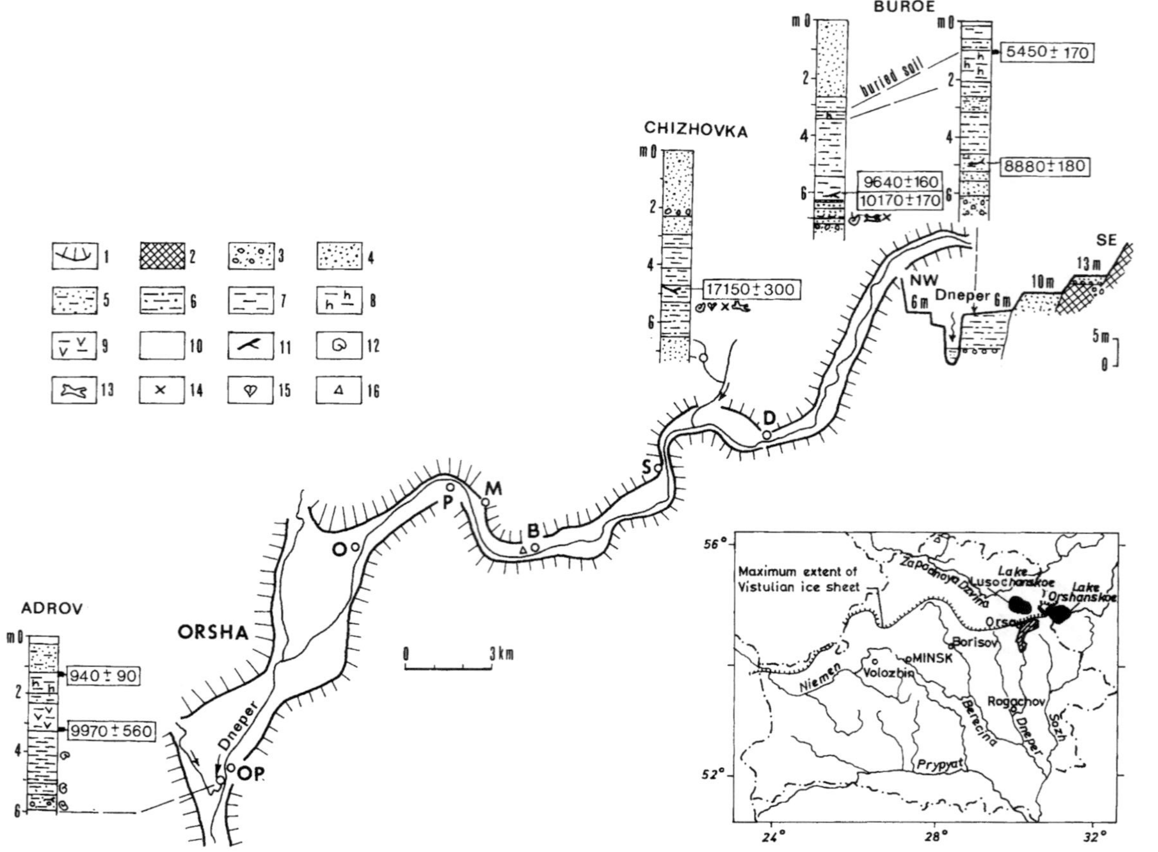

Figure 10.1   Sketch of the Dneper valley between Orsha and Adrov with study sites (by T. Kalicki). Key: 1, valley side; 2, till; 3, sand and gravel; 4, sand; 5, silty sand; 6, sandy silt; 7, silt; 8, organic silt; 9, peaty silt; 10, soil; 11, organic detritus; 12, malacofauna; 13, bones; 14, ostracodes; 15, insects; 16, archaeological site. Study sites: B, Berestenevo; D, Dubrovno; M, Mitkovshchina; O, Orsha; OP, Orsha Promyshlennaya; P, Pashino; S, Strazhevo

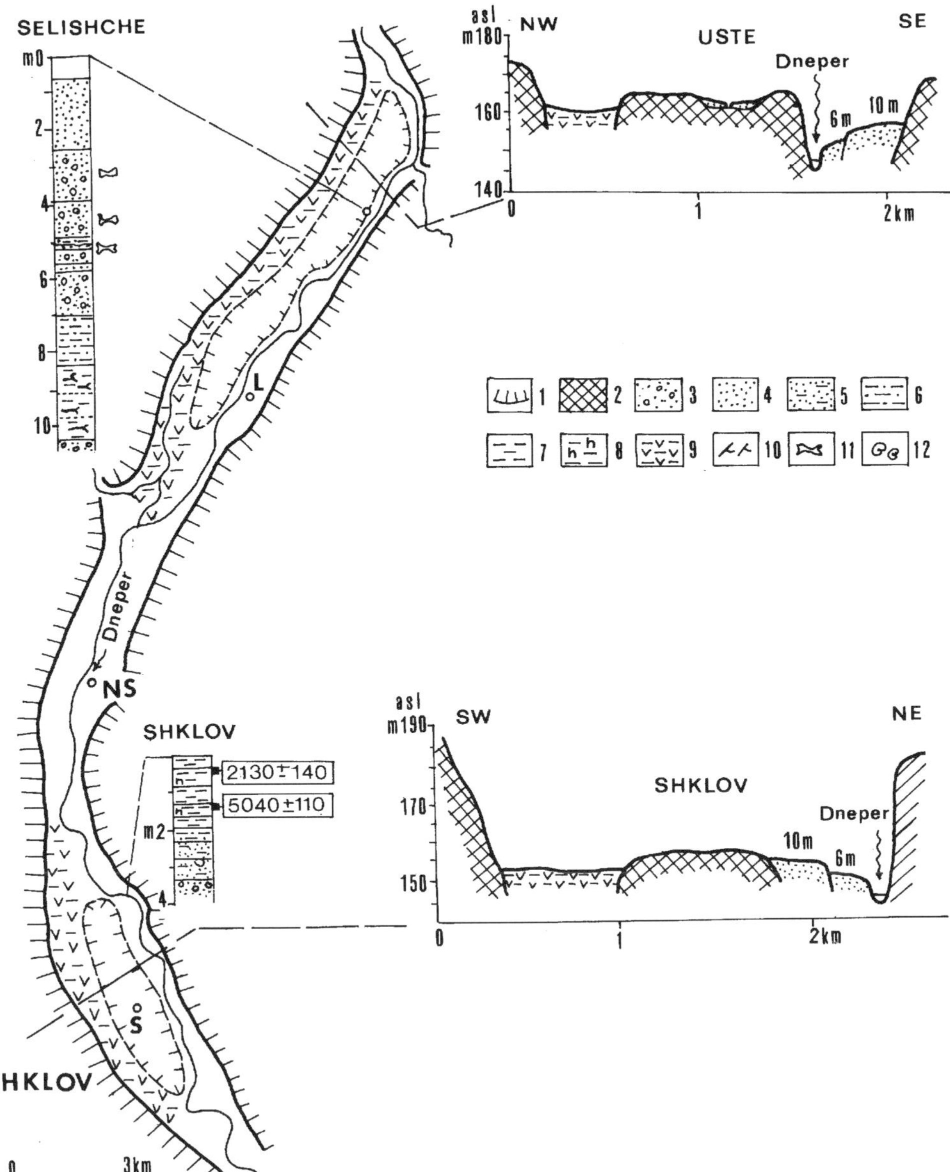

Figure 10.2   Sketch of the Dneper valley between Adrov and Shklov with study sites (by T. Kalicki). Key: 1, valley side; 2, till; 3, sand and gravel; 4, sand; 5, silty sand; 6, sandy silt; 7, silt; 8, organic silt; 9, peaty silt; 10, organic detritus; 11, bones; 12, malacofauna. Study sites: L, Levki; NS, Novye Stayki; S, Shklov

STRUCTURE AND AGE OF TERRACES AND FLOOD PLAINS

In spite of the fact that studies in the Dneper valley were carried out in the 19th century, the opinions on quantity and height of terraces in the study area are very varied and contradictory (Kalicki and Sanko, 1992). We described terraces 13 m high, 10 m high and 6 m high (submerged only during catastrophic floods) and the present flood plain is 4 m high.

The 13 m terrace is preserved in the form of narrow (20–200 m), separate strips on both banks of the Dneper (Figure 10.1). The surface of the terrace is even, rarely diversified by dunes or thermokarstic depressions (Buroe site). This terrace has an erosional socle (fluvioglacial gravels or tills) and the thickness of sandy alluvia with lag level in the bottom is very small (1.5–2.5 m).

The 10 m terrace is the best preserved overflood terrace in the Dneper valley. Its width oscillates between 50 and 100 m and on the surface was located an archaeological Mesolithic site (Figure 10.1). There are two erosive rectilinear 'dead branches' of the valley on this level (up to 1 km wide and 9–14 km long). They are both hanging valleys, parallel to the present-day Dneper valley and separated from this valley by moraine elevations (Figure 10.2). At present they are covered by vast peatbogs (Kalicki and Sanko, 1992). The accumulative terrace is composed of sandy gravel channel sediments with a thickness of 2.5 m (Buroe and Berestenevo sites) to more than 10 m (Selishche site) (Figure 10.2). At the Dubrovno and Mitkovshchina sites, cryogenic involutions and pseudomorphoses were reported in the alluvia of this terrace (Sanko, 1987). These sites prove that alluvia were accreted in two phases, and the accumulation took place in a cold climate. It is not unlikely that in both sites we are dealing with a buried terrace (Older Pleniglacial?) covered by younger sediments. The malacofauna, found in the sandy gravel alluvia, is characteristic of a cold climate (Mitkovshchina site). The bones of large tundra mammals (mammoth, horse, reindeer, hairy rhinoceros, musk-ox) and rodents (mainly lemmings and field-voles) were found in the sandy gravel sediments of the Selishche site (Kalinovski, 1983; Figure 10.2). The palynological profile of the sediments from this site reveals a small content of pollen (64% grasses) and a distinct blend of redeposited pollens (Sanko, 1987). Also, studies of palaeobotanical macrofossils indicate the occurrence of two complexes, one periglacial and the other redeposited (Velichkevich, 1982). In the Chizhovka valley, a small tributary of the Dneper, the sandy gravel alluvia of this terrace cover the organic sediments dated 17 150 ± 300 ^{14}C yr BP (Figure 10.1). Also, the palaeobotanical (pollen and macrofossils) and palaeontological (bones, insects, ostracods) findings confirm that the alluvia were deposited during exceptionally severe, periglacial climatic conditions (Sanko, 1987; Kalicki and Sanko, 1992).

The 6 m terrace is 100–200 m wide and stretches along the whole examined section of the Dneper valley. Characteristic features in the structure of this terrace are very thick (up to several metres) silty overbank deposits with buried soils covering the sandy channel alluvia (Figure 10.1). The bones of early Holocene fauna (mainly field-voles) with an admixture of tundra fauna were found in sands at Buroe and Pashino sites (Kalinovski, 1983). The studies of ostracods from the bottom of silts at the Buroe site gave similar results, because the species characteristic of a cold climate prevail (Zubovich, 1981). Datings of wood found in the bottom (depth about 6 m) of overbank deposits at the Buroe site gave dates of 10 170 ± 170 and 9640 ± 160 ^{14}C yr BP (Sanko, 1987). Also, according to the pollen diagram, these silts were accumulated during the Preboreal (Kalicki and

Sanko, 1992). A new dating of a wood from the middle of the silt deposits (depth 4.5–5.0 m) provided an age of 8880 ± 180 ^{14}C yr BP (IGSB–293). The top of the buried soil (depth 1.0–2.5 m) at this site was dated 5450 ± 170 ^{14}C yr BP (Kalicki, 1995) (Figure 10.1). Subfossil molluscs (with 21 land snails and 29 water species) have been found in the lower part of overbank sediments at a few sites (Kalicki and Sanko, 1992). A new analysis of malacofauna at the Adrov profile is shown as an example (Figure 10.3). The malacofauna have been found below 3.25 m depth in silty or sandy muds (mean grain size diameter, $Mz = 5.0$–5.9ϕ). The number of taxa in the samples fluctuates from 17 to 25 and the number of specimens from 361 to 2246. The stagnophile species form the basis of the freshwater fauna. The euryecological *Pisidium henslowanum* dominates in this group and in the fauna generally and reaches 50.6% (45–88% in other profiles). The main components of mollusc assemblages are also the water species (*Valvata piscinalis, Bithynia tentaculata, Gyraulus laevis, Armiger crista*), reophile (*Pisidium nitidum, P. amnicum*) and snails of temporary water bodies (*Valvata cristata, Lymnaea truncatula*). The open habitat species (*Vallonia pulchella, V. costata*) dominate in land malacofauna at Adrov and other sites. The presence of periglacial species (*Columella columella, Vallonia tenuilabris*) and one forest species (*Discus ruderatus*) allow us to connect the lowest part of sediments with the Late Vistulian and the upper part with the Early Holocene. A weak modification of the mollusc assemblages in the profiles testifies to the small differentiation and high rate of flood plain sedimentation in that period. The middle part of the Adrov profile (1.2–3.25 m depth) is composed of organic silty muds and peaty silts ($Mz = 5.9$–6.4ϕ). The bottom of this member was dated 9970 ± 560 ^{14}C yr BP and the top 940 ± 90 ^{14}C yr BP (Kalicki and Sanko, 1992). The upper part of this profile is composed of silty sands ($Mz = 4.1$–5.0ϕ) (Figure 10.1). The 6 m terrace is an accumulative plain, which began to be formed at the end of the Vistulian. Initially, numerous depressions with water (channels of braided river?) were filled with silty sands and sandy silts (Buroe and Adrov sites). In the next stage the overbank deposits accreted relatively quickly on the waterlogged flood plain with numerous small water basins. At the start of the Holocene we see a distinct change of the character of flood plain (Adrov and Pashino sites). It probably resulted from the high rate of accretion of the overbank deposits with simultaneous incision of the river and change to a dry climate in the early Holocene. During the Holocene the 6 m terrace functioned as a flood plain and silty muds were deposited on it. About 1000 years BP, the grain size of these overbank deposits changed from silty muds to sandy silty muds.

The present-day 4 m flood plain is a narrow (a few to several metres) strip (Figure 10.2). At the Shklov site, the present flood plain is composed of silty muds, about 2 m in thickness, with two buried soils. The lower buried soil gave an age of 5040 ± 110 ^{14}C yr BP (Kalicki and Sanko, 1992) and the upper one 2130 ± 140 ^{14}C yr BP (Kalicki, 1995). The accumulative flood plain was formed during the Holocene after incision of the 6 m terrace.

EVOLUTION OF THE DNEPER VALLEY

During the Quaternary, the Dneper has always flowed to the Black Sea (Goretski, 1970). However, during the maximum of the last glaciation (Orsha stage), in the upper reach of the Dneper valley (somewhat upstream from the examined section), dammed lake Orshanskoe was formed at the front of the ice sheet. Its formation indicates that the earlier

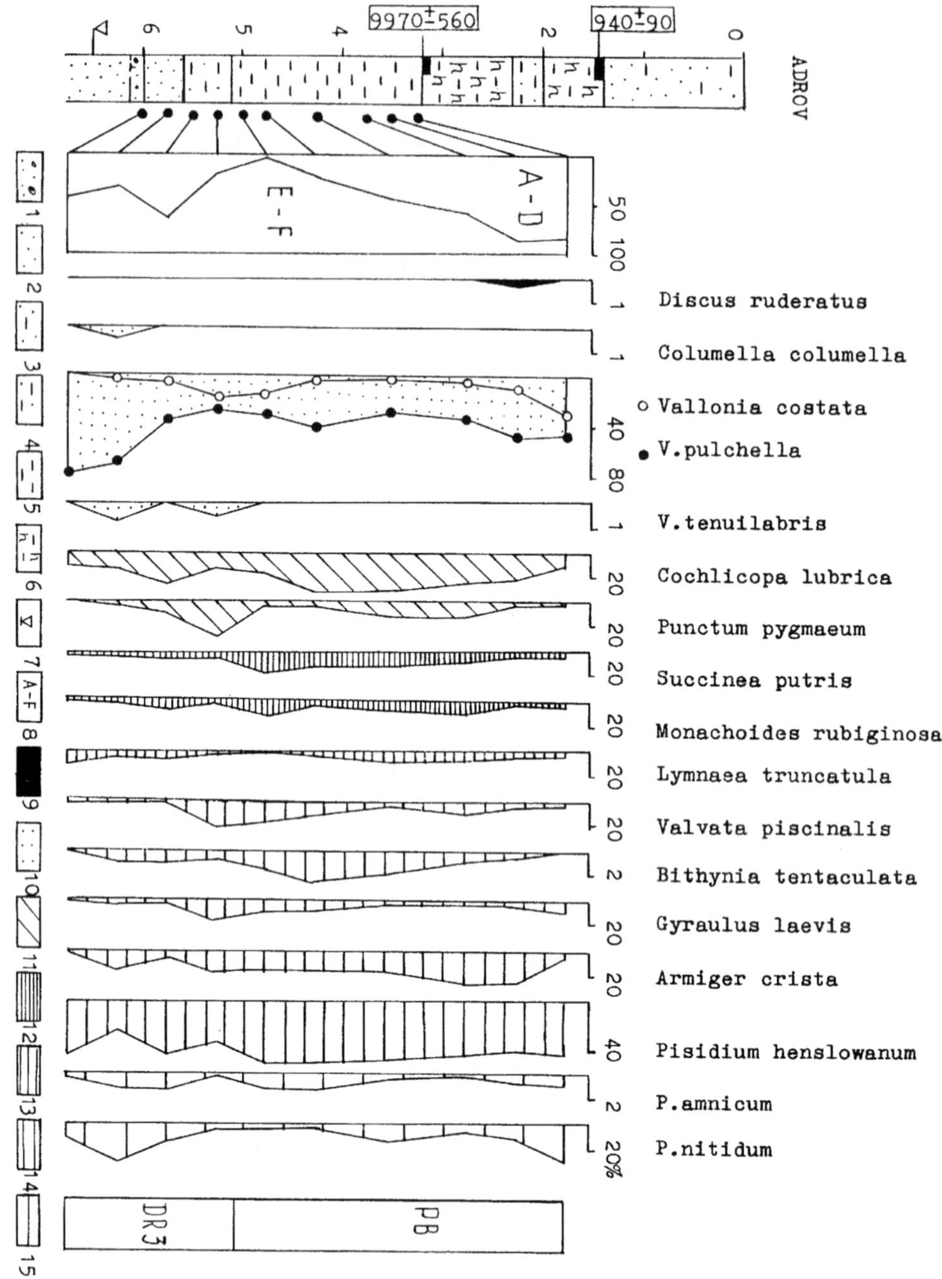

Figure 10.3   Malacological diagram of Adrov profile on 6 m high flood plain (by A. F. Sanko). Key: 1, silty sand with gravels; 2, sand; 3, silty sand; 4, sandy silt; 5, silt; 6, organic silt; 7, water level of river; 8, ecological groups of malacofauna according to Alexandrowicz (1987): A–D, land species; E–F, fresh-water species. Malacofauna: 9, forest (group I); 10, open ground (group V); 11, mesophile (group VII); 12, higrophile (group IX); 13, episodic reservoir (group X); 14, stagnophile (group XI); 15, reophile (group XII)

position of the European watershed was different (Kvasov, 1976). Glaciation, formation of the lake and the flow of proglacial waters into the Dneper valley caused the dissection of the watershed and its shift to the north.

In the proximity of the ice front, increased discharge was caused by proglacial waters which intensified erosion in the valley. During that time the Dneper valley upstream of Orsha became slightly sinuous with incised meanders, and near Uste and Shklov the valley was divided into two branches. Dneper, having multiplied discharges, probably flowed in two branches of the valley. After the proglacial water supply ceased, the river incised and abandoned one of the branches. Further evolution took place only in the active valley, where formation of the lower terraces occurred, whereas the abandoned branches were covered by peatbogs (Figure 10.4).

Destruction of older terraces in the examined section is also probably connected with the intensification of lateral erosion. The rest of these older terraces are the buried alluvia separated from the younger ones by a cryoturbation level (Kalicki and Sanko, 1992). The 10 m terrace locally has this type of structure: two members of alluvia in superposition (Dubrovno and Mitkovshchina sites; Figure 10.4). The alluvia are separated by cryoturbation levels (Kalicki and Sanko, 1992). Older Vistulian terraces are preserved in the valley only at a remarkable distance from the maximum extent of the Vistulian ice sheet. The middle Vistulian alluvia of the second overflood terrace, which date from 30 000 to 46 000 years BP, were described from the 18–20 m terrace in the region of Rogachov, which is almost 200 km from the ice sheet front (Arslanov et al., 1971; Zimenkov and Kuznetsov, 1985).

Only young Vistulian terraces are preserved in the study reach. A similar structure is common for all of them. They are composed of facially non-differentiated, sandy or sandy gravelly sediments deposited by braided rivers. As the research in the Dneper valley shows, terraces were formed after the last glacial maximum and they are connected with

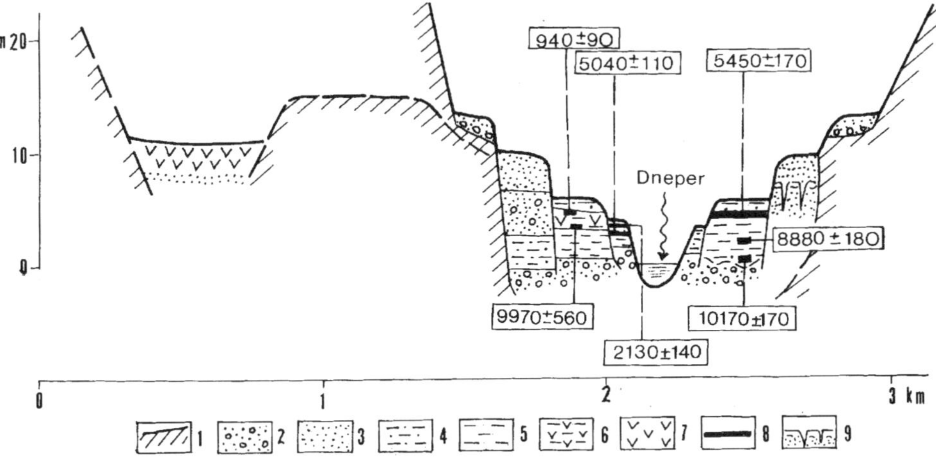

Figure 10.4 Schematic cross-section of the Dneper valley between Orsha and Shklov (by T. Kalicki). Key: 1, sediments older than Vistulian; 2, sand with gravels; 3, sand; 4, sandy silt and silty sand; 5, silt; 6, peaty silt; 7, peat; 8, buried soil; 9, fossil ice wedges and cryoturbations

the recessional stages of the ice sheet. The 13 m terrace dovetails with the highest level of the Lusochanskoe glacial dammed lake. Alluvia from the 10 m terrace sediments were dated from 17 150 ± 300 ^{14}C yr BP (Kalicki and Sanko, 1992). The lack of a loess cover on these terraces is evidence that they must have been active accumulative plains until the end of the loess deposition about 15 000 years BP (Voznyachuk, 1973). On the other hand, about 10 000 years ago the 10 m terrace must have been an overflood terrace or have flooded very rarely since the Mesolithic population settled on it (Berestenovo site) (Figure 10.1). We meet a similar situation in other river valleys of the Russian Plain, where Mesolithic sites are found on first overflood terraces, for instance in the valley of Kama (Baler, 1957; Ivanova, 1982). So far, the link between the young terraces occurring in the examined section of the valley and the older ones preserved farther from the front of the ice sheet has not been investigated.

In the Late Glacial a change in the river pattern took place. This change is distinctly indicated by facial diversity of the alluvia of the 6 m terrace. The alluvial plain of the braided Dneper was dissected as a result of concentration of channels and changed itself into a flood plain on which an intensive sedimentation of overbank deposits began. On the 6 m terrace of the Dneper a depression on top of channel deposits, which are probably traces of braided beds, is filled with fine sands or sandy silts (Buroe site: 10 170 ± 170 ^{14}C yr BP) (Kalicki and Sanko, 1992; Figure 10.4).

The postglacial evolution of the Dneper valley was connected with pronounced uplift (Kalicki and Sanko, 1992). The Dneper was incising and formed a very narrow valley, with narrow strips of terraces and flood plains. The horizontal channel stability conditioned by the incision of the river led during the whole Holocene to the flood plain being built by vertical accretion of overbank deposits (Figure 10.4).

Dating has established the periods of increased river activity (Starkel, 1983; Kalicki, 1991b). A periodic increase of flood frequency, leading to an increase in mud sedimentation rate, caused fossilization of soils, which developed during the Early and Middle Holocene (since 9970 ± 560 ^{14}C yr BP) on the flood plains (Figure 10.4). Depending on the magnitude of the floods, fossilization of soils occurred on a lower (5040 ± 110 and 2130 ± 140 ^{14}C yr BP) or on a higher level (5450 ± 170 and 940 ± 90 ^{14}C yr BP) of the flood plain (Kalicki and Sanko, 1992).

Although the first human sites in the examined valley were established during the Mesolithic, distinct anthropogenic changes have been observed only since historical times, where fertile loess soils began to be intensively cultivated after these regions were occupied by Kiev Russia in the middle of the 9th century. The deforestation led to an increase in flood frequency and magnitude during periods of cooler and wetter climate (Zernitskaya and Kozharinov, 1988). As a consequence, fossilization of the soil on the higher level of the flood plain took place about 940 ± 90 ^{14}C yr BP, which was covered by sandy muds. The change of grain size composition of the muds was the result of intensified erosion on cultivated regions. Silty muds ($Mz = 5.9$–6.4ϕ) deposited during most of the Holocene changed into sandy silty muds ($Mz = 4.1$–5.0ϕ) during the last 1000 years (Kalicki and Sanko, 1992).

In the Dneper valley, the higher flood plain level, which was overlapped in some places by more than 1 m of sandy silty muds during the last millennium, is flooded only during catastrophic floods. At present very characteristic muds, not observed in older sediments, are deposited on the lower flood plain level of the Dneper. They are composed of alternating thin laminae of sands and silts. Their thickness reaches about 2.5 m in some places.

DISCUSSION AND CONCLUSION

During the last glaciation the examined valley section was dominated by proglacial waters flowing to the south. A similar situation is described in the neighbouring catchment of the Berezina (Kalicki, 1991a), Volga (Kvasov, 1987), as well as on the plains of North America (Baker, 1983). On the Russian Plain, both in the catchments of the Dneper and the Volga (Kvasov, 1987), river captures are characteristic in the upper courses. These river captures occurred by overflow of water to the south from lakes dammed at the front of the ice sheet. This caused the shift of the post-Vistulian watershed to the north. However, in the examined valley there are no traces of catastrophic floods connected with rapid outflow of waters from dammed lakes, analogous to the ones described in North America (Baker, 1983).

In the examined valley section only Young Vistulian terraces occur. High meltwater discharges with simultaneous overloading of rivers did not favour bed erosion. Braided rivers formed alluvial plains and, at the same time, in the narrow sections they widened the valley, cutting an erosional terrace that in the Dneper river is at 13 m above the present channel bed. However, at the same time, destruction of older terraces occurred mainly by lateral erosion. These older terraces are only preserved several hundred kilometres from the ice sheet front or, locally, as a buried alluvial body covered by younger sediments. Young terraces with a similar type of alluvium are also found in analogous positions in the Volga catchment (Kvasov, 1987) and in North America (Baker, 1983).

Applying Dury's (1970) classification, the Dneper is an underfit river. The valley reach upstream of Orsha represents the c1 variant and in some parts also resembles the Osage type (c3). The Dneper valley between Orsha and Shklov is divided twice into two branches which functioned simultaneously during periods of high discharges. After the decrease of discharges only one of the branches was active. Therefore it is suggested that a new type is introduced into Dury's classification: the Dneper type. This type of underfitting is manifested by the existence of 'abandoned branches of a valley'.

Tectonic uplift had a remarkable influence on the postglacial evolution of the examined valley. The incising Dneper formed a narrow valley. Natural channelization of the river caused the evolution to be limited to vertical accretion by overbank deposition in the narrow flood plains along the channel.

In the Dneper valley, traces of intensified river activity occurred at 5500–5000, 2100 and 1000 yr BP. Their manifestation is by fossilization of soils on the flood plains of the Dneper. Already in earlier works (Kalicki, 1991a; Kalicki and Sanko, 1992) a general uniformity of these phases in the Russian Plain and in Poland has been established. These phases occurred in periods with a cool and humid climate according to Starkel (1983) and Starkel et al. (1996). As far as older phases are concerned, it is agreed by these authors that they are climatically conditioned. Subatlantic phases, however, were often connected with widespread human settlements such as the anthropogenic level from the 10th to 11th century in the Russian Plain (Chotinski and Starkel, 1982). Yet, simultaneously in this period a distinct phase of cooler and wetter climate occurs which marked itself in the river valleys in Belarus and the Russian Plain, among others, by changes of sedimentation type on flood plains and by buried soils (Zolotokrylin et al., 1986; Klimanov and Serebrannaya, 1986; Aleksandrovski et al., 1987; Zernitskaya and Kozharinov, 1988; Kalicki, 1991a; Kalicki et al., 1997). Nevertheless, research in the valleys of the Berezina and

Dneper shows that this level is marked by an intensification of river activity, both in a natural river basin and in a basin strongly modified by human impact (Kalicki, 1996a,b). It proves the overlap of climatic changes and human activity in the subatlantic. Human activity caused an increase in the rate of overbank sedimentation (fossilization of soil on the higher flood plain level) and a change of grain size composition of muds in the Dneper valley.

ACKNOWLEDGEMENTS

We wish to thank Dr N. Mikhailov from the Institute of Geological Sciences, National Academy of Sciences of Belarus, for a new radiocarbon dating, and two unknown reviewers for comments.

REFERENCES

Aleksandrovski, A.L., Glasko, M.P. and Folomeyev, B.A., 1987. Issledovaniya pogrebennykh poymennykh pochv kak geokhronologicheski urovney vtoroi poloviny golotsena. *Byulletin Komissii po Izucheniyu Chetvertichnego Perioda*, **56**, 123–128.

Alexandrowicz, S.W., 1987. Analiza malakologiczna w badaniach osadów czwartorzędowych. *Kwartalnik AGH, Geologia*, **13**, 1–2.

Arslanov, H.A., Voznyachuk, L.N., Velichkevich, F.J., Makhnach, N.A., Kalechyts, E.G. and Petrov, G.S., 1971. Paleogeografia Byelorussii v ranniye fazy formirovaniya srednevaldayskikh generatsi alluviya vtoroi nadpoimennoi terrasy Dnepra. *Doklady AN SSSR*, **200**(6), 141–144.

Baker, V.R., 1983. Late Pleistocene fluvial systems. In H. E. Wright (ed.), *Late Quaternary Environments of the United States*, University of Minnesota Press, Minnesota, 115–129.

Baler, O.N., 1957. Khronologiya formirovaniya rechnykh terras na Urale v arkheologicheskom osveshchenii. *Trudy Komissii po Izucheniyu Chetvertichnego Perioda*, **13**, 307–314.

Chotinski, N.A. and Starkel, L., 1982. Naturalne i antropogeniczne poziomy graniczne w osadach holoceńskich Polski i centralnej części Niziny Rosyjskiej. *Przegląd Geograficzny*, **54**(3), 201–218.

Dury, G.H., 1970. General theory of meandering valleys and underfit streams. In G. H. Dury (ed.), *Rivers and River Terraces*, Macmillan, London, 264–275.

Goretski, G.I., 1970. *Alluvialnaja letopis Velikogo Pra-Dnepra*, Izdatelstvo 'Nauka', Moscow.

Ivanova, I.K., 1982. Iskopayemyy chelovek i yego kultura. In E. V. Shantser (ed.), *Stratigrafiya SSSR, Chetvertichnaya sistema I*, Nedra, Moscow, 382–412.

Kalicki, T., 1991a. Budowa teras i wiek równiny zalewowej Berezyny koło Borysowa (Białorus). *Przeglad Geograficzny*, **63**(3–4), 363–377.

Kalicki, T., 1991b. The evolution of the Vistula river valley between Cracow and Niepolomice in late Vistulian and Holocene times. In L. Starkel (ed.), *Evolution of the Vistula River Valley During the Last 15 000 Years, Part IV*, Geographical Studies, Special Issue 6, 11–37.

Kalicki, T., 1995. Lateglacial and Holocene evolution of some river valleys in Byelorussia. In B. Frenzel (ed.), *European River Activity and Climatic Change During the Lateglacial and Early Holocene*, Special Issue: ESF Project European Palaeoclimate and Man 9. Paläoklimaforschung, 14, 89–100.

Kalicki, T., 1996a. Climatic or anthropogenic alluviation in Central European valleys during the Holocene. In J. Branson, A.G. Brown and K.J. Gregory (eds), *Global Continental Changes: the Context of Palaeohydrology*, Geological Society, London, Special Publication 115, 205–215.

Kalicki, T., 1996b. Phases of increased river activity during the last 3500 years. In L. Starkel and T. Kalicki (eds), *Evolution of the Vistula River Valley During the Last 15 000 Years, Part VI*, Geographical Studies, Special Issue 9, 94–101.

Kalicki, T. and Sanko, A.F., 1992. Genesis and age of the terraces of Dneper between Orsha and Shklov (Byelorussia). *Geographia Polonica*, **60**, 151–174.

Kalicki, T., Sanko, A.F., Zernicka, V.P. and Litvinjuk, G.I., 1997. Ewolucja doliny Dźwiny na

Nizinie Suraskiej w późnym glacjale i holocenie. In T. Kalicki (ed.), Badania ewolucji dolin rzecznych na Białorusi-I. *Dokumentacja Geograficzna*, **6**, 13–52.

Kalinovski, P.F., 1983. *Teriofauna pozdnogo antropogena i golotsena Belorussii*, Nauka i Tekhnika, Minsk.

Klimanov, V.A. and Serebrannaya, T.A., 1986. Izmeneniya rastitelnosti i klimata na Srednerusskoi Vozvyshennosti v golotsene. *Izvestiya Akademii Nauk SSSR, Geografiya*, **2**, 93–102.

Kvasov, D.D., 1976. Paleogidrologiya Vostochnoy Evropy v valdayskoye vremiya. In G. P. Kalinin and R. K. Klige (eds), *Problemy paleogidrologii*, Nauka, Moscow, 260–266.

Kvasov, D.D., 1987. The Late Quaternary history of the Volga river. *Palaeohydrology of the Temperate Zone I, Rivers and Lakes*, Inst. Geologii Akademia Nauk Estonskoy SSR, Tallin, 43–55.

Lewin, J. and Brindle, B.J., 1977. Confined meanders. In K. J. Gregory (ed.), *River Channel Changes*, Macmillan, Belfast, 221–233.

Sanko, A.F., 1987. *Neopleystotsen severo-vostochnoi Belorussii i smezhnykh rayonov RFSSR*, Nauka i Tekhnika, Minsk.

Starkel, L., 1983. The reflection of hydrologic changes in the fluvial environment of the temperate zone during the last 15 000 years. In K.J. Gregory (ed.), *Background to Palaeohydrology*, Wiley, Chichester, 213–237.

Starkel, L., Kalicki, T., Krąpiec, M., Soja, R., Gębica, P., Czyżowska, E., 1996. Hydrological changes of valley floor in the upper Vistula basin during Late Vistulian and Holocene. In L. Starkel and T. Kalicki (eds), *Evolution of the Vistula River Valley During the Last 15 000 Years, Part VI*, Geographical Studies, Special Issue 9, 7–128.

Velichkevich, F.J., 1982. *Pleistocenovyye flory lednikovykh oblastey Vostochno-Evropeyskoi ravniny*, Nauka i Tekhnika, Minsk.

Voznyachuk, L.N., 1973. K stratigrafii i paleogeografii neopleistotsena Belorussii i smezhnykh territorii. In *Problemy paleogeorafii antropogena Belorussii*, 'Nauka i Technika', Minsk, 45–75.

Zernitskaya, V.P. and Kozharinov, A.V., 1988. Paleofitokhorologicheskiye aspekty territorii Belorussii. In J. M. Punning (ed.), *Izotopno-geokhemicheskiye issledovaniya v Pribaltike i Belorussii*, Inst. Geologii Akademia Nauk Estonskoy SSR, Tallin, 77–85.

Zimenkov, O.I. and Kuznetsov, W.A., 1985. Vremiya formirovaniya alluviya nadpoimennykh terras i poimy rek Belorussii. In A.W. Matveev, M.A. Valchik and O.N. Shpakov (eds), *Geologiya i gidrogeologiya keynozoya Belorussii*, Nauka i Tekhnika, Minsk, 127–132.

Zolotokrylin, A.N., Krenke, A.N., Liyakhov, M.E., Popova, V.V. and Chernavskaya, M.M., 1986. Kolebaniya klimata evropeyskoy chastii SSSR v istoricheskom proshlom. *Izvestiya Akademii Nauk SSSR, Geografiya*, **1**, 26–36.

Zubovich, S.F., 1981. Pervyye svedeniya ob iskopayemykh ostrakodakh iz obnazheniya u d. Buroe. *Geologicheskiye issledovaniya keynozoya Belorussii*, Nauka i Tekhnika, Minsk, 40–44.

PART III

Palaeohydrology and Environmental Changes

Palaeoenvironmental Changes in the Last Glacial Maximum around the Wakasa Bay Area, Central Japan

SHINJI MIYAMOTO

Department of Geoecology, Krasnoyarsk State University, Russia

YOSHINORI YASUDA

AND

HIROYUKI KITAGAWA

International Research Center for Japanese Studies, Kyoto, Japan

INTRODUCTION

A number of palaeobotanical studies of the Last Glacial Maximum in Japan have been carried out on the basis of fossil pollen assemblages. Flora, vegetation and palaeoenvironments during the Last Glacial Maximum in Japan have been reconstructed by some researchers. For example, Tsukada (1974, 1984, 1985), Yasuda (1978) and Nasu (1980) made different vegetation maps of the Japanese islands around the Last Glacial Maximum; Furutani (1979) constructed a palaeovegetation map of the Kinki district; Shoma and Tsuji (1987) compiled a fossil plant list which comprised 52 pollen assemblages and 34 plant macrofossil assemblages using the Aira-Tn ash as a time marker for this period. However, many aspects of flora and vegetation in the Last Glacial Maximum are not yet fully clarified. Another necessary palaeobotanical study of the Last Glacial period is one that focuses on the relationship between geological events and vegetation changes. In particular, the effect of the fall of widespread tephra like the Aira-Tn ash on the vegetation should be discussed. These studies discussing the vegetation in relation to sudden environmental change have been entirely lacking.

The Aira-Tn ash is often observed in sediments of the Last Glacial period. This ash originated from the Aira Caldera in southern Kyushu at about the beginning of the Last Glacial Maximum, and covered almost all the Japanese islands and surrounding seas (Machida and Arai, 1976, 1983). In the Kinki district, the thickness of the Aira-Tn ash is

Palaeohydrology and Environmental Change. Edited by G. Benito, V. R. Baker and K. J. Gregory
© 1998 John Wiley & Sons Ltd.

about 20 cm. It can be used as a good time marker for contemporaneous comparisons in different places of plant fossil assemblages. Such a huge eruption must have played an important role in the vegetation history. Tsuji and Kosugi (1991) pointed out that the Aira-Tn ash fall caused a rise in pH level and changed the sedimentary water environment from acid to alkaline around the Kanto district, in the eastern part of Japan.

The aim of this study is to reconstruct the vegetation changes during the fall of the Aira-Tn ash, based on fossil pollen assemblages. The cause of the vegetation changes which occurred before and after the ash fall is discussed in relation to climatic changes and influence of the ash fall.

GENERAL CHARACTERISTICS OF THE STUDY REGION

There are three moors in the eastern part of the Tsuruga Plain (Figure 11.1). They are surrounded by hilly lands with an elevation of less than 200 m, and are called Uchi-Ikemi, Naka-Ikemi and Yoza-Ikemi from the north to the south, respectively. The Naka-Ikemi Moor site is located at lat. 35°39′N and long. 136°05′E, in the eastern part of the Wakasa Bay region. This moor is regarded as a waste-filled valley created by tectonic subsidence at the eastern part of the Tsuruga Plain (Okada, 1978). The marshes and lakes of this region have provided the most suitable conditions for palynological study because the continuous sediments since the Last Glacial period have been well preserved there (Yasuda, 1982), and recently many drilling projects have been conducted (Takemura et al., 1994).

The modern upland vegetation around the Naka-Ikemi Moor is characterized by evergreen broad-leaved forest mainly composed of *Machlus thunbergii*, *Castanopsis* var. *sieboldii*, *Celtis sinensis* and *Cleyera japonica*. The presence of *Pinus densiflora* and *Castanea serrata* due to human impact characterizes the upland vegetation as secondary forest.

Now, most of the moor is occupied by paddy rice fields, and buried tree stumps are scattered on the ground surface. Two buried stumps were identified as *Cryptomeria japonica* (Ueda and Tsuji, 1994). The abandoned rice fields of the moor are mainly dominated by Cyperaceae and Gramineae.

METHODS

Sampling

Sediment samples were taken by drilling at the centre of Naka-Ikemi Moor. The following samplers were used from top to bottom: piston sampler, Denison sampler and core pack sampler.

Radiocarbon dating

Five levels of the Naka-Ikemi Moor (NKM) samples which fall within the time span of the Last Glacial Maximum at 13.9 m, 17.5 m, 17.7 m, 20.0 m and 23.2 m in depth, were analysed. The samples were measured at the laboratory of the International Research Center for Japanese Studies using the 1220 QUANTULUS liquid scintillation counter of Pharmacia Wallac Ltd, UK (Miyamoto et al., 1995).

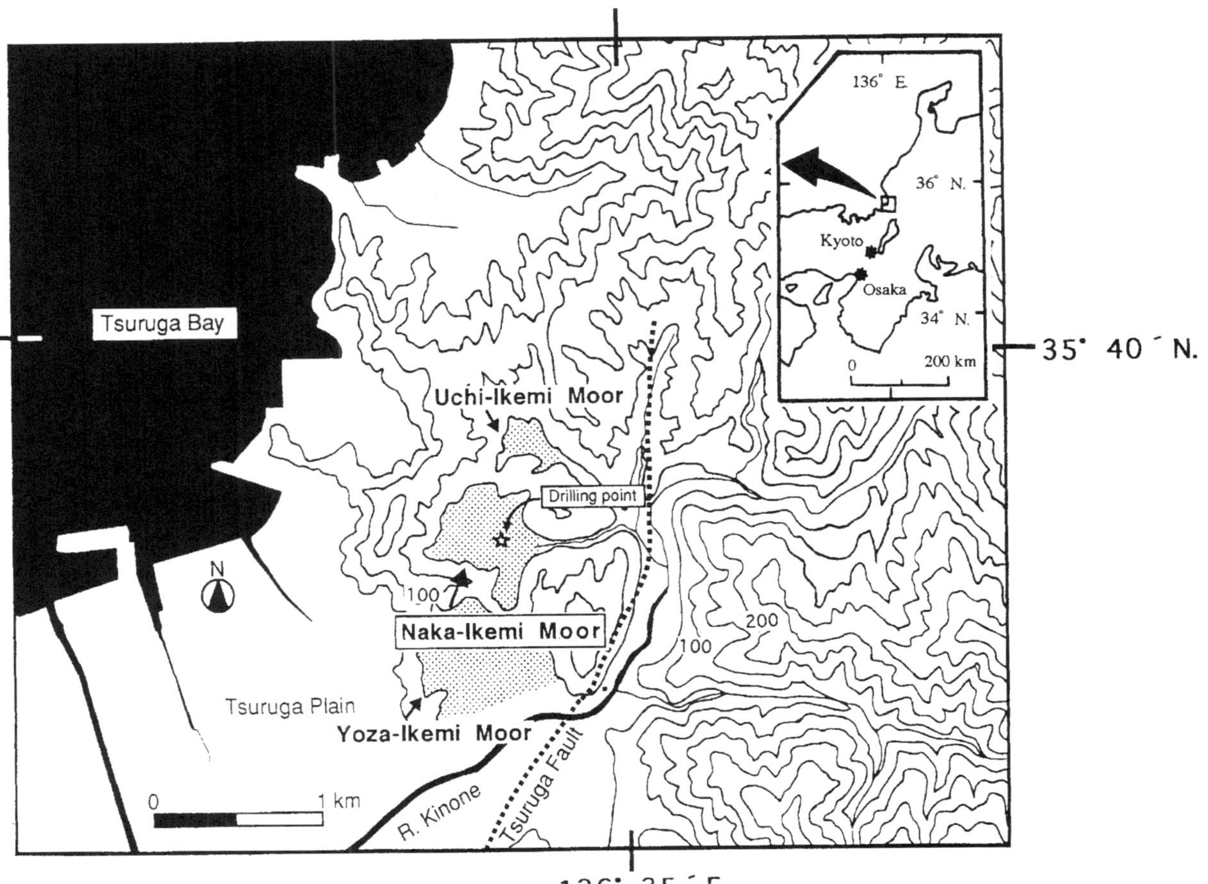

Figure 11.1   Location of Naka-Ikemi Moor and the drilling point, Fukui Prefecture, Central Japan

Bulk density

The bulk density for all of the 20 cm interval samples was measured. All 3 cm^3 samples were taken using a syringe and were stored in glass vials. After this process the samples were dried at 105 °C for 48 h in an oven and then weighed on an electronic balance.

Ignition loss

After the measurement of bulk density, the samples were burned at 450 °C for 48 h in an electronic heat furnace (ADVANTEC, KM-280) and then ignition loss was measured with an electronic balance. The weight ratio was calculated on the total dry weight.

Pollen analysis

Pollen and spore fossils were extracted by the following procedures. Samples of 2 g were taken at 20 cm intervals and were treated with 10% KOH, sieved through a 0.5 mm mesh, decanted to remove organic macromaterials, treated with wash and $ZnCl_2$ (gravity 2.0), dehydrated with acetic acid, and treated for 2.5 min by the acetolysis method. The residue was saturated in 50% glycerin, and was mounted on a glass slide. All pollen and spores on each slide were counted. A mass of pollen grains was counted, as single grain occurrences were illustrated in a pollen diagram. The percentages of pollen taxa and spore types were calculated based on the total arboreal pollen counts excluding *Alnus*.

Identification of volcanic ash

Volcanic glass fragments extracted from the washed samples were determined from the measurements of refractive index using the RIMS 86 at the laboratory of the Tokyo Metropolitan University.

STRATIGRAPHY AND CHRONOLOGY

The core, from the top to 29 m depth, is mainly composed of peat with intercalated plant macrofossils. The lower layers, between 29 m and 39 m, consist of silt and peat layers. The lowest part from 39 m to 41 m is composed of peaty clay (Figure 11.2).

Volcanic ashes were observed at three levels: 5.28 to 5.29 m, 17.56 to 17.67 m, and 26.93 to 26.97 m. They are Kikai-Akahoya ash (K-Ah), Aira-Tn ash (AT) and Daisen Kurayoshi pumice (DKP) from the highest to the lowest (Figures 11.2 and 11.3), respectively. The results of the refractive index measurement of volcanic glasses and hornblende were cited from Miyamoto et al. (1995).

The section of core sediments from 10 to 25 m in depth corresponds to the Last Glacial period because the oldest and youngest radiocarbon ages of these layers were 40 270/-1257 + 1471 ^{14}C yr BP (JAS-190, Table 11.1), and 11 717 ± 190 ^{14}C yr BP (JAS-194), respectively (Table 11.1). The radiocarbon ages of the upper and lower horizons of the Aira-Tn ash layer are 23 468/-588 + 635 ^{14}C yr BP (17.5 m, JAS-198) and 24 892/-531 + 569 ^{14}C yr BP (17.5 m, JAS-191), respectively. Therefore, the eruption age of the Aira-Tn ash does not conflict with the estimated age of *c.* 24 000 yr BP (Matsumoto et al., 1987). Based on the above dating, the deposits from 10 to 25 m in depth were treated in the

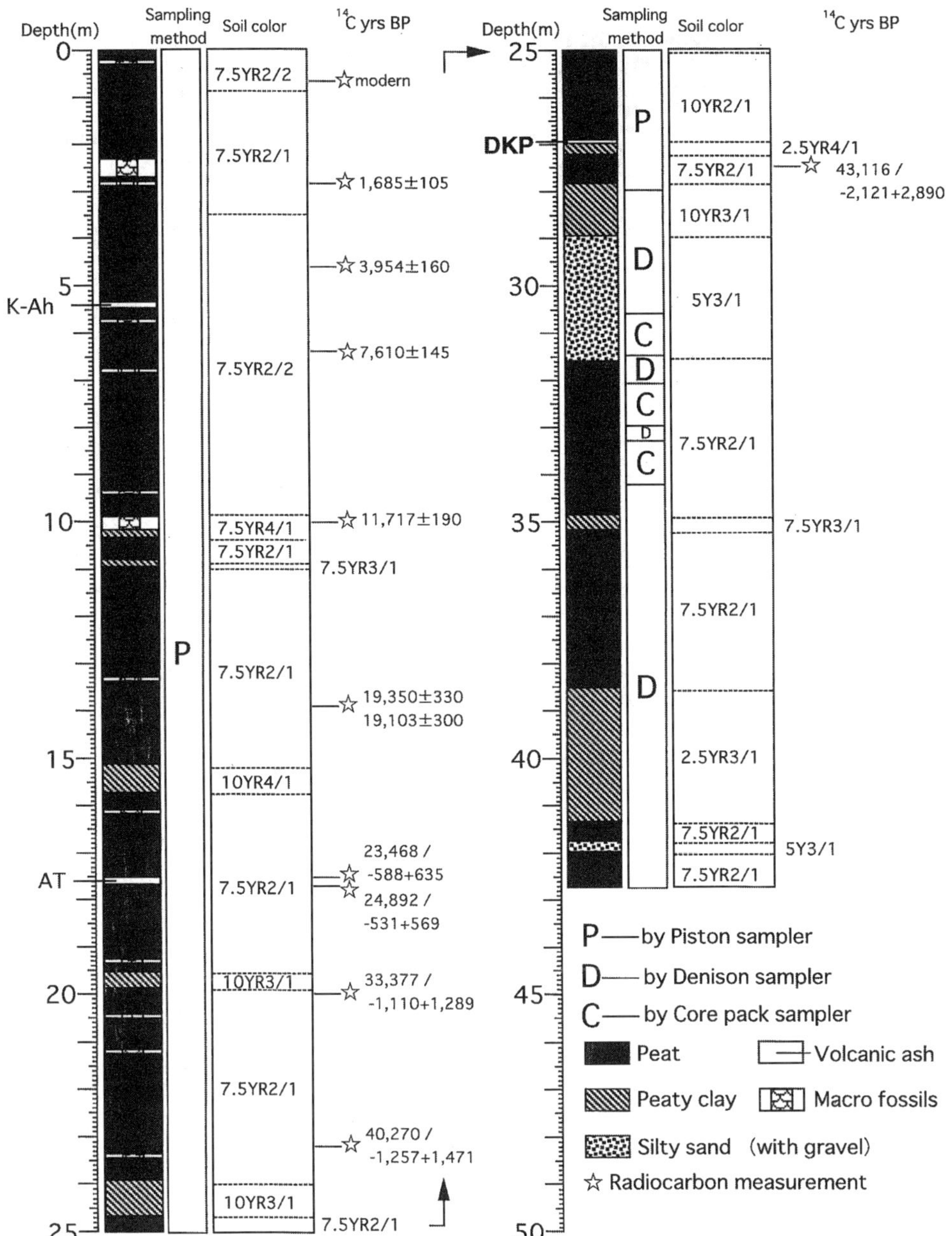

Figure 11.2   Sedimentary facies of core samples from Naka-Ikemi Moor (from Miyamoto et al., 1995)

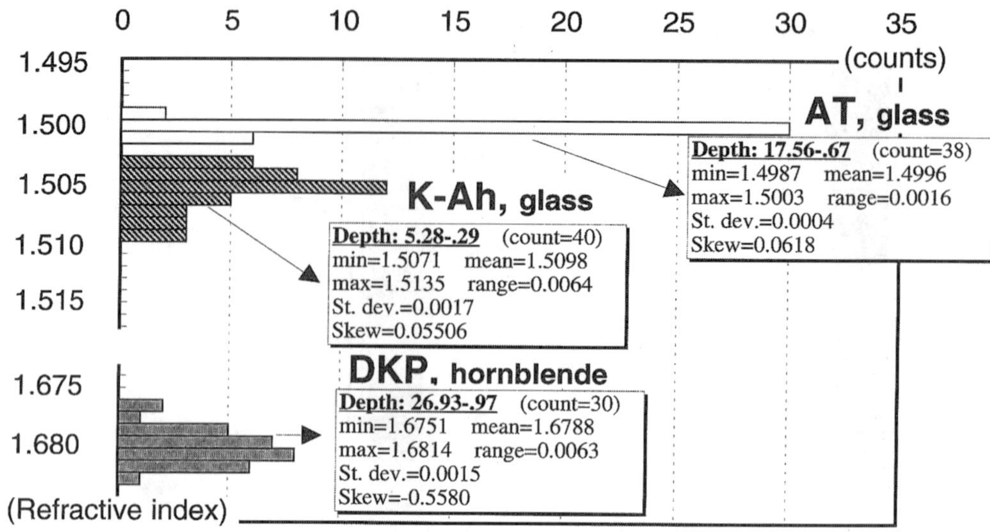

Figure 11.3   Refractive index histogram of the volcanic glass and hornblende of Naka-Ikemi Moor (from Miyamoto et al., 1995)

Table 11.1   Radiocarbon dates of Naka-Ikemi Moor

Code no.*	Sample no.	Material	Depth (m)	Age (^{14}C yr BP)
JAS-189	NKM2-3	Peat	0.65	Modern (1.07 $\pm$ 0.2 pMC)
JAS-196	NKM4-B	Wood	2.80	1685 $\pm$ 105
JAS-188	NKM7-3/4	Peat	4.60	3954 $\pm$ 160
JAS-195	NKM9-5	Peat	6.35	7610 $\pm$ 145
JAS-194	NKM13-12	Peat	10.00	11 717 $\pm$ 190
JAS-193	NKM18-19	Peat	13.90	19 350 $\pm$ 330
JAS-197	NKM18-9	Wood	13.90	19 103 $\pm$ 300
JAS-198	NKM23-AT.U	Peat	17.50	23 468/-588 + 635
JAS-191	NKM23-AT.L	Peat	17.70	24 892/-531 + 569
JAS-192	NKM25-17/18	Peat	20.00	33 377/-1110 + 1289
JAS-190	NKM29-14/15	Peat	23.20	40 270/-1257 + 1471
JAS-199	NKM35-8/9	Peat	27.50	43 116/-2121 + 2890

* JAS is the laboratory code number of the International Research Center for Japanese Studies. pMC: percentage modern ^{14}C

following parts: the bulk density is about 0.1 to 0.3 in the peaty sediments, and ignition loss increases upwards from about 20 to 60%, from 25 to 10 m in depth.

POLLEN SPECTRA

Ninety-five taxa were recognized in the fossil pollen and spores (Figures 11.4 to 11.6). In all samples studied, arboreal pollen including *Tsuga, Pinus* subgen. *Haploxylon, Betula, Alnus* and *Quercus* subgen. *Lepidobalanus* were common. Non-arboreal pollen included

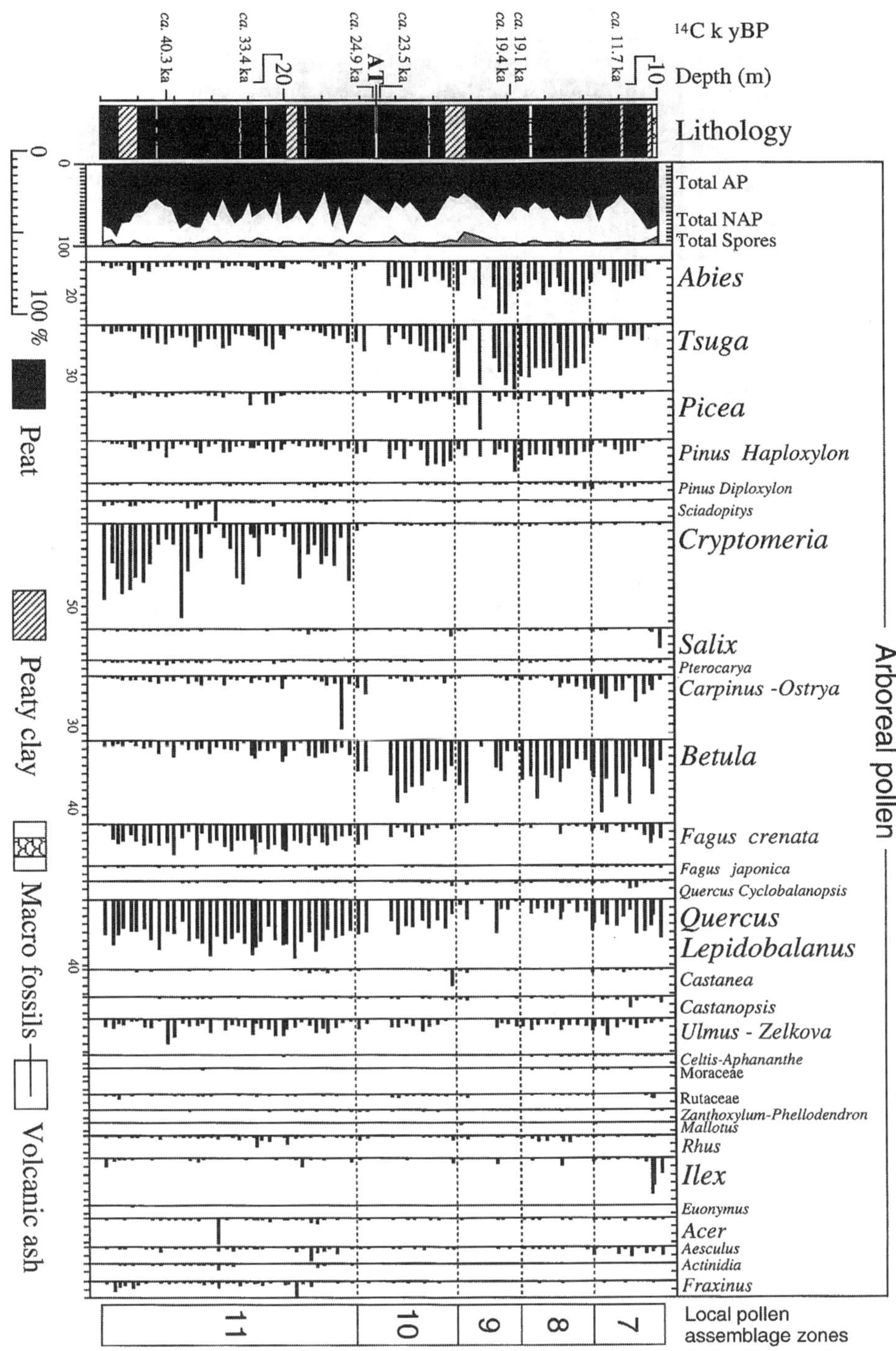

Figure 11.4　Major arboreal pollen diagram of Naka-Ikemi Moor

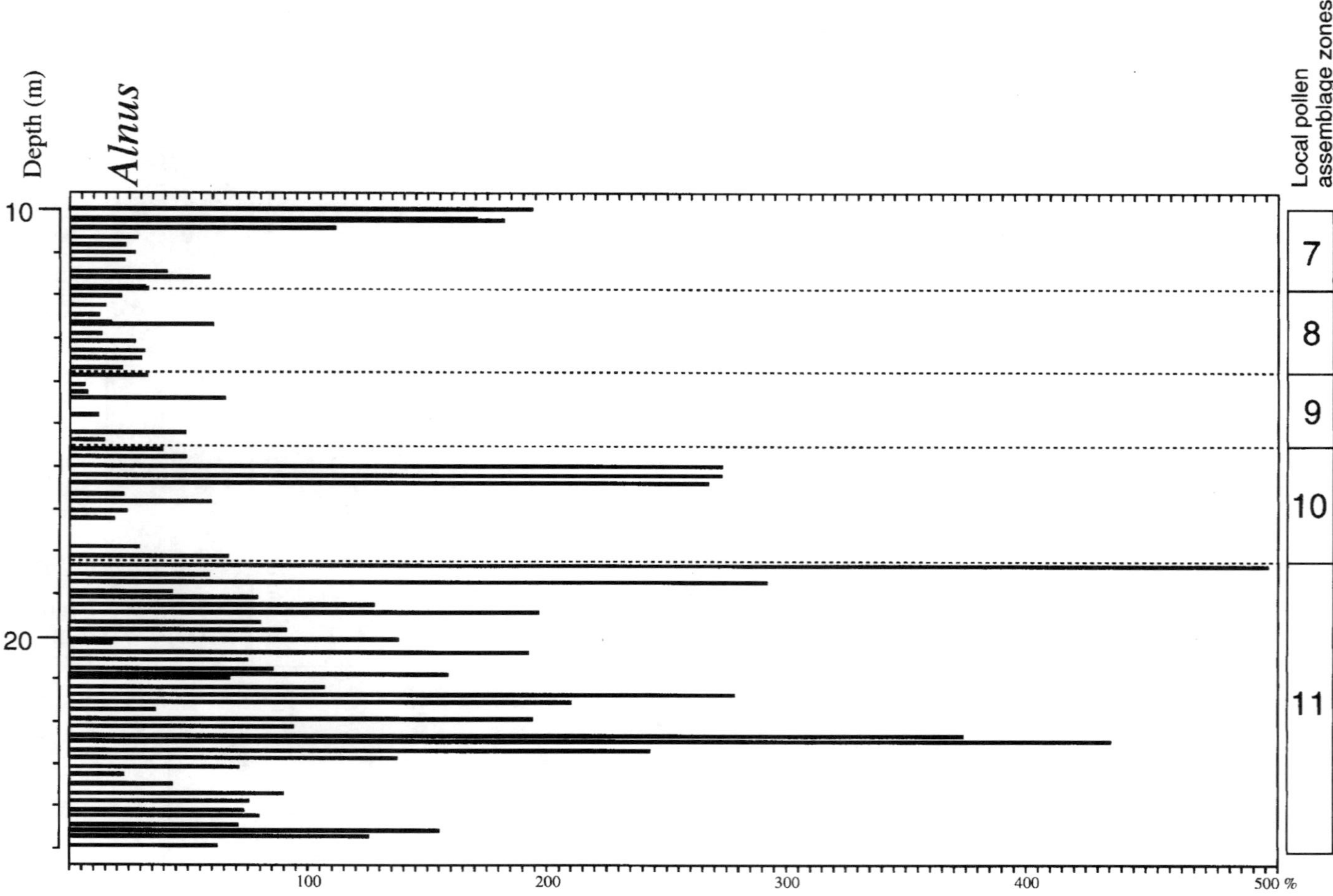

Figure 11.5   *Alnus* pollen diagram of Naka-Ikemi Moor

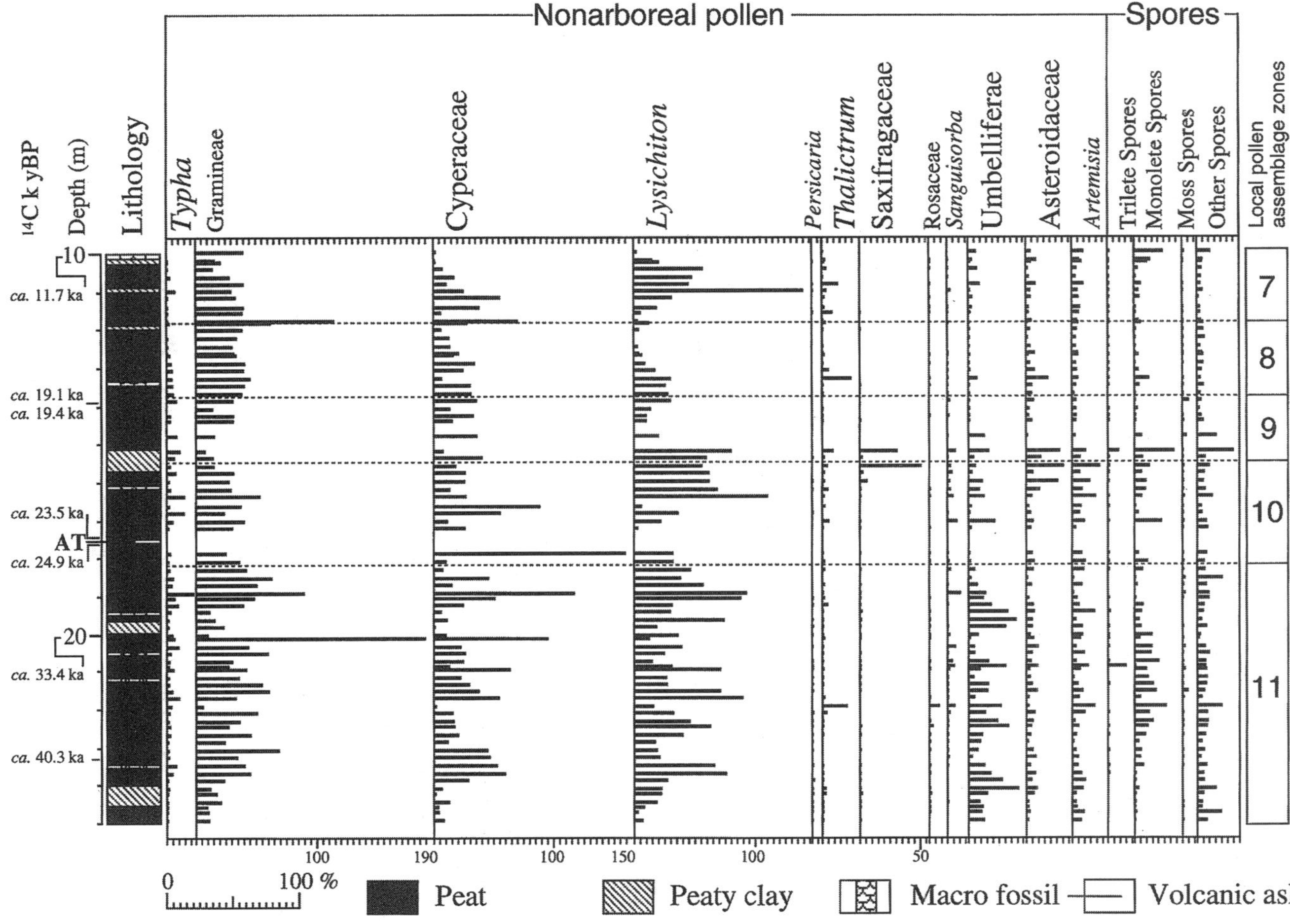

Figure 11.6   Major non-arboreal pollen diagram of Naka-Ikemi Moor

Gramineae, Cyperaceae and *Lysichiton*. Five local pollen assemblage zones from lowest to highest, NKM-11 to NKM-7, were established mainly on the basis of arboreal pollen spectra.

(1) **Zone NKM-11**. *Cryptomeria* occupies about 30% of the total arboreal pollen, *Tsuga* about 10%, and *Abies*, *Picea* and *Pinus* subgen. *Haploxylon* about 5%. *Fagus crenata* and *Quercus* subgen. *Lepidobalanus* are about 15 to 20%. The end of this zone is distinguished by a rapid decrease in *Cryptomeria* pollen.

(2) **Zone NKM-10**. This zone is characterized by an increase in coniferous and *Betula* pollen. Coniferous pollen begins to increase from the lower part of this zone. *Abies* increases upwards, from about 5 to 15%. *Tsuga* and *Abies* also increase, from about 10 to 16% and from 1 to 15%, respectively. *Pinus* subgen. *Haploxylon* increases up to 15% at the middle. In contrast, *Cryptomeria* suddenly decreases to about 1%. *Quercus* subgen. *Lepidobalanus*, *Betula*, *Fagus crenata* and *Ulmus-Zelkova* are common, at 10 to 20%. Gramineae and Cyperaceae occupy more 30%.

(3) **Zone NKM-9**. This zone is distinguished from the top of NKM-10 by a rapid increase in *Abies*, *Tsuga*, and *Picea* pollen, and decreases in *Betula* and *Fagus crenata*. At the top of this zone *Abies* increases up to about 30%, and *Tsuga* indicates about 40%. *Alnus* decreases to about 20%.

(4) **Zone NKM-8**. This zone is characterized by an increase in deciduous pollen, and decrease in coniferous pollen. *Tsuga* decreases upwards, from about 40 to 25%. *Quercus* subgen. *Lepidobalanus*, *Carpinus-Ostrya* and *Fagus crenata* are common, at about 10%. On the other hand, Gramineae stays at about 30%, and Cyperaceae and *Lysichiton* decrease to about 5%.

(5) **Zone NKM-7**. This zone is characterized by a rapid decrease in coniferous pollen, and increase in deciduous pollen. Coniferous pollen decreases to about 10%. In deciduous pollen, *Carpinus-Ostrya* increases to about 15%, and *Betula* increases to reach 38% maximum. *Ilex* rapidly increases to about 20% in the upper part of this zone. *Lysichiton* shows the highest percentage at the middle of this zone, about 140%.

Pollen of several taxa were occasionally found as pollen masses. Pollen masses of *Alnus*, Gramineae and Cyperaceae are common in all samples, and pollen masses of *Betula* and *Lysichiton* were observed.

DISCUSSION

Vegetation and climatic changes during the Last Glacial period

In this section, we discuss the vegetation changes based on the local pollen assemblage zones of NKM-11 to NKM-7. The local pollen assemblage zones from NKM-11 to NKM-7 indicate that the vegetation around Naka-Ikemi Moor during the Last Glacial Maximum was mixed vegetation of coniferous and broad-leaved deciduous trees in the upland, and wetland vegetation and grasslands in the lowland. The type of past vegetation can be reconstructed from the modern vegetation. Wetland vegetation is distributed in and around the moor, and upland vegetation is distributed in the montane regions around the moor. Wetland vegetation included stands of *Alnus* in a grassland of Gramineae, Cyperaceae, *Lysichiton*, *Thalictrum*, *Sanguisorba*, Umbelliferae and Compositae. On account of the buried stumps of *Cryptomeria japonica* observed in the moor,

C. japonica is regarded as mainly distributed in and around the wetland. Upland vegetation included mixed forests of coniferous and deciduous broad-leaved trees, such as *Abies, Tsuga, Picea, Pinus* subgen. *Haploxylon, Cryptomeria, Quercus* subgen. *Lepidobalanus, Carpinus-Ostrya* and *Ulmus-Zelkova.*

The ages of pollen zones were estimated from the radiocarbon age and the eruption age of the Aira-Tn ash. The bulk density indicates that a continuous record is preserved in this period.

Zone NKM-11 (c. 40–25 ka)

Vegetation of this zone was characterized by a spread of cool temperate deciduous broad-leaved trees such as *Betula, Fagus crenata, Quercus* subgen. *Lepidobalanus* and *Ulmus-Zelkova,* and *Cryptomeria* both in wetland and upland. Tsukada (1967) suggested that *Cryptomeria japonica* is related to precipitation, being restricted to areas with 1800 to 3000 mm annual total rainfall. Upland forests were mainly composed of cool temperate deciduous trees and coniferous trees forming a mixed forest vegetation. Wetland vegetation included stands of *Alnus,* Gramineae, Cyperaceae, *Lysichiton* and Umbelliferae. This pollen assemblage suggests that cool and wet climatic environments prevailed in this period. This zone is correlated with the MG-1 to FG-2 zones of Lake Mikata (Yasuda, 1982), which corresponds to the Interstadial of the Pleniglacial period (Isotope Stage 3).

Zone NKM-10 (c. 25–20 ka)

This zone was marked by an increase in sub-boreal coniferous trees and cool temperate deciduous trees. This period was characterized by mixed forest of sub-boreal coniferous trees and temperate broad-leaved deciduous trees (mainly *Betula*). Cyperaceae and *Lysichiton* dominated in the wetland vegetation. A sudden decline of *Cryptomeria* occurred at *c.* 25 000 yr BP, just before the fall of Aira-Tn ash. This *Cryptomeria* pollen signal is a reflection of the climatic decline, that is, the decrease of temperature and precipitation from the Interstadial (Isotope Stage 3) to the Last Glacial Maximum (Isotope Stage 2).

Zone NKM-9 (c. 20–17 ka)

This zone was characterized by increases of sub-boreal coniferous trees such as *Abies, Tsuga, Picea* and *Pinus* subgen. *Haploxylon.* These sub-boreal coniferous trees occurred with cool temperate deciduous trees (mainly *Betula* and *Quercus* subgen. *Lepidobalanus*) on the hill slopes around the moor. This reconstructed vegetation corresponds to zone IW-5 of the Mikata Lowland (Takahara and Takeoka, 1992) and zone FG-5 of Lake Mikata (Yasuda, 1982) of the western part of the Wakasa Bay region, both of which have been correlated with the Last Glacial Maximum.

From the palaeovegetation maps, the vegetation of this area is reconstructed as mixed conifer and deciduous broad-leaf forest (Nasu, 1980; Tsukada, 1985), and mixed forest of sub-alpine coniferous and temperate deciduous broad-leaved trees (Yasuda, 1978). The reconstructed vegetation types of this area are consistent with the result of the present study.

Tsukada (1980) suggested that the Wakasa Bay region was one of the forest refuges of *Cryptomeria japonica* around the Last Glacial Maximum. However, only very few

Cryptomeria pollen were observed in this study during the Last Glacial Maximum. *Cryptomeria* pollen did not appear successively since *c.* 25 ka. *Cryptomeria* pollen reaches less than 5% since *c.* 25 ka BP at the Naka-Ikemi Moor site and does not increase until the latest stage of the Last Glacial period. It is clear that *Cryptomeria japonica* was not distributed around Naka-Ikemi Moor, located in the eastern part of the Wakasa Bay region.

Zone NKM-8 (c. 17–13 ka)

In this zone, cool temperate deciduous trees (mainly *Betula*) began to increase with the decrease in sub-boreal coniferous trees. It suggests a warmer and wetter climate than that of the Last Glacial Maximum.

Zone NKM-7 (c. 13–11 ka)

After the previous zone NKM-8, sub-boreal coniferous trees rapidly decreased, while cool temperate broad-leaved deciduous trees increased. In particular, *Betula* dominated in this period, as a pioneer tree. The rapid decline in sub-boreal coniferous trees and the rise in *Betula* indicate the successive extension of broad-leaved trees replacing the coniferous trees. This was a reflection of a transitional period in respect to climatic oscillations. The increase in the ignition loss indicates increased litter fall from cool temperate deciduous trees such as *Betula*.

Vegetation and climate prior to the fall of Aira-Tn ash

At the top of zone NKM-11, the abundance of *Cryptomeria* and broad-leaved trees like *Fagus crenata* and *Quercus* subgen. *Lepidobalanus* indicates warm and wet climatic environments. From the upper part of NKM-11 (about 20 m in depth; *c.* 33 000 yr BP) to the lower part of NKM-9, *Abies*, *Tsuga* and *Pinus* subgen. *Haploxylon* began to increase and *Cryptomeria* rapidly decreased before the fall of the Aira-Tn ash (*c.* 25 000 yr BP). These pollen signals indicate that the climate became cooler prior to the fall of the Aira-Tn ash. Similar climatic changes at this period are also recognized in other palaeobotanical studies in the Japanese islands (Nakamura, 1973; Yasuda, 1987; Tsuji and Kosugi, 1991).

In the wetland, Gramineae, *Lysichiton* and Umbelliferae pollen decrease in the zone NKM-4 (*c.* 25 ka). This reflects the drier soil conditions and reduction of wetland forests around the study site. A vegetational change similar to this also occurred at the Mikata Lowland in the eastern part of the Wakasa Bay region (Takahara and Takeoka, 1992).

Vegetation and climate after the fall of Aira-Tn ash

Just above the Aira-Tn ash, most pollen taxa decreased or vanished. This change implies that the ash fall seriously destroyed the vegetation. After the Aira-Tn ash, the recovery of vegetation began with sub-boreal coniferous trees such as *Tsuga* and *Pinus* subgen. *Haploxylon*. This tendency may reflect the continuous climate cooling during this period. It is also emphasized that the climate during this period was dry, as suggested by the decline in *Alnus*.

There are, however, no other published pollen records which indicate the destruction of vegetation by the fall of Aira-Tn ash, except the Itai-Teragatani Site, the Sasayama Basin, Hyogo Prefecture (Ooi et al., 1990). This suggests that the recovery of vegetation would have been so rapid that the vegetation destruction was not recorded in sediments in other sites.

The ash fall caused severe environmental changes such as widespread devastation of land (Tsuji and Kosugi, 1991), but its direct effect on the vegetation needs to be further clarified in relation to other environmental factors.

ACKNOWLEDGEMENTS

This chapter is a part of the master's thesis of the first author (Shinji Miyamoto) submitted to the Department of Geography, Tokyo Metropolitan University. The authors are grateful to Hiroshi Kadomura, Professor Emeritus of Tokyo Metropolitan University, Professors Nobuyuki Hori and Shuji Iwata, Associate Professor Hitoshi Fukusawa, Assistant Professor Sadao Takaoka of Tokyo Metropolitan University, and Professor J.R. Flenley of Massey University for their many valuable comments.

This study was financially supported in part by a Grant-in-Aid of the Japanese Ministry of Education, Science and Culture, Project No. A0201, 'Vegetational and Climatic Changes in Relation to the Rise and Fall of Civilizations (led. by Professor Yoshinori Yasuda)'.

REFERENCES

Furutani, M., 1979. Studies on the forest history in the Osaka area since Würm Glacial Age in Japan. *Quaternary Research* (Japan), **18**, 121–141 (in Japanese with English abstract).

Machida, H. and Arai, H., 1976. Kouiiki ni Bunpu suru Kazanbai, Aira-Tn Kazanbai no Hakkenn to sono Igi (A widespread tephra: discovery of the Aira-Tn ash and its significance). *Kagaku (Sci. J.)*, **46**, 339–347 (in Japanese).

Machida, H. and Arai, H., 1983. Wide spread Late Quaternary tephras in Japan with special reference to Archaeology. *Quaternary Research* (Japan), **22**, 133–148 (in Japanese with English abstract).

Matsumoto, E., Maeda, Y., Takemura, K. and Nishida, S., 1987. New radiocarbon age of Aira-Tn ash (AT). *Quaternary Research* (Japan), **26**, 79–83 (in Japanese with English abstract).

Miyamoto, S., Yasuda, Y. and Hiroyuki, K., 1995. Sedimentary facies and chronology of core samples from the Naka-Ikemi Moor-Sedimentary environment for the past 50 ka. *Journal of Geography* (Japan), **104**, 865–873 (in Japanese with English abstract).

Nakamura, J., 1973. Palynological aspects of the Late Pleistocene Japan. *Quaternary Research* (Japan), **12**, 29–37 (in Japanese with English abstract).

Nasu, T., 1980. Urumu Hyouki Saiseiki no Koshokusei (Palaeovegetation at Wurm Glacial Maximum). *Daishiki*, **20**, 55–63 (in Japanese).

Okada, A., 1978. Structure of the waste-filled valleys and associated crustal movements at the eastern part of the Tsuruga Plain, North of Lake Biwa. *Paleolimnology of Lake Biwa and the Japanese Pleistocene*, **6**, 66–80.

Ooi, N., Minaki, M. and Noshiro, S., 1990. Vegetation changes around the Last Glacial Maximum and effects of the Aira-Tn ash, at the Itai-Teragatani Site, Central Japan. *Ecological Research*, **5**, 81–91.

Shoma, K. and Tsuji, S., 1987. Vegetation. In Japan Association for Quaternary Research (eds), *Nihon Daiyonki Chizu Kaisetsu (Explanatory Text for Quaternary Maps of Japan)*, University of Tokyo Press, 80–86 (in Japanese).

Takahara, H. and Takeoka, S., 1992. Vegetation history since the Last Glacial period in the Mikata Lowland, the Sea of Japan area, Western Japan. *Ecological Research*, **7**, 371–386.

Takemura, K., Kitagawa, H., Hayashida, A. and Yasuda, Y., 1994. Sedimental facies and chronology of core samples from the Lake Mikata, Lake Suigetsu and Mikata Lowland. Sedimentary environment since the Last Interglacial. *Journal of Geography* (Japan), **103**, 233–242 (in Japanese with English abstract).

Tsuji, S. and Kosugi, M., 1991. Influence of Aira-Tn ash (AT) eruption on ecosystem. *Quaternary Research* (Japan), **30**, 419–426 (in Japanese with English abstract).

Tsukada, M., 1967. Vegetation and climate around 10,000 B. P. in central Japan. *American Journal of Sciences*, **265**, 562–585.

Tsukada, M., 1974. *Koseitaigaku II, Ouyou Ron* (*Paleoecology II, Synthesis*), Kyoritsu Shuppan (in Japanese).

Tsukada, M., 1980. *Sugi no Rekishi, kako Ichimann Gosennenn Kann* (The History of *Cryptomeria japonica*: for the past 15 ka). *Kagaku* (*Sci. J.*), **50**, 538–546 (in Japanese).

Tsukada, M., 1984. A vegetation map in the Japanese Archipelago approximately 20,000 years BP. *Japan Journal of Ecology*, **34**, 203–208 (in Japanese with English abstract).

Tsukada, M., 1985. Map of vegetation during the Last Glacial Maximum in Japan. *Quaternary Research* (Japan), **23**, 369–381 (in Japanese with English abstract).

Ueda, Y. and Tsuji, S., 1994. The buried forests and their radiocarbon dating at Naka-Ikemi in Tsuruga, along the Wakasa Bay. *Japan Journal of Historical Botany*, **2**, 29–30 (in Japanese).

Yasuda, Y., 1978. Prehistoric environments in Japan. Palynological approach. *Science Reports of Tohoku University, 7th series* (*Geography*), **28**, 117–281.

Yasuda, Y., 1982. Pollen analytical study of the sediment from the Lake Mikata in Fukui Prefecture, Central Japan. Especially on the fluctuation of precipitation since the Last Glacial Age on the side of Sea of Japan. *Quaternary Research* (Japan), **21**, 255–271 (in Japanese with English abstract).

Yasuda, Y., 1987. The cold climate of the Last Glacial Age in Japan – comparison with southern Europe. *Quaternary Research* (Japan), **25**, 277–294 (in Japanese with English abstract).

CHAPTER 12

Late Pleistocene and Holocene River Development in Mediterranean Steepland Environments, Southwest Crete, Greece

G. S. MAAS, M. G. MACKLIN

AND

M. J. KIRKBY

School of Geography, University of Leeds, UK

INTRODUCTION

The Mediterranean is characterised by a seasonal precipitation regime, with cool moist winters and hot dry summers. Much of the region has a semi-arid climate with annual precipitation below 500 mm, though increasing to over 2000 mm in mountainous terrain. Its mountainous rivers typically have high gradients and boulder- or cobble-bed channels, often incised into older alluvial or colluvial fills or bedrock (Macklin et al., 1995). The 'flashy' nature of many of these streams is in response to the highly seasonal precipitation regime, with intense frontal storms enhanced by orographic effects. This results in infrequent floods with very high rates of sediment transport. Flood events in these environments range in nature from water floods through to hyperconcentrated and debris flows (*sensu* Costa, 1984).

Previous studies concerned with the effects of Late Quaternary environmental change on mountain river systems within the Mediterranean have generally concentrated on alluvial fan environments. They are represented by examples from southeast Spain (Harvey, 1984, 1988, 1996), southern Turkey (Roberts, 1983, 1995), the Dead Sea, Israel (Frostick and Reid, 1989) and southwest Crete, Greece (Nemec and Postma, 1993) and indicate aggradation and dissection sequences associated with a change in sediment supply and runoff regime in response to Late Quaternary climate change onto which the effects of local and regional tectonics have been superimposed.

One key question that these studies, however, have rarely addressed is how the upper and middle reaches of steepland systems have responded to environmental change during the late Holocene, in particular the effects of the climatic deterioration associated with the

Little Ice Age (Grove, 1988), and more recent short-term changes in precipitation. In addition, the degree to which human activity has affected geomorphic processes in small, mountainous catchments in the eastern Mediterranean has been little researched. Previous studies in the region have tended to concentrate on relatively large catchments where the history of settlement is better documented (e.g. Wagstaff, 1981; Pope and van Andel, 1984; van Andel et al., 1986; Gutiérrez Elorza and Peña Monné, 1990; Lewin et al., 1991).

To address these issues a geomorphological study has been undertaken within a small (6.6 km^2) steepland catchment in the Lefka Ori mountains of southwestern Crete, Greece. This steep gradient, coarse-grained river system flows into the Omalos basin, an internally drained karstic depression. It exhibits a suite of valley floor alluvial and colluvial deposits and an extensive alluvial fan. The major aims of this study were to:

(1) identify and date valley floor alluvial units within the Omalos catchment;
(2) evaluate the depositional environments of individual units and relate these to major Late Pleistocene and Holocene environmental controls.

STUDY AREA

The Omalos basin is situated 20 km southwest of Chania, 23°54′30″E and 35°19′00″N (Figure 12.1), within the Lefka Ori mountains which form a series of east–west trending horsts and grabens. The region is underlain by Upper Triassic crystalline limestone (IGSR, 1969). Western Crete as a whole has undergone major episodes of tectonic activity during the Holocene, the most recent of which at 1530 ± 40 yr BP resulted in uplift of between 4 and 8.5 m (Pirazzolli et al., 1982). The effect of tectonics upon the Omalos basin in terms of river base-level change are, however, insignificant.

Crete has a typical Mediterranean climate, characterised by a highly seasonal precipitation regime, with mild wet winters and long dry summers. The nearest weather station at Souda (20 km to the northwest of Omalos) provides precipitation data for the period 1958 to 1992, with a mean annual total of 678 mm. The altitude and topography around Omalos increase the total annual precipitation, much of which falls as snow in the winter and lies on the upper northern slopes until May.

Investigations have focused on an unnamed ephemeral gravel-bed channel that drains the southern part of the Omalos basin (Figure 12.2). With a catchment area of 6.6 km^2 and relative relief of 909 m, it rises in extensive scree-covered headwaters at a maximum altitude of 1984 m, and then passes downstream into a short, narrow bedrock reach. Below this the valley floor widens and contains a suite of river terraces, covering more than half of the valley floor. At the mountain front the stream has built an alluvial fan which is currently entrenching. Channel gradients range from about 0.4 mm^{-1} in the upper part of the basin to about 0.08 mm^{-1} in the lower reaches.

Vegetation within the catchment is dominated by sparse phrygana (*Anthylieto-poterietum*) on the lower slopes, while at higher altitudes cypress (*Cupressus sempervirens*) and oak (*Quercus calliprinos*) shrubs are beginning to colonise extensive relict scree slopes. The lower slopes of the alluvial fan and the Omalos basin have been terraced for cultivation, with grazing generally restricted to the upper part of the alluvial fan.

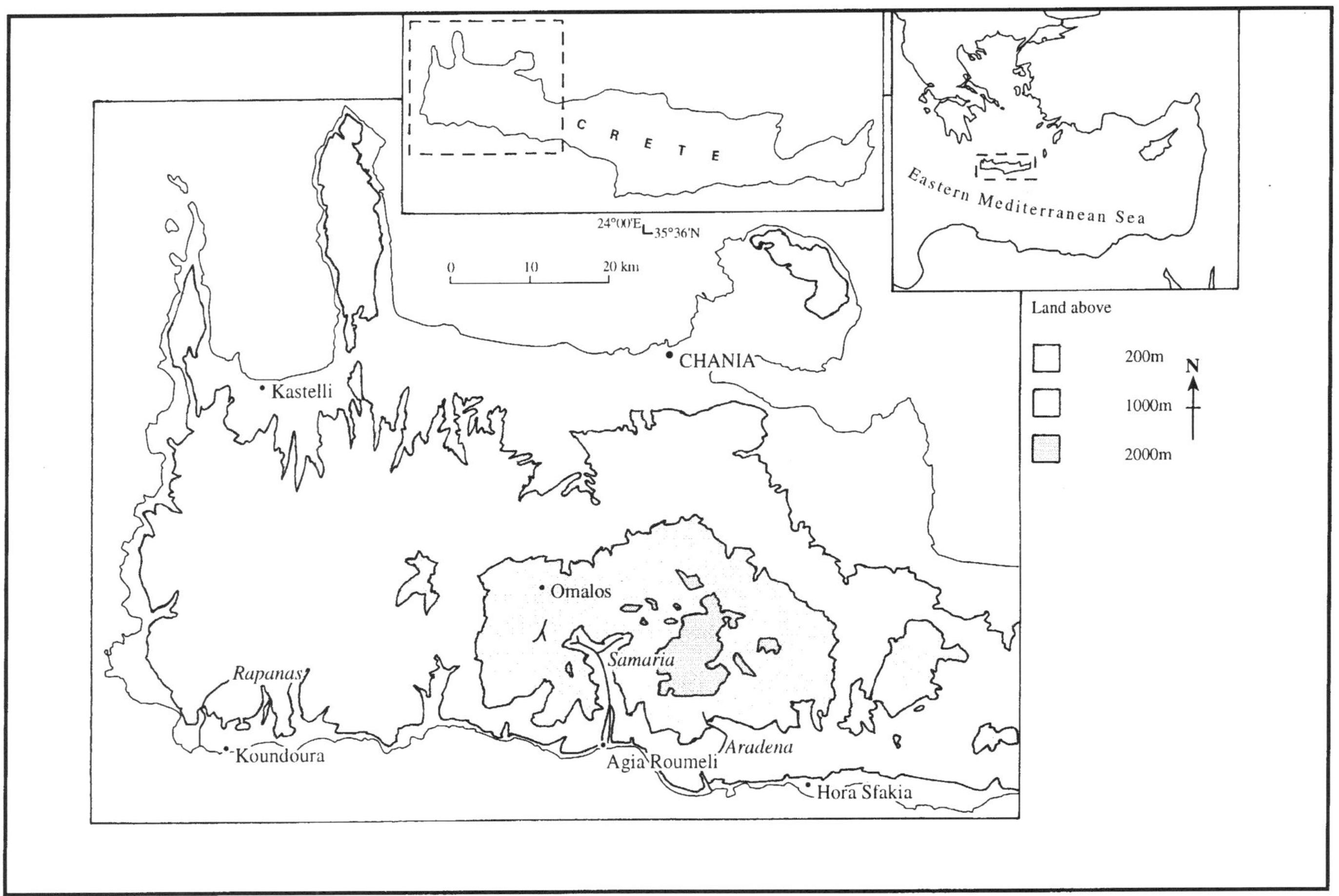

Figure 12.1   Location of Crete, Greece, within the eastern Mediterranean and the position of the Omalos basin within the Lefka Ori mountains of southwestern Crete

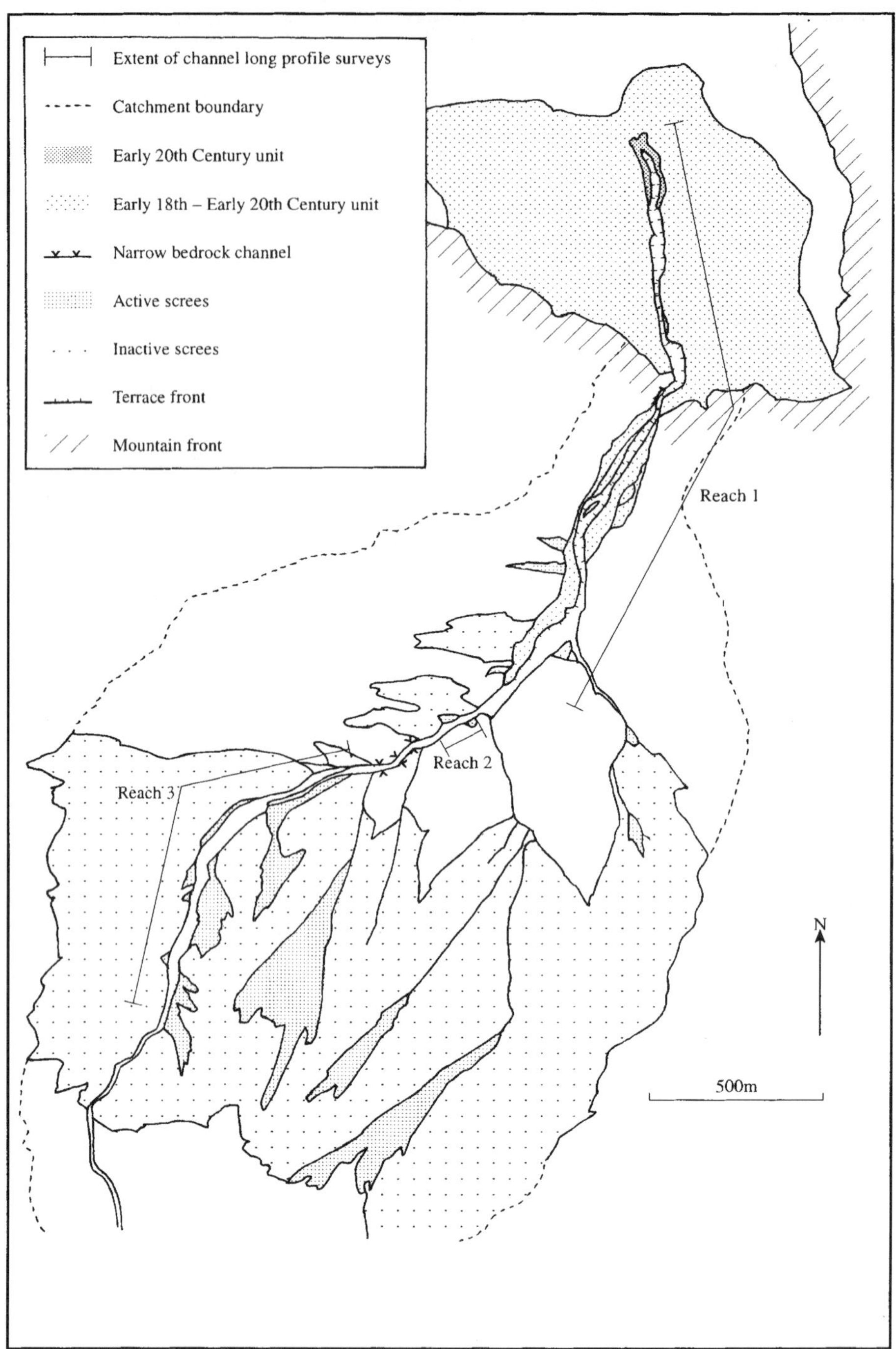

Figure 12.2   Major geomorphological features within the study stream and location of surveyed reaches

METHODS OF INVESTIGATION

The major alluvial fills within three study reaches (Figure 12.2), were mapped using EDM (electronic distance meter) survey. The sedimentological characteristics of each unit were recorded using standard techniques that included measuring the *b*-axis of ten of the largest boulders on the terrace surfaces to estimate palaeocompetence. Stream bank sections of terrace units were logged to determine their depositional environment.

Five dating techniques have been used to establish a chronology for river erosion and aggradation within the valley floor.

(1) Lichenometry, using the mean of the five largest thalli of a white crustose species *Aspicilia Calcarea*, commonly found on boulders exposed at the surface of river terraces. Lichen growth rates were calibrated by measuring the five largest lichens on dated substrates in western Crete (Figure 12.3).
(2) Tree ring analysis of cypress and oak both growing on the surface of river terraces and also buried by deposits.
(3) Air photographs from 1945, 1968 and 1989 were used to determine the age of recent channel deposits.
(4) Uranium series dating of calcretes to provide a minimum age of deposition.
(5) ^{14}C dating of wood and charcoal incorporated within alluvial units. In this chapter, the 2σ (95% probability) calibration result of the radiocarbon age is used in the alluvial geochronology interpretation.

ALLUVIAL DEPOSITS IN THE OMALOS BASIN

Alluvial terrace sequence

Six major alluvial units (including the present channel bed) were identified within the

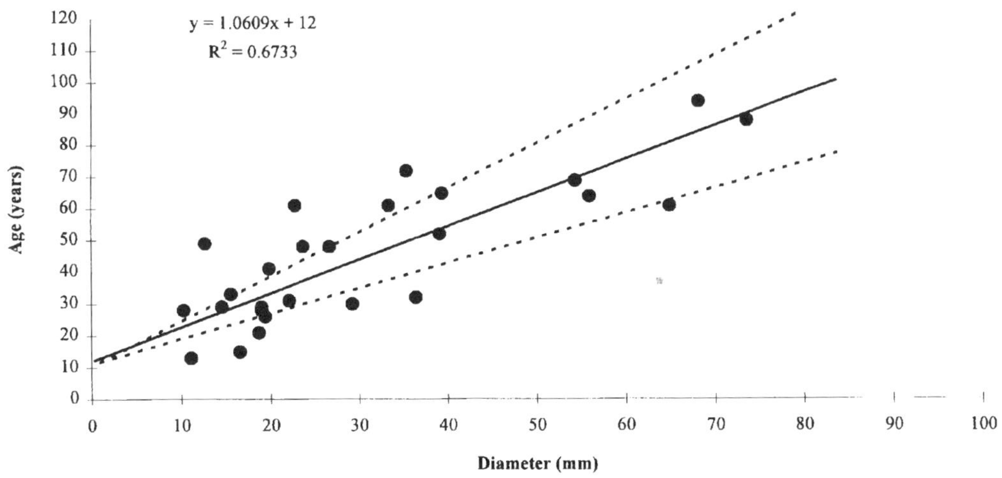

Figure 12.3 Correlation between lichen thallus diameter (*Aspicilia calcarea*) and age, significant at 99%. Upper and lower quartile envelope (dotted lines) provides maximum and minimum age estimates

catchment and their ages, height relationships to the present channel and mean largest clast size are shown in Table 12.1. The surface of the most extensive unit, U5, lies up to 9.2 m above the present channel and can be traced continuously downstream from the top of reach 1 (Figure 12.4c), grading to the alluvial fan (Figure 12.4d). U2 and U3 are minor units found only at the top end of reach 1 (Figure 12.4c), with the latter unit having a steep depositional front and a surface height of 4.5 m above the present channel. Unit U4, located within the fan-head channel, has a maximum thickness of 1.2 m and its surface converges downstream towards the present channel (Figure 12.4d). The surface of unit U6 similarly converges downstream towards the channel bed (Figure 12.4d). Long-profiles of unit surfaces indicate that maximum channel incision has occurred in reach 2 and the upstream end of reach 1, decreasing downstream and towards the apex of the alluvial fan.

Sedimentology of alluvial fills

In reach 2 unit U6 consists of subangular to subrounded cobbles with a matrix of highly reddened calcrete. No internal bedding or imbrication exists in the upstream exposures of this unit, and its poor sorting reflects its deposition close to the sediment source area, with limited reworking. This unit filled the entire valley floor of the narrow bedrock section (Figure 12.2) to a depth of more than 7.2 m. Exposed within the fan-head trench the unit displays the same calcrete development as found in reach 2; however clasts are smaller (mean of the ten largest boulders (MTLB) = 180 mm) and more rounded, suggesting a greater degree of transportation.

Unit U6 fines upwards into silty sands containing occasional friable pebbles in discrete layers. A well-developed red soil is evident in the upper part of this unit and in the top 0.8 m of the profile, charcoal and fragments of pottery (undatable) were recovered.

Unit U5 covers a large proportion of the valley floor in reach 1 and also the alluvial fan downstream (Figure 12.2). In reach 1, the upper part of this unit is 3.4 m thick, consisting of imbricated boulders (MTLB = 684 mm) with a silt–sand matrix. It has a convex cross-valley floor surface profile representing a series of accretionary flood events with maximum traction forces concentrated in the channel centre, diminishing towards the valley sides (Costa, 1984). At the transition between the mountain front and the alluvial fan, U5 consists of cobble-sized clasts (MTLB = 167 mm) with a silt–sand matrix that display a series of fining-upward sequences. Larger clasts are imbricated with long axes orientated parallel to the present channel.

Unit U4 in reach 1 (MTLB = 183 mm) is confined within the channel cut into the alluvial fan and has sedimentological characteristics similar to upstream sections of U5. At the most distal end of the channel the unit splays onto the fan surface in a series of lobes several clast diameters thick.

Both units U3 and U2 are poorly sorted and weakly imbricated and exhibit sedimentary features characteristic of rapid deposition (e.g. Pierson, 1980; Ackroyd and Blakeley, 1984). Unit U3 is a large, mid-channel lobe with a steep depositional front and surface 4.5 m above the channel bed (Figure 12.4c). The largest clasts (MTLB = 342 mm) are located towards the front and base of this deposit, decreasing in size upstream and towards its surface.

Table 12.1   Age, height, thickness and clast size of six alluvial units

Unit number	Alluvial unit age (all dates obtained from reach 1 unless otherwise indicated)	Maximum height of unit surface above channel (m)	Maximum observed unit thickness (m)	Mean largest clast size (mm) ($n = 10$)
U1	Present channel			162
U2	AD 1968–1989 (d)	3.2	3.2	412
U3	AD 1968–1989 (d)	4.5	4.5	342
U4	AD 1947 + / − 11 (a)	3.1	1.2	183
	AD 1871 (c)			
U5	AD 1927 + / − 19 (a)	9.2	3.4	167 (downstream)
	AD 1883 (c)			684 (upstream)
	cal AD 1690–1735 (b)			
U6	cal AD 1285–1470 (b)	2.6	2.6	silt–sand (soil)
	cal 3535–3960 BC (b)			
	16.6 + / − 3.8 ka (e)	1.6	1.6	180 (calcrete, reach 1)
	14.4 + / − 3.6 ka (e), reach 2	7.2	7.2	324 (calcrete, reach 2)

Data techniques used: (a) lichenometry; (b) ^{14}C; (c) dendrochronology; (d) air photograph; (e) Uranium-series.

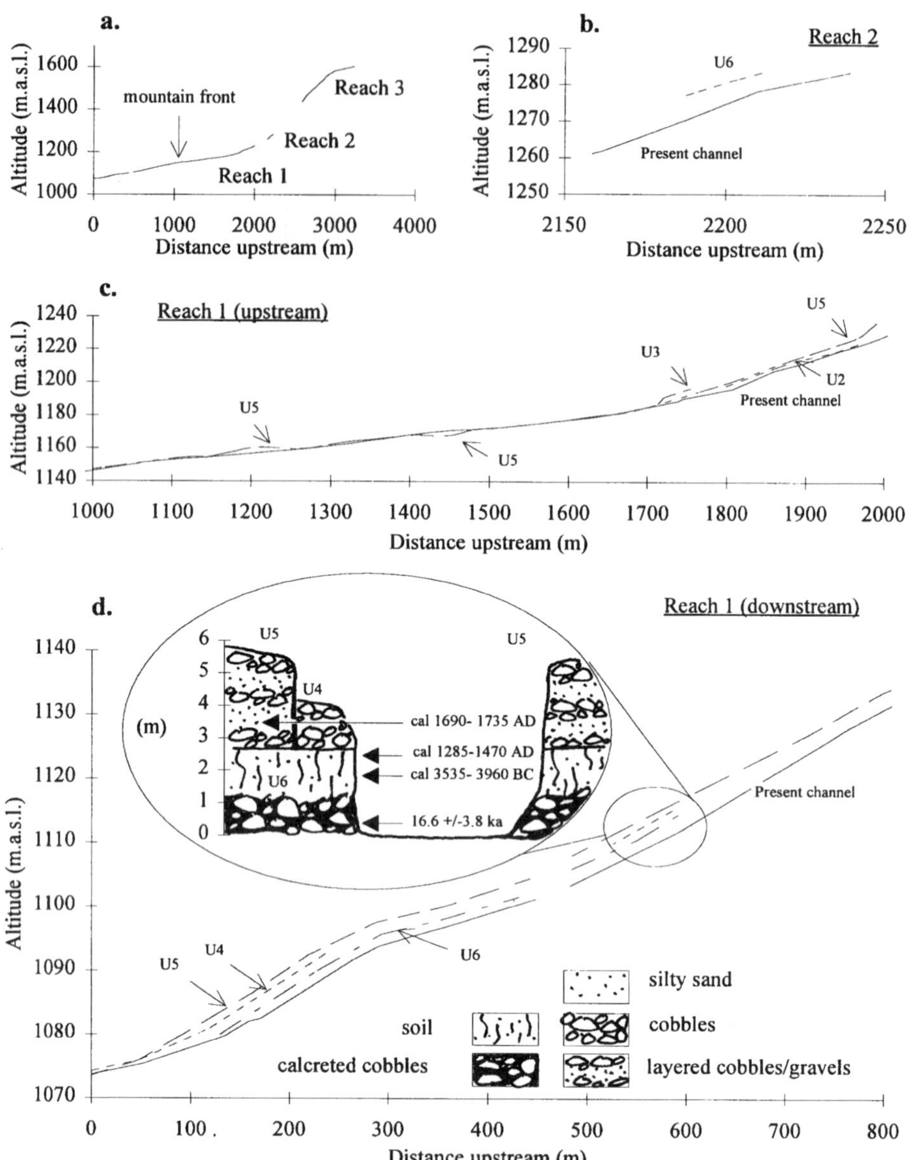

Figure 12.4   (a) Surveyed reaches of the study stream, showing long-profile gradients of the present channel surface. (b) Reach 2 long-profile gradient of present channel and alluvial unit surface. (c) Upper 1000 m of reach 1, showing long-profile gradients of present channel and alluvial unit surfaces. (d) Lower 800 m of reach 1, showing longitudinal profiles of alluvial unit surfaces and schematic cross-section of alluvial sequence, indicating relative height of dated material

Alluvial geochronology

Two calcrete samples taken from the top of the gravel horizon of U6 gave U/Th dates of 14.1 $\pm$ 3.6 and 16.6 $\pm$ 3.8 ka for reach 2 and reach 1 (Figure 12.4d), respectively, and provide a minimum age for the deposition of this unit. Two charcoal layers within the upper part of the soil developed within U6 were ^{14}C dated. The upper sample (depth of 0.3 m) was dated AD 1285 to 1470 cal (560 $\pm$ 80 ^{14}C yr BP; Beta-85403), while the lower sample (depth of 0.8 m) gave a date of 3535 to 3960 cal BC (5130 $\pm$ 90 ^{14}C yr BP; Beta-97872). A buried tree stump 2.5 m below the surface of U5 was ^{14}C dated at AD 1690 to 1735 cal (10 $\pm$ 60 ^{14}C yr BP; Beta-80842) (Figure 12.4d), though lichens and tree ring dates from the surface of this unit indicate that deposition probably ceased around the turn of the present century. Unit U4 was deposited some time after the late 19th century as it buried a tree that dates from AD 1871. Units U3 and U2, on the basis of air photographs, date to between AD 1968 and 1989; however, the absence of lichens on the surfaces of these units suggests an age closer to AD 1989.

RIVER RESPONSE IN THE OMALOS BASIN TO LATE PLEISTOCENE AND HOLOCENE ENVIRONMENTAL CHANGE

Alluvial and soil sequences in the Omalos catchment relate to periods of increased runoff and sediment load punctuated by phases of catchment stability and reduced sediment supply leading to valley floor incision. To interpret the controls of these it is important to understand how this small, steepland system has responded to changes in major environmental parameters and, in particular, to assess the relative importance of natural and anthropogenic factors.

The time scale of this study encompasses several global- and regional-scale changes in climate, the most notable of which occurred at the Last Glacial Maximum (18 000 yr BP) to Holocene transition (Birman, 1968; Gat and Magaritz, 1980; Prentice, et al., 1992; Pérez-Obiol and Julià, 1994), and more recently during the Little Ice Age (c. AD 1450 to 1850; Grove, 1988).

The record and nature of human interference in the Omalos catchment is, however, very poorly documented. Deforestation and variation in grazing intensity are likely to have been the only significant land-use change, but can only be inferred from the history of land-use change on adjacent mountain ranges in Crete. Deforestation has been documented on the mountain of Psilorites (2400 m altitude and 50 km east of Omalos; Lyrintzis and Papanastasis, 1995) where vegetation clearance began during the Neolithic and culminated in Minoan times (3000–1000 BC). Since this period the forests of Psilorites have experienced minor fluctuations in size and density depending upon human pressure, though the timing and extent of these fluctuations are uncertain. For example, a 17th century Scottish traveller reported that Psilorites was 'over clad even to the top with cypress trees' (Meiggs, 1982), where at present it is treeless. At present it is unknown whether deforestation occurred simultaneously in the Omalos catchment, and therefore only very tentative speculations can be made with respect to the effect of human activity on river development.

Unit U6 represents a major phase of channel aggradation and alluvial fan formation, which on the basis of U-series dated calcrete occurred prior to 10.8 to 20.4 ka, probably during the Last Glacial Maximum. Prentice et al. (1992) have proposed that climate during the Last Glacial Maximum in the Mediterranean region was characterised by cold winters and pronounced summer drought with intense winter precipitation and high runoff. Global circulation models produced by COHMAP (1988) indicate that increased cloudiness and reduced evaporation in summer, coupled with lower winter temperatures, resulted in increased seasonality of precipitation, with an increase in winter and decrease in summer precipitation. This has been shown to be a critical control of sediment yield and runoff in Mediterranean environments (Kirkby and Neale, 1987). Such conditions would have promoted hillslope stripping and channel aggradation.

Soil development on U6 is likely to have commenced at the same time as the formation of calcrete within this unit, at the end of the Pleistocene, and ended with gravel deposition on the terrace surface some time after the 15th century. The two dated charcoal samples found within the top 0.8 m of the soil horizon appear to have been incorporated within the upper part of the profile during periods of minor sediment reworking and influxes of sediment from adjacent hillslopes or channel deposits. The older phase (3535 to 3960 cal BC) could relate to deforestation during the Neolithic period which may have occurred contemporaneously with deforestation documented by Lyrintzis and Papanastasis (1995) on Mount Psilorites.

Unit U5 represents a significant change in river behaviour with a switch from low rates of river sediment transport and channel stability to valley floor and alluvial fan aggradation. The dramatic increase in sediment supply and calibre is likely to have been in response to an increase in both hillslope erosion and flood magnitude during the latter half of the Little Ice Age (Grove, 1988). There is growing evidence for widespread river metamorphosis throughout northern, western and central Europe in response to climatic deterioration during this period (Rumsby and Macklin, 1996). The Little Ice Age in the Mediterranean, however, has been less researched though evidence from a number of sources, including documentary, geomorphological and dendrochronological (e.g. Birman, 1968; Lamb, 1977; Serre-Bachet et al., 1992), suggest a significant reduction in mean annual temperatures and increased mean annual precipitation. Venetian records of trade with Crete between AD 1548 and 1648 (Grove and Conterio, 1995) indicate an increase in winter snowfall and generally more extreme and extensive rain storms, caused by more frequent blocking high pressure systems over northern Europe and southerly diversion of depression tracks.

Aggradation of U4 occurred between AD 1871 to 1947 (Table 12.1) and coincides with an increase in precipitation in the eastern Mediterranean around the turn of the century (Kutiel et al., 1996), (Figure 12.5a). Runs of relatively wet years observed at four stations in the eastern Mediterranean tend to be associated with the dominance of prevailing westerly winds from the adjacent Mediterranean Sea, compared to dryer sequences of years which are associated with easterly winds of continental origin. This wet phase is also recorded in Iraklion (northern Crete) during the 1920s and 1930s (Figure 12.5b).

Units U3 and U2, the youngest deposits in the Omalos catchment (dated between AD 1968 and 1989), appear to relate to single flood events. The large volume of sediment and calibre of U3 suggests it was the result of an exceptionally large flow event. Local observers have reported that major floods in the area are generally produced by particularly intense periods of rainfall in the autumn and winter.

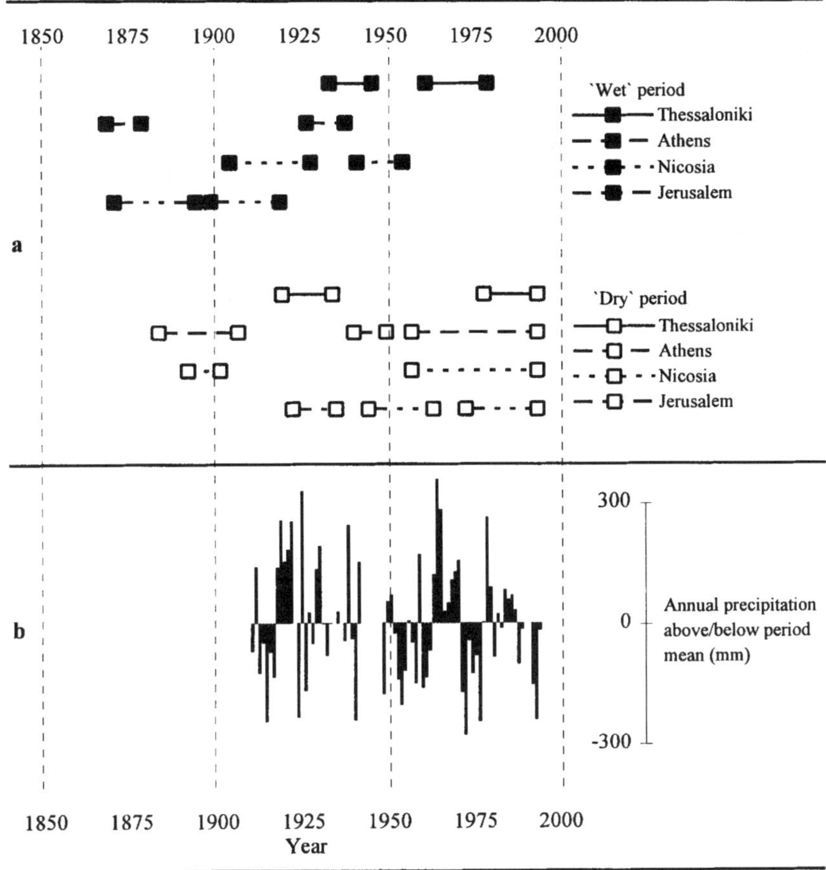

Figure 12.5 (a) The timing of 'wet' and 'dry' sequence of years in the eastern Mediterranean during the late 19th and 20th centuries. Modified from Kutiel et al. (1996). (b) Precipitation data from Iraklion, Crete, showing annual precipitation totals above and below the period mean (1910 to 1991). Data supplied by J. Palutikof, University of East Anglia

CONCLUSIONS

Geomorphological investigations of alluvial deposits in the Omalos system, southwest Crete, Greece, have revealed a quite detailed record of river response to Late Pleistocene and Holocene environmental change. Six major alluvial units are present in the river system, ranging in age from the Last Glacial Maximum to the present century. The oldest unit (U6), dated some time before 14.4 to 16.6 ka, seems likely to have been deposited during a cold period of the Late Pleistocene when reduced temperatures and increased effective precipitation accelerated hillslope erosion and sediment delivery to the channel. This was followed by an extended phase of valley floor stability and soil development perhaps lasting until the end of the 15th century. This was followed by renewed valley floor sedimentation during the latter part of the Little Ice Age (c. AD 1450 to 1850) which continued until the 1930s. More recently, between AD 1968 and 1989, two exceptionally large flood events have affected Omalos, resulting in further incision in the headwaters of the catchment and extensive gravel deposition in the middle and lower reaches of the river.

This study has demonstrated that historical changes in climate, including those that occurred during the Little Ice Age, have profoundly affected river behaviour in Mediterranean mountain environments. However, because land-use histories are presently rather poorly documented in many Mediterranean mountain basins, what is less clear is the importance of human interference in promoting historic and earlier periods of river instability. To answer these questions, future research will need to focus on steepland basins where land-use change is both temporally and spatially well constrained.

ACKNOWLEDGEMENTS

This study was funded by the School of Geography, University of Leeds, as part of the Superslugs research initiative. We would like to thank Martin Cookson, Ian Dyer, Lauren Morrison, Sophie Ryder and Caius Simmons for assistance in the collection of field data. Dr Stuart Black (Environmental Science Division, Lancaster University) provided the uranium series dates and assistance in the field. This work forms part of G.S.M.'s doctoral thesis, in preparation.

REFERENCES

Ackroyd, P. and Blakely, R.J., 1984. En masse debris transport in a mountain stream. *Earth Surface Processes and Landforms*, **9**, 307–320.

Birman, J.H., 1968. Glacial reconnaissance in Turkey. *Geological Society of America Bulletin*, **79**, 1009–1026.

COHMAP Members, 1988. Climate changes of the last 18,000 yrs: Observations and model simulations. *Science*, **241**, 1043–1052.

Costa, J.E. 1984. Physical geomorphology of debris flow. In J.E. Costa and P.J. Freisher (eds), *Developments and Applications of Geomorphology*, Springer Verlag, Berlin, 268–317.

Frostick, L.E. and Reid, I., 1989. Climatic versus tectonic controls of fan sequences: Lessons from the Dead Sea, Israel. *Journal of the Geological Society*, **146**, 527–538.

Gat, J.R. and Magaritz, M., 1980. Climatic variations in the eastern Mediterranean sea area. *Naturwissenschaften*, **67**, 80–87.

Grove, J.M., 1988. *The Little Ice Age*, Methuen, London.

Grove, J.M. and Conterio, A., 1995. The climate of Crete in the sixteenth and seventeenth centuries. *Climatic Change*, **30**, 223–247.

Gutiérrez Elorza, M. and Peña Monné, J.L., 1990. Upper Holocene climatic change and geomorphological processes on slopes and infilled valleys from archaeological dating (N.E. Spain). In A.C. Imeson and R.S. De Groot (eds), *Landscape Ecological Impact of Climatic Change on the Mediterranean Region*, Universities of Wegeningen, Utrecht and Amsterdam, 1–18.

Harvey, A.M., 1984. Aggradation and dissection sequences on Spanish alluvial fans: Influence on morphological development. *Catena*, **11**, 289–304.

Harvey, A.M., 1988. Controls of alluvial fan development: the alluvial fans of the Sierra de Carrascoy, Murcia, Spain. *Catena Supplement*, **13**, 123–137.

Harvey, A.M., 1996. The role of the alluvial fans in the mountain fluvial systems of southeast Spain: Implications of climatic change. *Earth Surface Processes and Landforms*, **21**, 543–553.

IGSR, 1969. *The Geological Map of Greece: Alikianou Sheet*, Institute for Geology and Subsurface Research.

Kirkby, M.J. and Neale, R.H., 1987. A soil erosion model incorporating seasonal factors. In: V. Gardiner, (ed.), *International Geomorphology 1986, Part II*, Wiley, Chichester, 189–210.

Kutiel, H., Maheras, P. and Guika, S., 1996. Circulation and extreme rainfall conditions in the eastern Mediterranean during the last century. *International Journal of Climatology*, **16**, 73–92.

Lamb, H.H., 1977. *Climate: Present, Past and Future*, Methuen, London.

Lewin, J., Macklin, M.G. and Woodward, J.C., 1991. Late Quaternary fluvial sedimentation in the Voidomatis basin, Epirus, N.W. Greece. *Quaternary Research*, **35**, 103–115.

Lyrintzis, G. and Papanastasis, V., 1995. Human activities and their impact on land degradation – Psilorites mountain in Crete: A historical perspective. *Land Degradation and Rehabilitation*, **6**, 79–93.

Macklin, M.G., Lewin, J. and Woodward, J.C., 1995. Quaternary fluvial systems in the Mediterranean basin. In J. Lewin, M.G. Macklin and J.C. Woodward (eds), *Mediterranean Quaternary River Environment*, Balkema, Rotterdam, 1–25.

Meiggs, R., 1982. *Trees and Timber in the Ancient Mediterranean World*, Clarendon Press, Oxford.

Nemec, W. and Postma, G., 1993. Quaternary alluvial fans in S.W. Crete: Sedimentary processes and geomorphic evolution. In M. Marzo and C. Puigdefábregas (eds), *Alluvial Sedimentation*, International Association of Sedimentologists, Special Publication 17, 235–276.

Pérez-Obiol, R. and Julià, R., 1994. Climatic change on the Iberian peninsula recorded in a 30 000-yr pollen record from Lake Banyoles. *Quaternary Research*, **41**, 91–98.

Pierson, T.C., 1980. Erosion and deposition by debris flows at Mt Thomas, North Canterbury, New Zealand. *Earth Surface Processes*, **5**, 227–247.

Pirazzoli, P.A., Thommeret, J., Thommeret, Y., Laborel, J. and Montaggioni, L.F., 1982. Crustal block movements from Holocene shorelines: Crete and Antikythira (Greece). *Tectonophysics*, **86**, 27–43.

Pope, K. and van Andel, T.H., 1984. Late Quaternary alluviation and soil formation in the southern Argolid: Its history, causes and archaeological implications. *Journal of Archaeological Science*, **11**, 281–306.

Prentice, C.I., Guiot, J. and Harrison, S.P., 1992. Mediterranean vegetation, lake levels and palaeoclimate at the Last Glacial Maximum. *Nature*, **360**, 658–660.

Roberts, N., 1983. Age, palaeoenvironments, and climatic significance of Late Pleistocene Konya Lake, Turkey. *Quaternary Research*, **19**, 154–171.

Roberts, N., 1995. Climatic forcing of alluvial fan regimes during the Late Quaternary in Konya basin, south central Turkey. In J. Lewin, M.G. Macklin and J.C. Woodward (eds), *Mediterranean Quaternary River Environments*, Balkema, Rotterdam, 205–217.

Rumsby, B.T. and Macklin, M.G., 1996. River response to the last neoglacial (the 'Little Ice Age') in northern, western and central Europe. In J. Branson, A.G. Brown and K.J. Gregory (eds), *Global Continental Changes: the Context of Palaeohydrology*, Geological Society, London, Special Publication 115, 217–233.

Serre-Bachet, F., Guiot, J. and Tessier, L., 1992. Dendroclimatic evidence from southwestern Europe and northwestern Africa. In R.S. Bradley and P.D. Jones (eds), *Climate since A.D. 1500*, Routledge, London, 349–365.

van Andel, T.H., Runnels, C.N. and Pope, K.O., 1986. Five thousand years of land use and abuse in the southern Argolid, Greece. *Hesperia*, **55**, 103–128.

Wagstaff, J.M., 1981. Buried assumptions: some problems in the interpretation of the 'Younger Fill' raised by recent data from Greece. *Journal of Archaeological Science*, **8**, 167–177.

CHAPTER 13

Anthropogenic Influence on the Development of the Holocene Terraces of the River Lippe, Germany

JÜRGEN HERGET

Institute of Geography, Ruhr Universität Bochum, Germany

INTRODUCTION

During studies of the development of the Lippe valley system, the Holocene terraces of the river were one of the main issues (Liedtke and Herget, 1996; Herget, 1997). The Lippe valley lies in northwestern Germany, close to the border with The Netherlands. The river originates from a karstic spring at the town of Bad Lippspringe along the southern border of the Westfalian Bight and flows west to the lower Rhine at Wesel (Figure 13.1). The lower part of the valley is situated north of the industrial zone Ruhrgebiet with Essen and Dortmund, two of the bigger cities in the region north of Cologne. Two Holocene terraces in the valley system are known as the Inselterrasse and the Aue. Detailed studies of these terraces did not exist until our project. Whereas the Inselterrasse is a local feature, the Aue or Auenterrasse is the first and lowest terrace level above the river channel, i.e. floodplain. But it is important to keep in mind that the floodplain might consist of several morphological terraces.

The development of the Lippe valley bottom is a special case of anthropogenic influence on river development and might be an interesting addition to studies of the human impact on palaeohydrology (Brown, 1996; Gregory, 1995; Kalicki, 1996). A well-known example of anthropogenic influence on valley bottom development is the upper Rhine valley between Basel and Mannheim. Its development has several parallels with the Lippe valley. In the 19th century several meanders in the upper Rhine valley were cut, and secondary channels blocked, for the drainage of the floodplain. These river modifications led to an incision of several metres and changed the channel in the valley section from an anastomosing to a meandering pattern (Bensing, 1966; Gallusser and Schenker, 1992). The incision resulted from the concentration of the discharge into a single channel following the closure of the entrance of the other channels by small dams. At the Isteiner Schwelle locality, the former cliffs are now higher in comparison to the mean water level and their surface stays dry most of the year (Unterseher, 1992). This chapter describes the characteristics of the Inselterrasse and Aue and an explanation is given for the development of the atypical characteristics of the valley bottom.

Palaeohydrology and Environmental Change. Edited by G. Benito, V. R. Baker and K. J. Gregory

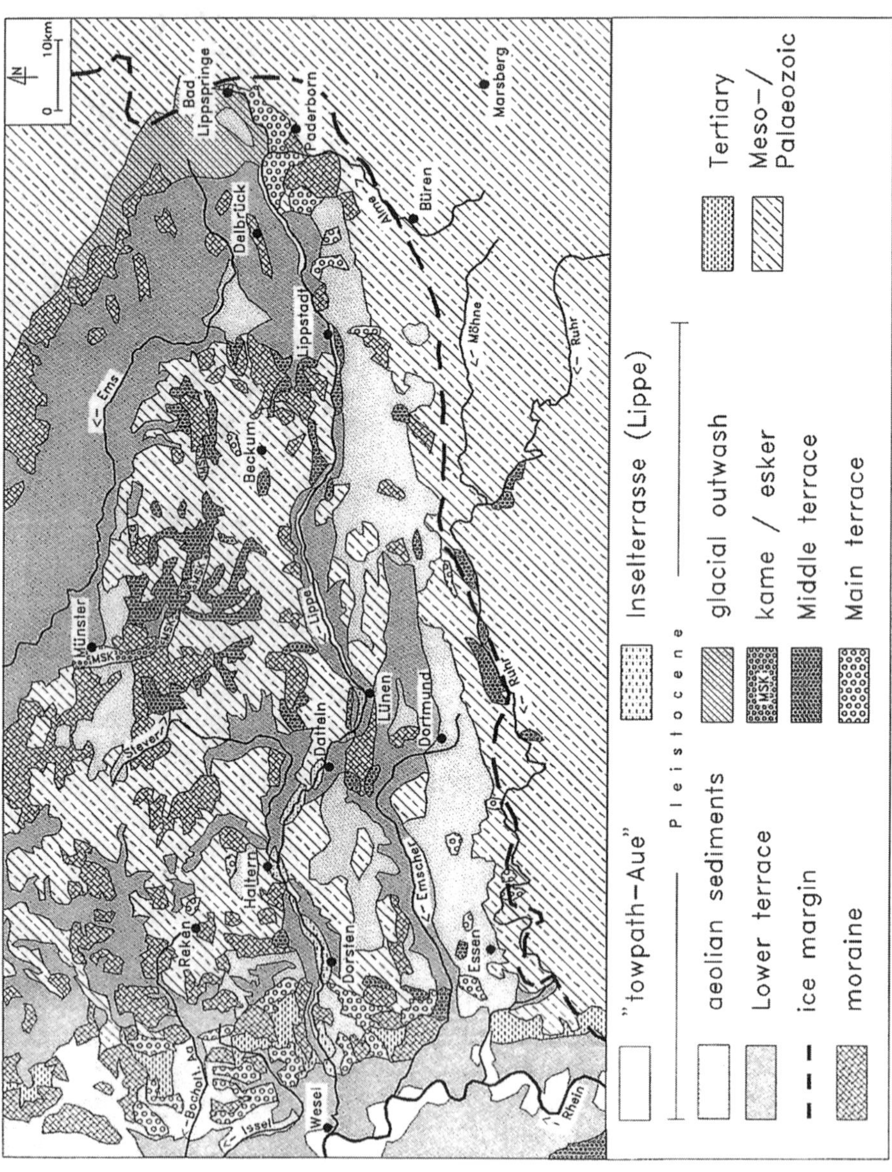

Figure 13.1 Simplified Quaternary sketch map of the Lippe catchment (modified from Deutloff (1976) and Speetzen (1986))

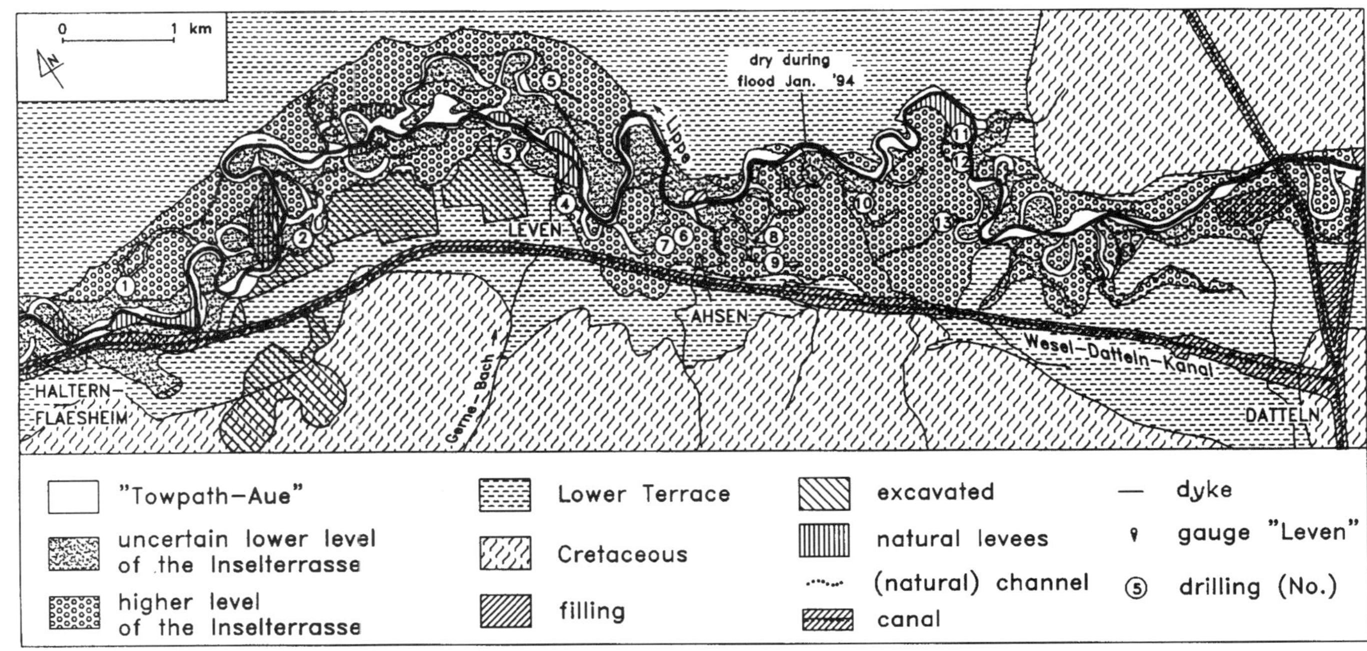

Figure 13.2   Holocene valley bottom terraces in the lower Lippe valley between Datteln and Haltern-Flaesheim. Numbers within circles refer to Table 13.1

GENERAL CHARACTERISTICS OF THE HOLOCENE TERRACES

Inselterrasse

The Inselterrasse is a terrace level located between the Weichselian Lower Terrace and the Aue. The name Inselterrasse ('island terrace') refers to the separation of the terrace level into several islands by abandoned channels (Udluft, 1933). As a morphological terrace, the Inselterrasse only exists between the area around Lünen (north of Dortmund) and the lower terraces of the Rhine at the mouth of the Lippe (Figure 13.1). In the headwaters the unconformity between the floodplain sediments and the sediments of the Weichselian Lower Terrace is seen as the counterpart of the Inselterrasse (Skupin, 1982, 1983). In the headwaters the Inselterrasse is developed as a denudation terrace consisting of a single broad level and, according to Schirmer (1995), may be characterised as a monoplain terrace. In the lower reaches the segregated islands of the Inselterrasse form a mosaic or loop terrace pattern (Figure 13.2) following the classification of Schirmer (1995).

Aeolian sediments (loess, coversand, dunes) and the Lower Terrace are of Weichselian age (115–10 ka). The ice margin, moraine, glacial outwash and kame/esker features belong to the Saalian ice age (300–127 ka), when coverage by the Scandinavian iceshield occurred. The Middle and Main terraces were thought to belong to the Saalian and older ice ages, but may also be seen as glacio-fluvial sediments deposited by a temporary drainage system at the advancing ice margin during the Saalian glaciation (Herget, 1997).

In the area of Lünen the Inselterrasse and Aue diverge reaching a maximum vertical difference of 3 m around Haltern. In the lower valley the terraces converge again. In detail the Inselterrasse can be differentiated into a higher and a lower level (Figure 13.2) as originally recognised by Müller (1950). The area around the gauge of Leven is suggested as the type locality for the division of the Inselterrasse. There the two levels of the Inselterrasse are close to each other and are clearly developed with a maximum difference in altitude of 0.5 m. However, because of small-scale relief on the terraces, their levels are not everywhere as distinct as shown in Figure 13.2. Natural levees of different height, extent and age cover the Aue and the lower Inselterrasse and make it difficult to recognise the different terrace levels in some places. Because of these difficulties the lower level of the Inselterrasse must be regarded as uncertain.

The abandoned channels also lie at different levels. Their outlets are orientated to different terrace levels, but many have no clearly visible inlet. The origin of this situation is related to the recent sediment dynamics of floods. Since even the higher level of the Inselterrasse can be reached by modern floods, the Lippe deposits of sand and silt accumulate on the edge of the floodplain at the inlets and outlets of abandoned channels. The outlets of the former river channels are cleared by the outflowing water during the falling limb of the hydrograph. Consequently the outlets still convey flow on the falling limb of a flood while the inlets are filled gradually by sediments. In some sections only small hollows mark the former inlet, whereas the channel form is clear at the outlet.

Owing to the lack of dating, only estimates of the age of the Inselterrasse were possible until this study. Several authors agree on an early Holocene age (Preboreal–Boreal) of the Inselterrasse (Arnold, 1960, 1977; Braun, 1975; Skupin, 1982, 1983). In the headwaters south of Delbrück, Skupin (1982) found a peat layer with a pollen spectrum typical of the Atlantic period (7500–4500 yr BP). Therefore the underlying unconformity between the sediments of the Weichselian Lower Terrace, and the floodplain sediments as counterpart of the Inselterrasse, must be older.

An excavation in Dorsten revealed radiocarbon ages in the cross-bedded sands of the Inselterrasse of between 8100 ± 150 yr BP and AD 775–980. This demonstrates that the accumulation of the Inselterrasse did not end before the Atlantic period, but continued until historic times (Speetzen and Lanser, unpublished data).

In the valley section between Datteln and Haltern-Flaesheim (Figure 13.2) several drillings and exploratory excavations were carried out for investigation of the terrace sediments. Dating by radiocarbon and dendrochronology of terrace sediments and channel fills yields ages that vary from before 8000 yr BC to after 300 yr BP (Table 13.1).

Table 13.1 Datings of channel fills and terrace sediments in the lower Lippe valley between Datteln and Haltern-Flaesheim (exposure numbers refer to Figure 13.2). The radiocarbon datings (conventional radiometric analysis) were carried out by BETA-Analytics Inc. (Miami), M. Geyh (Hannover) and B. Kromer (Heidelberg) by using a half-life of 5568 years, a two sigma level of error (95% probability) and dendrochronological calibrations (regional with respect to northern hemisphere scale). B. Schmidt (Cologne) examined the tree-ring ages by using the north German oak calendar

Drilling/ exposure no.	Terrace Sediment character	Material Depth (cm)	Laboratory sample no.	Dating method Age (^{14}C yr BP) $\pm 1\sigma$	Calibrated age (2σ)
1	Lower Terrace Terrace sediment	Humic loam 345–375	Beta 87309	^{14}C (conv.) 24 940 $\pm$ 200	
2	Lower Inselterrasse Channel fill	Humic loam 160–190	Beta 87312	^{14}C (conv.) 2910 $\pm$ 50	1260–930 BC
3	Inselterrasse Terrace sediment	Humic sand 255–260	Beta 87311	^{14}C (conv.) 2960 $\pm$ 50	1305–1005 BC
4	Higher Inselterrasse Terrace sediment	Wood 190	Geyh 20071	^{14}C (conv.) 40 $\pm$ 80	AD 1700–1955
5	Lower Terrace Terrace sediment	Humic loam 340–370	Beta 87310	^{14}C (conv.) 26 090 $\pm$ 1040	
6	Higher Inselterrasse Channel fill	Wood: 240			
		Peat: 300–330	Geyh 20069 Geyh 20070	^{14}C (conv.) 1715 $\pm$ 65 ^{14}C (conv.) 2475 $\pm$ 45	AD 250–415 765–420 BC
7	Higher Inselterrasse Fossil tree	Fossil oak: 450	Schmidt	Dendrochronology	Dendrochronology: 732–621 BC
			Geyh 20507	^{14}C (cal.) 2440 $\pm$ 50	760–405 BC
8	Lower Inselterrasse Channel fill	Wood: 120–130	Beta 87317	^{14}C (conv.) 3540 $\pm$ 50	1975–1735 BC
9	Lower Inselterrasse Channel fill	Wood: 120–125	Beta 87318	^{14}C (conv.) 9100 $\pm$ 70	8230–8005 BC
10	Lower Inselterrasse Channel fill	Humic loam 185–190	Beta 87319	^{14}C (conv.) 3530 $\pm$ 70	2025–1675 BC
11	Aue Terrace sediment	Wood: 50	Beta 87322	^{14}C (conv.) 560 $\pm$ 100	AD 1275–1505 AD 1595–1620
12	Higher Inselterrasse Channel fill	Wood: 210	Beta 87320	^{14}C (conv.) 1870 $\pm$ 60	AD 25–265 AD 290–320
13	Higher Inselterrasse Channel fill	Wood: 225–235	Beta 87321	^{14}C (conv.) 6180 $\pm$ 80	5270–4920 BC

Two dates from sediments of Weichselian age (nos 1 and 5) show that the sediments of the Lower Terrace underlie the Inselterrasse and Aue (Figure 13.3).

Owing to the lack of exposures, only limited information about the stratigraphy of the valley bottom is known. The sandy, gravel-free sediments of the Inselterrasse are cross-bedded. The cross-bedded gravels and sands of the Lower Terrace were deposited by a braided river Lippe during the Weichselian. Cretaceous marl underlies the Quaternary sediments. Typically the sediments of the higher level of the Inselterrasse are about 2 m in thickness, locally rising to 7 m. Similar variations occur with the Weichselian Lower Terrace with a thickness that varies between 2 m and more than 20 m. Locally, these sediments are missing completely. These fluctuations of thickness might be seen as an indication of young or recent natural vertical movements (Herget, 1997). The maximum historical floodstage lies above the higher level of the Inselterrasse.

Aue

As shown in Figure 13.1, in the headwaters the Aue occupies wide sections of the valley bottom. In the lower reaches the Aue typically consists of a small narrow strip running parallel to the river channel (Udluft, 1933). The map of the valley bottom between Datteln and Haltern-Flaesheim (Figure 13.2) shows that some sections of the Aue appear wide due

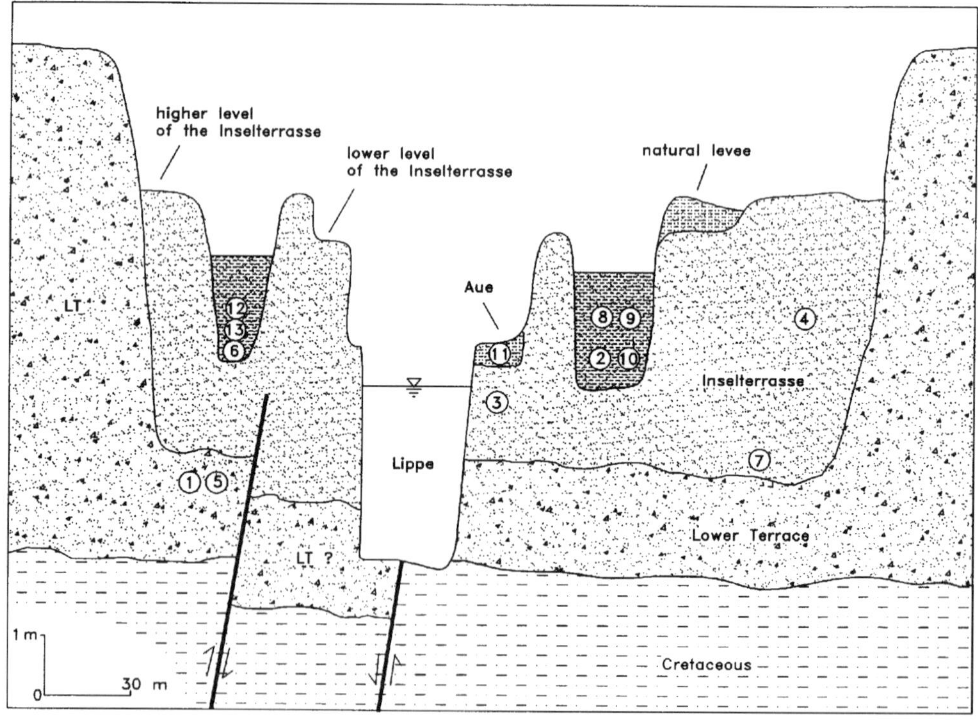

Figure 13.3   Schematic cross-section of the valley bottom in the lower Lippe valley. (The 1 km broad valley bottom is drawn to a smaller scale, compare with Figure 13.2.) Numbers within circles refer to Table 13.1

to flood erosion. Locally the accumulation of natural levees reaches the level of the (lower) Inselterrasse. The sediments consist of fine to medium sands with a higher silt content than the sands of the Inselterrasse. The age of the sediments is expected to be younger than the Atlantic period and includes recent redeposition. A post-Atlantic age is supported by pollen analysis from the headwaters (Skupin, 1982, 1983; Hiss 1989) and radiocarbon dating from the lower reaches (Figure 13.2, nos 3 and 11; Table 13.1). Several dated samples from the higher level of the Inselterrasse are younger than the samples from the Aue.

During and after a flood event in January 1994, an extension of the floodplain was mapped (Herget, 1997). As shown in Figure 13.2, several sections of the lowest terrace of the Aue were above the flood level while large areas of the lower level of the Inselterrasse were flooded. In historical times even the higher level of the Inselterrasse was frequently flooded (Krakhecken 1939). Geomorphological mapping of the valley bottom confirmed the terrace differentiations because of sharp terrace edges with heights up to 1.5 m although only a relative differentiation of the terraces is possible. It is not possible to make a stratigraphical differentiation of the Holocene terraces in the lower Lippe valley only by comparing the levels of the terrace surfaces.

The atypical characteristics of the Holocene valley bottom of the Lippe valley can be summarised as follows.

(a) The Inselterrasse only exists in the lower Lippe valley west of Lünen.
(b) In some places the Inselterrasse consists of two levels, which are not always clearly separated from each other.
(c) The level of the Inselterrasse diverges and converges from the Aue.
(d) In the lower reaches the Aue consists typically only of a small strip parallel to the river channel.
(e) The Aue lies partly above the average flood level whereas sections of the Inselterrasse are periodically flooded. In historical times the Inselterrasse was frequently flooded.
(f) The sediments of the Inselterrasse and the fills of the abandoned channels accumulated during the entire Holocene.

DEVELOPMENT OF THE HOLOCENE VALLEY BOTTOM

The characteristics of the Holocene valley bottom are atypical for valley bottoms in central Europe (Schirmer, 1995) and can be explained as a result of anthropogenic influences. Specifically, two models are advanced (Figure 13.4). The first model assumes a natural anastomosing pattern and cites Roman dams as the first anthropogenic influence. The other assumes a natural meandering pattern in which case human modification began in medieval times.

(1.1) Under natural conditions the Lippe was an anastomosing river with discharge distributed in several channels. The valley bottom consisted of a single broad level. This is supported by the dimension and position of some abandoned channels in the lower reaches of the Lippe valley (Figure 13.2). Some channels are too narrow and shallow to convey the mean discharge. In the easily erodible sandy sediments several channels might have developed with individual meandering patterns.
(1.2) During their campaign against German tribes the Roman conquerors used the river

for their supply transport (Eckoldt, 1980; Kühlborn, 1995; Morel, 1987). For the improvement of navigation on the river, the Romans may have dammed single channels. If so, the discharge would have been concentrated in one single channel that would subsequently deepen and broaden. This channel shows the typical pattern of a meandering river. The Romans would only have needed small dykes on local tributary forks to accomplish this. The dammed channels would gradually fill naturally by sediments during floods, and some abandoned channels would lose their inlets. If proven, these first steps of river engineering nearly 2000 years ago during the Roman campaign would be some of the earliest examples of river management in Europe (Gregory, 1995).

Actually there is no archaeological evidence for these activities by the Roman conquerors. But the historical sources do not present any detailed information about the Roman activities during their campaign, except their defeat by the German tribes (Kühlborn, 1995). Experts on Roman archaeology of the area do not expect to find any facts regarding river management but think it was achievable by the Roman military (J.S. Kühlborn, pers. comm.). Further studies on the navigation on the Lippe by the Romans, especially their logistics and engineering, are in preparation.

(2.1) In its natural condition with a meandering river channel pattern, the Aue consists of several small channels that carry discharge only during flood events. The Lippe actively meandered on the floodplain, eroding into the meanders and forming avulsions during floods. There is not enough information available for a comparison of the phases of activity and stability with other river catchments in Europe (Schirmer, 1995; Starkel, 1996) but the sediments of the terraces and the channel fills accumulated during the entire Holocene. This is revealed by the radiocarbon datings of Speetzen and Lanser (unpublished) from the excavation at Dorsten.

(2.2) Several meanders were artificially cut to shorten the navigation route. This meander-cutting resulted in incision because of the steeper channel gradient. A new towpath had to be built in the sections of the avulsions.

In the 12th century shipbuilding started in Dorsten and from 1332 the first written records about a navigation toll at Dorsten are known (Koppe, 1992). To move the ships upstream a towpath was built next to the river channel at variable heights. Long periods of low water in the broad and shallow river channel with its sandbars have always been a problem for navigation. Sluices had to be built for passing the cliffs in the river channel. Detailed historical evidence about activities for the improvement of navigation on the river exists only from the early 19th century and later times (Koppe, 1986; Strotkötter, 1896; Vollmer, 1993). Only from the area around Haltern are the first meander cuttings between AD 1000–1700 reported (Strichling, 1932). But, since navigation on the river is documented, the existence of a towpath is an aspect that does not need further evidence, even during the Roman campaign.

Owing to later meander cutting, the incision amounted to 3 m in a period of 100 years and continues today (Vollmer, 1993).

(2.3) The river channel experienced anthropogenic lateral sedimentation. Owing to aggradation a higher water level was needed for navigation on the river. Sediments from sections with steep embankments and the natural levees were used to narrow the river channel (Vollmer, 1993). This narrowing resulted in another incision pulse and again a towpath was constructed. The meander cutting continued.

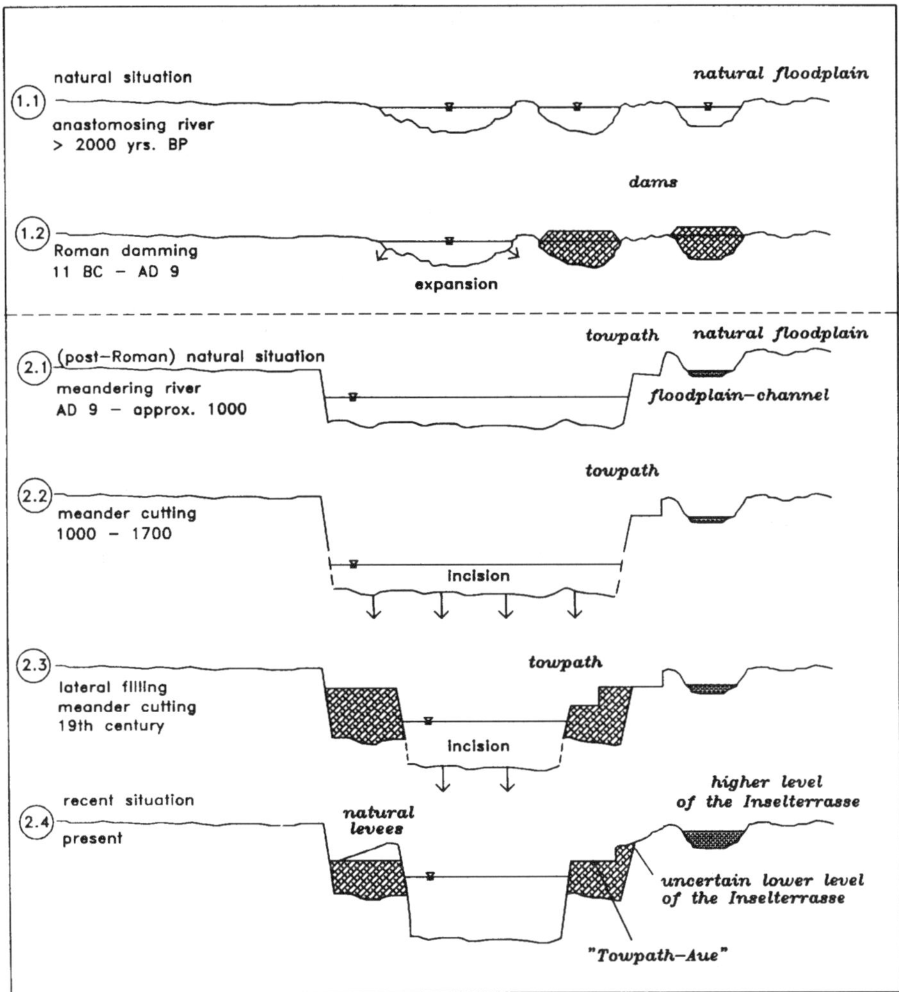

Figure 13.4   Hypothesised development of the Holocene terraces in the lower Lippe valley

(2.4) The towpath is expanded due to flood erosion. The natural floodplain is presently
       building a terrace level (higher level of the Inselterrasse) between the river channel
       and the Weichselian Lower Terrace. As a result of incision, the distance between the
       mean water level and the floodplain has increased, but natural levees and erosional
       scars make it difficult to identify the different terraces.

The natural single floodplain terrace of phase 1.1/2.1 is the origin of the higher level of the
modern Inselterrasse. The lower level of the Inselterrasse is developed from the towpath
constructed during phase 2.2. The Aue as lowest terrace level above the river channel
might be named 'Towpath-Aue' as it developed from the towpath which was built up
during phase 2.3 and is still developing as a narrow strip parallel to the river channel. The

variable height of the towpath is the explanation for its non-continuous occurrence above the mean floodlevel.

THE NATURAL SITUATION: ANASTOMOSING OR MEANDERING RIVER?

To determine whether the former natural river pattern was anastomosing or meandering, the dimensions of the abandoned channels were compared to the present-day discharge. The objective is to estimate whether the abandoned channels are large enough to carry the flow in a single channel. If this is the case the river pattern was meandering. However, if they are too small, more than one channel must have been in use at the same time and the river was anastomosing. The assumption for this conclusion is that the discharge of the period concerned was roughly similar to the present discharge.

To calculate the cross-sectional area the dimensions of the channel at location 10 are determined by interpretation of drillings. In this location the channel fill of humic loam can clearly be separated from the sands of the Inselterrasse (Figure 13.5). This river section was chosen for the age of the channel fill (2025–1675 BC) to ensure that anthropogenic influence can be excluded for this period. Additionally there are no significant differences between palaeohydrology and the modern fluvial regime (e.g. flood frequency and sediment dynamics) in northwestern Germany (Klostermann, 1992; Starkel, 1996).

The channel fill consists of humic loam which is covered by younger flood sediments.

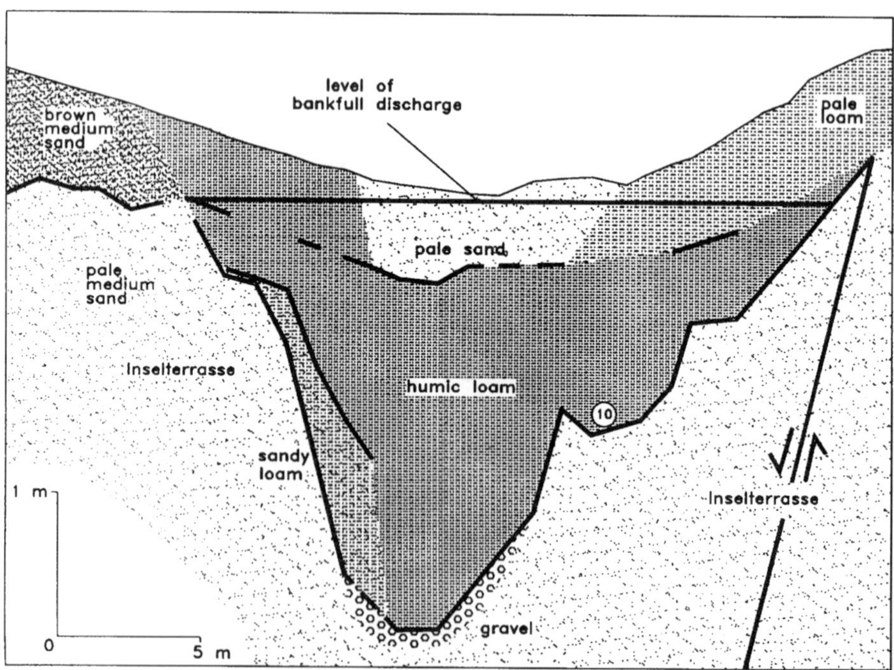

Figure 13.5   Cross-section of the abandoned channel at location no. 10 (compare with Figure 13.2)

Because of technical problems with the drilling equipment and a high groundwater level, the sample for radiocarbon dating could not be taken at the deepest part of the channel fill due to the risk of contamination by younger sediments and carbon from higher sections of the profile. Therefore the abandonment of the channel is probably older than the radiocarbon date.

The Manning formula (Morisawa, 1985) is used to estimate the bankfull discharge Q_b of the abandoned channel:

$$Q_b = \frac{1}{n} A_b R^{2/3} S^{1/2}$$

The cross-sectional area A_b of bankfull discharge can be derived from the profile by separating the younger sand cover from the channel fill and is calculated as $26.12\,\mathrm{m}^2$. The hydraulic radius R is the quotient of the cross-sectional area A_b and the bankfull wetter perimeter P ($= 29.46\,\mathrm{m}$). The slope S ($= \Delta h/\Delta l$) was estimated by the difference in altitude ($\Delta h = 2.22\,\mathrm{m}$) and distance ($\Delta l = 206\,\mathrm{m}$) of the surface of the channel fill between two cross-sections. The average surface of the channel fill sediments might not be the correct input data to estimate S, but there are no other available data. The steep slope of 1% might be caused by recent tectonic movements. The roughness coefficient n was taken from Morisawa (1985). By assuming a natural clean and sinuous river channel with some pools and riffles it can be estimated at 0.04. Using these data the mean velocity in the channel at bankfull discharge was calculated as $v = 2.40\,\mathrm{m\ s}^{-1}$ and the bankfull discharge as $Q_b = 62.6\,\mathrm{m}^3\,\mathrm{s}^{-1}$.

The recurrence interval of bankfull discharge appears to be in the range of one to two years, in special cases up to four years (Gregory and Walling, 1973; Knighton, 1984; Leopold et al., 1964). Therefore the discharge of a modern flood with a recurrence interval of two years can be compared with the bankfull discharge of the abandoned channel. It is still assumed that the palaeohydrology of the period around 4000 yr BP is comparable to the present. But, owing to the incision of the modern river channel, the comparison with its bankfull discharge is pointless.

The discharge of a modern flood with a recurrence interval of two years is about $160\,\mathrm{m}^3\,\mathrm{s}^{-1}$ (Vollmer, 1995). This corresponds to more than twice the bankfull discharge of the abandoned channel. Therefore more than one channel must have been in use in the period around 4000 yr BP and the channel pattern was anastomosing.

Further studies on other abandoned channels of different ages in several sections of the Lippe valley have to be carried out to validate this result. For this reason the anastomosing channel pattern in the lower Lippe valley can actually only be interpreted as probable and not as a definite fact.

CONCLUSION

The Inselterrasse developed only in the lower reaches of the Lippe valley because there was more navigation engineering and channel alteration (Koppe, 1992; Krakhecken, 1939; Wiel, 1970). The incision of the river channel has uncovered cliffs of harder units of the underlying Cretaceous marls. These cliffs are the foundation for the sluices in several places along the river. In the 19th century the navigation was more intensive in the lower reaches than in the headwaters. This led to more channel alteration and consequently deeper incision. The cliffs, however, stopped the headcut incision from moving upstream.

Meanders were also cut in the headwaters (Vollmer, 1993), but owing to the stable Cretaceous marls in the valley bottom no additional towpath was needed. Therefore there was no anthropogenic mechanism for the development of a 'Towpath-Aue'. The river incised with steep and straight slopes and the natural floodplain survived. The unconformity between the sediments of the lower terrace and the floodplain sediments is not an equivalent situation for the Inselterrasse in the lower reaches (Skupin, 1982).

The Rhine as receiving stream was not influenced by the improvements of the navigation in the Lippe valley. Therefore the incision decreases in the direction of the mouth of the Lippe, and the Holocene terraces converge again to the single level of the natural floodplain.

By using the Manning formula, first approximations revealed that the bankfull discharge of an abandoned channel is significantly less than the comparable modern flood with a recurrence interval of two years. Therefore the natural channel pattern of the Lippe was probably anastomosing.

The Holocene bottom of the Lippe valley is not comparable with other valley bottoms in central Europe because its development is dominated by anthropogenic influence which resulted in individual characteristics.

ACKNOWLEDGEMENTS

I thank H. Liedtke, E. Bremer, G. Eggenstein, S. Harnischmacher, J.-S. Kühlborn, K.N. Thome and H. Zepp for important discussions and comments. Financial support for the studies on the development of the Lippe valley was kindly provided by the 'Deutsche Forschungsgemeinschaft' (German Research Foundation) (Li 86/21-1).

REFERENCES

Arnold, H., 1960. Geologische Karte. In Geologisches Landesamt NRW (ed.), *Übersichtskarte von Nordrhein-Westfalen 1:100.000 – Erläuterungen zu Blatt Münster C 4310*, Geologisches Landesamt, Krefeld, 9–134.

Arnold, H., 1977. *Geologische Karte von Nordrhein-Westfalen 1:100.000 – Erläuterungen zu Blatt C 4314 Gütersloh*, Geologisches Landesamt, Krefeld.

Bensing, W., 1966. Gewässerkundliche Probleme beim Ausbau des Oberrheins. *Deutsche Gewässerkundliche Mitteilungen*, **10**, 85–101.

Braun, F.J., 1975. Quartär. In F.J. Braun and A. Thiermann (eds), *Geologische Karte von Nordrhein-Westfalen 1:100.000 – Erläuterungen zu Blatt C 4306 Recklinghausen*, Geologisches Landesamt, Krefeld, 94–122.

Brown, A.G., 1996. Human dimensions of palaeohydrological change. In J. Branson, A.G. Brown and K.J. Gregory (eds), *Global Continental Change – the Context of Palaeohydrology*, Geological Society, London, Special Publication 115, 57–72.

Deutloff, O., 1976. Geologie. In Akademie für Raumforschung und Landesplanung und Ministerpräsident des Landes NRW (eds), *Deutscher Planungsatlas Bd. I – NRW – Lieferung 8: Geologie*, Schroedel, Hannover.

Eckoldt, M., 1980. *Schiffahrt auf kleinen Flüssen Mitteleuropas in Römerzeit und Mittelalter*, Schriften des deutschen Schiffahrtsmuseums 14, Stalling, Oldenburg.

Gallusser, W.A. and Schenker, A. (eds), 1992. *Die Auen am Oberrhein – Ausmaß und Perspektiven des Landschaftswandels am südlichen und mittleren Oberrhein seit 1800; eine umweltdidaktische Aufarbeitung*, Birkhäuser, Basel.

Gregory, K.J., 1995. Human activity and palaeohydrology. In K.J. Gregory, L. Starkel and V.R. Baker (eds), *Global Continental Palaeohydrology*, Wiley, Chichester, 151–172.

Gregory, K.J. and Walling, D.E., 1973. *Drainage Basin Form and Process – A Geomorphological Approach*, Arnold, London.

Herget, J., 1997. *Die Flußentwicklung der Lippe*, Bochumer Geographische Arbeiten 62, Bochum.

Hiss, M., 1989. *Geologische Karte von Nordrhein-Westfalen 1:25.000 – Erläuterungen zu Blatt 4417 Büren*, Geologisches Landesamt, Krefeld.

Kalicki, T., 1996. Climatic or anthropogenic alluvation in Central European valleys during the Holocene? In J. Branson, A.G. Brown and K.J. Gregory (eds), *Global Continental Change – the Context of Palaeohydrology*, Geological Society, London, Special Publication 115, 205–215.

Klostermann, J., 1992. *Das Quartär der Niederrheinischen Bucht – Ablagerungen der letzten Eiszeit am Niederrhein*, Geologisches Landesamt, Krefeld.

Knighton, D., 1984. *Fluvial Forms and Processes*, Arnold, London.

Koppe, W., 1986. *Die Lippeschiffer – Zur Geschichte eines ausgestorbenen Berufes*, Vestische Zeitschrift **84/85**, 241–258.

Koppe, W., 1992. Die Bedeutung der schiffbaren Lippe für die Wirtschafts- und Sozialgeschichte der Stadt Dorsten. *Heimatkalender der Herrlichkeit Lembeck und Dorsten*, **51**, 34–40.

Krakhecken, M., 1939. *Die Lippe*, Arbeiten der geographischen Kommission 2, Münster.

Kühlborn, J.-S., (ed.), 1995. *Germaniam pacavi – Germanien habe ich befriedet – Archäologische Stätten augusteischer Okkupation*, Westfälisches Museum für Archäologie, Münster.

Leopold, L.B., Wolman, M.G. and Miller, J.P., 1964. *Fluvial Processes in Geomorphology*, Freeman, San Francisco.

Liedtke, H. and Herget, J., 1996. *Forschungsprojekt 'Lippeterrassen' Li 86 / 21–1*, Bochum (unpublished final report for the Deutsche Forschungsgemeinschaft).

Morel, J.-M.A.W., 1987. Frührömische Schiffshäuser in Haltern, Hofestatt. *Ausgrabungen und Funde in Westfalen-Lippe*, **5**, 221–249.

Morisawa, M., 1985. *Rivers – Form and Process*, Longman, London.

Müller, H., 1950. Die Halterner Talung. *Westfälische Geographische Studien*, **3**, Münster.

Schirmer, W., 1995. Valley bottoms in the late Quaternary. *Zeitschrift für Geomorphogie N.F. Supplement-Band*, **100**, 27–51.

Skupin, K., 1982. *Geologische Karte von Nordrhein-Westfalen 1:25.000 – Erläuterungen zu Blatt 4218 Paderborn*, Geologisches Landesamt, Krefeld.

Skupin, K., 1983. *Geologische Karte von Nordrhein-Westfalen 1:25.000 – Erläuterungen zu Blatt 4217 Delbrück*, Geologisches Landesamt, Krefeld.

Speetzen, E., 1986. Das Eiszeitalter in Westfalen. *Alt- und mittelsteinzeitliche Fundplätze in Westfalen 6, Teil 1*, Westfälisches Museum für Archäologie, Münster.

Starkel, L., 1996. The upper Vistula catchment on the background of changes in the fluvial systems in Europe and in the temperate zone. In L. Starkel (ed.), *Evolution of the Vistula River Valley during the Last 15000 Years – part VI*, Polish Academy of Sciences, Wroclaw, Geographical Studies Special Issue 9, 102–110.

Strichling, A., 1932. Aus dem Leben der Lippe. *Vestischer Kalender*, **10**, Recklinghausen, 12–27.

Strotkötter, G., 1896. *Die Lippeschiffahrt im neunzehnten Jahrhundert*, Münster.

Udluft, H., 1933. Das Diluvium des Lippetals zwischen Lünen und Wesel und einiger angrenzender Gebiete. *Jahrbuch der preußischen geologischen Landesanstalt*, **54**, Berlin, 37–57.

Unterseher, E., 1992. Istein – vom Fischerdorf zum Rebbau- und Bergbaudorf. In W.A. Gallusser and A. Schenker (eds), *Die Auen am Oberrhein – Ausmaß und Perspektiven des Landschaftswandels am südlichen und mittleren Oberrhein seit 1800; eine umweltdidaktische Aufarbeitung*, Birkhäuser, Basel, 92–102.

Vollmer, A., 1993. *Lippeauenprogramm Abschnitt Lippstadt – Lippborg – Wasserwirtschaftlicher Fachbeitrag zum ökologischen Gutachten*, Geseke (unpublished report for the Staatliches Amt für Wasser und Abfallwirtschaft Lippstadt).

Vollmer, A., 1995. Wasserwirtschaftliche Untersuchung zum Lippeauenprogramm Lippborg bis Wesel (unpublished report for the Lippeverband).

Wiel, P., 1970. *Wirtschaftsgeschichte des Ruhrgebietes – Tatsachen und Zahlen*, Siedlungsverband Ruhrkohlenbezirk, Essen.

CHAPTER 14

Floods and Quaternary Sedimentation Style in a Bedrock-controlled Reach of the Bergantes River, Ebro Basin, Northeast Spain

ALMA LÓPEZ-AVILÉS, PHILIP J. ASHWORTH AND
MARK G. MACKLIN

School of Geography, University of Leeds, UK

INTRODUCTION

The Ebro basin, northeast Spain (Figure 14.1), straddles an important ecotone in the transitional zone between the humid Pyrenees in the north and the eastern part of the Iberian Range, and contains some of the most arid regions in Spain such as Los Monegros and El Bajo Aragón (Pueyo-Mur, 1978; Macklin et al., 1994). Downstream of the city of Zaragoza, many of the headwater catchments of the southern tributaries to the Ebro are characterised by steep gradients, sparse vegetation cover and highly folded and fractured sedimentary bedrock, which, together with the climatic regime and increasing demand of water supply for irrigation, make them particularly vulnerable to erosion and sensitive to anthropogenic activity (Davis, 1994).

Much work has been undertaken on the study of erosion and sedimentation patterns of Mediterranean basins (e.g. Lewin et al., 1995; Poesen and Hooke, 1997), and more specifically, on fluvial geomorphological processes (e.g. Sala, 1982), flow frequency and channel morphology (e.g. Conesa-García, 1995), and bedload transport (e.g. Batalla and Sala, 1995) on Spanish Mediterranean streams. In addition, work on the dynamics and sedimentology of the Ebro River and its delta (e.g. Maldonado, 1975; Nelson, 1990; Palanques et al., 1990; Guillén and Palanques, 1992) has recently been completed. However, little attention has been paid to structurally controlled bedrock channels in the Mediterranean uplands, despite the fact that the literature on bedrock channel morphology and roughness is available for other fluvial environments (e.g. Baker and Kochel, 1988; Wohl, 1992; Carling and Grodek, 1994; Wohl et al., 1994). Bedrock channels are common in the headwaters of the southern tributaries to the Ebro River as a result of the intensive tectonism experienced in the Iberian uplands (Aurell et al., 1990; Sánchez-Navarro et al., 1990). Strongly folded bedrock may influence drainage network development (Harvey and Wells, 1987; Collier et al., 1995), local channel morphology and migration (Macklin et al., 1995) and sediment supply (Woodward, 1995). In addition, the

---

Palaeohydrology and Environmental Change. Edited by G. Benito, V. R. Baker and K. J. Gregory
© 1998 John Wiley & Sons Ltd.

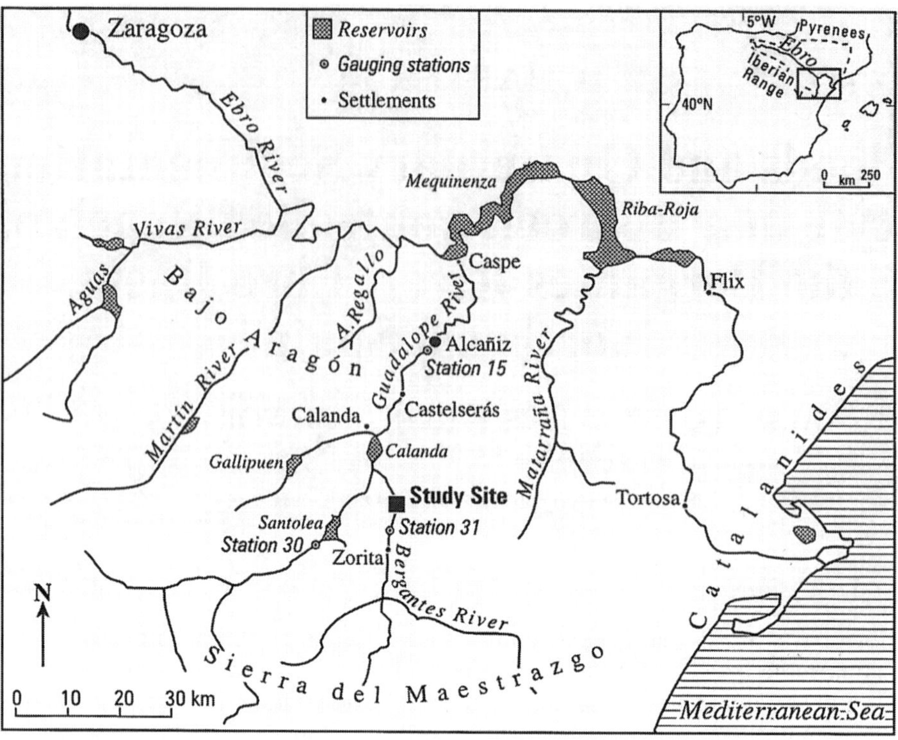

Figure 14.1   Location map of the Guadalope–Bergantes river system and other major southern tributaries of the Ebro River in the Bajo Aragón region. The gauging stations are referred to in the text

exposure of bedrock affects the vegetation density and hillslope sediment storage (Thornes, 1987; Bull, 1988). However, it remains unclear how all these different processes interact and control the long-term development of the valley floor morphology and alluvial stratigraphy. Understanding these interactions is important since these fragile environments are particularly sensitive to climate change (Woodward, 1995; Benito et al., 1996) and therefore their morphology and sedimentation style may be used to reconstruct the bioclimatic conditions that formerly prevailed in semi-arid regions (e.g. Macklin et al., 1994; Fuller et al., 1996).

This chapter describes the valley floor geomorphology and sedimentology associated with a structurally controlled bedrock river, the Bergantes (Figure 14.1), in the southern headwaters of the Ebro basin. Geochronological dating and investigation of the change in flood regime and anthropogenic activity in the region permit a reconstruction of the Late Quaternary sedimentation style that characterises these Mediterranean upland bedrock channels.

GEOLOGICAL SETTING

The Bergantes is an unregulated tributary of the Guadalope River which flows from the Sierra del Maestrazgo in the Iberian Range towards the semi-arid Bajo Aragón region

(Figure 14.1). The Iberian Range was a large sedimentation basin fully or partially occupied by the ocean in the marine transgression/regression episodes during the Jurassic and Early Cretaceous. In the transition towards the Tertiary, the entire Iberian Range was uplifted as a consequence of the compressional motion between the African and European cratons regarded as the 'Alpine Orogeny' (Smith and Woodcock, 1982).

The Bergantes and Guadalope headwaters are characterised by strongly folded and faulted Mesozoic limestone, dolomite, marl, conglomerates and Cretaceous calcareous sandstone, together with Tertiary detritic conglomerates of Oligocene age (IGME, 1978, 1981). The Alpine tectonism is superimposed on a regional pattern of tectonic fractures generated during the Palaeozoic, and is responsible for producing intense faulting and a series of anticlines and synclines that trend northwest–southeast, from Calanda (Figure 14.1) to the highest peaks of the Sierra del Maestrazgo (Aurell et al., 1990; Sánchez-Navarro et al., 1990). This folding and faulting and upland denudation supplied large volumes of sediment for piedmont mantle and alluvial fan formation during the Quaternary (Ibáñez et al., 1983; Peña-Monné et al., 1984).

Tectonism was still active in the Quaternary in some parts of the Sierra del Maestrazgo, which experienced extensional effects leading to fault initiation and alluvial fan enlargement (Simón-Gómez, 1990). Alluviation during the Pleistocene has given rise to a series of terraces of unconsolidated gravels up to 25 m thick, which are discernible along the Guadalope and Bergantes rivers as reported by Mensua and Ibáñez (1973), Macklin and Passmore (1995) and Fuller et al., (1996). These Pleistocene terraces supply sediment to the present channel when they are undercut during flood events.

BERGANTES STUDY AREA

A 2 km long study reach of the Bergantes was selected 35 km downstream of its source. The catchment area to the study reach is 1125 km² with a mean valley slope in the reach of 0.007. Mean grain-size in the study reach is approximately 55 mm, although the channel bed alternates between solid bedrock and boulders (b-axis up to 205 mm). Most of the Bergantes channel is incised into a meandering bedrock gorge along the Sierra del Maestrazgo. The valley widens at the confluence with the Guadalope (10 km downstream of the study site) and further downstream of the Calanda anticline and reservoir (14 km downstream of the study site, Figure 14.1), which marks the border between the Maestrazgo and Bajo Aragón regions.

The Guadalope hydrological regime is classified as Mediterranean and the Bajo Aragón region is characterised by semi-arid climatic conditions with precipitation in the range of 250–350 mm per year (Pueyo-Mur, 1978; Rodríguez-Vidal, 1982). However, the headwaters of the Bergantes receive marginally higher precipitation than the Guadalope catchment owing to their location on the eastern side of the Sierra del Maestrazgo, away from the rest of the rain-shadowed Ebro basin (e.g. mean annual precipitation was 620 mm at station 31 in Zorita, see Figure 14.1, between 1960 and 1970). Precipitation in the Bergantes catchment is irregularly distributed throughout the year, with both summer convective and autumn frontal storms causing large amounts of runoff. Most of the steep valley sides of the Bergantes catchment are covered by between 30 and 70% pine (*Pinus halepensis*) and associated Mediterranean bushes (*Quercus coccifera* and *Quercus ilex*). The degradation of *Quercus* natural woodlands has led to the development of the traditional maquis vegetation. Many of the agricultural terraces bordering the Bergantes study site are cultivated with vineyards and almond trees (Ministerio de Agricultura, 1977).

The Bergantes has been periodically gauged at station 31 (4 km upstream of the study reach, Figure 14.1) from 1931 until 1935, 1953 until 1971 and from 1990 onwards (data supplied by the Confederación Hidrográfica del Ebro). Figure 14.2a displays the hydrological characteristics of the Bergantes and Guadalope rivers for the period 1959–1969 recorded at station 31 and stations 30 (upstream of the Santolea reservoir) and 15 (Alcañiz, see Figure 14.1). Figure 14.2a shows that the upstream sections of the Guadalope and Bergantes have annual peaks in spring and autumn, respectively, and that the Bergantes River experiences higher discharges than the Guadalope headwaters, contributing to the increase of the discharge further downstream, as recorded for the Guadalope at station 15. This is corroborated by a good correlation between the annual maximum instantaneous discharge recorded on the same dates for stations 15 and 31 (Figure 14.2b). An annual maximum instantaneous flood discharge of 1560 m^3 s^{-1} was recorded for the Bergantes (station 31) in October 1967 (Figure 14.2b).

METHODS AND DATA COLLECTION

In order to isolate and describe the main sedimentary deposits on the Bergantes study site, the principal geomorphological units were defined using aerial photographs, field mapping and EDM (electronic distance meter) survey. A 1:20 000 aerial photograph was enlarged and the main river terraces, bedrock strata, overbank deposits, bars, channels and sharp breaks in slope superimposed to produce a map of the geomorphological units (Figure 14.3). Over 900 spot heights were then taken onto the geomorphological units using an EDM station and reduced to a common datum. A long-profile of the current channel thalweg and water surface and nine cross-sections perpendicular to the mean valley direction were also surveyed.

Major river terraces were dated using infra-red stimulated luminescence (IRSL). The general principles of luminescence dating are described in Bailiff (1992), Wintle (1993) and Duller (1996). Thermoluminescence dating has previously been used on alluvial terraces in eastern Spain by Fumanal and Carmona (1995), and optically stimulated luminescence (OSL) and IRSL techniques have been used to date alluvial units in semi-arid areas of northeast Spain by Macklin et al. (1994), Macklin and Passmore (1995) and Fuller et al. (1996). The IRSL technique has proved particularly suitable for dating alluvial sediment from semi-arid regions because the water content is essentially constant, improving the accuracy in the estimation of the radiation dose (Bailiff, 1992). IRSL dating determines the date of deposition and subsequent burial (or time of last exposure to light) of alluvial sediment by measuring the accumulated radiation within the buried sediment as luminescence emitted by either quartz or potassium feldspar grains. It must be assumed that all the alluvial sediment samples have been 'zeroed' and not re-exposed to light by subsequent scour or minor reworking. Three samples for IRSL were taken by inserting a 50 mm diameter light-tight PVC tube into a freshly exposed outcrop or pit face (locations in Figure 14.3). IRSL dates of alluvial sediments are complemented by 25 dendro-dates to ascertain a minimum age of deposition for vegetated gravel bars adjacent to the present-day channel.

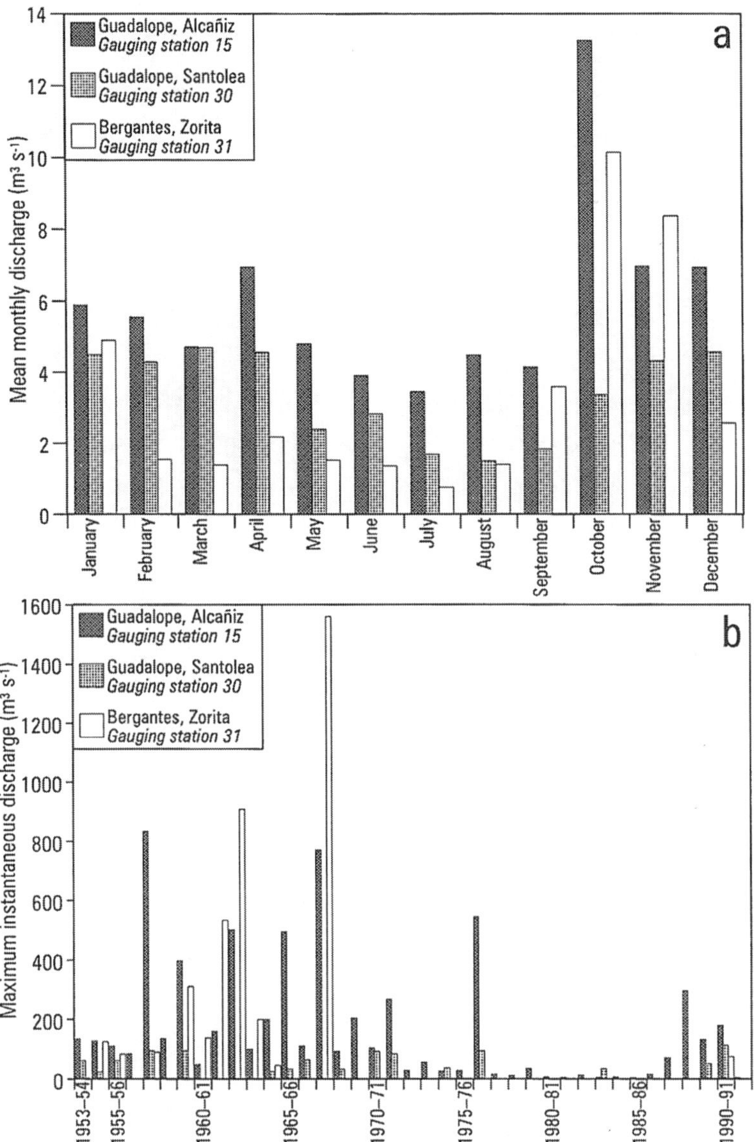

Figure 14.2 (a) Mean monthly discharge ($m^3\ s^{-1}$) recorded at station 15, station 30 and station 31 for the years 1959–1969. See Figure 14.1 for location of each gauging station. (b) Annual maximum instantaneous discharge ($m^3\ s^{-1}$) at station 15, and the corresponding annual maximum instantaneous discharge when registered on the same days at station 30 for the years 1953–1992, and station 31 between 1953–1971 and 1990–1992 (note the gauging record for station 31 was interrupted from 1971 until 1990, and that the hydrological years are from October to September, starting the data series for this graph in October 1953)

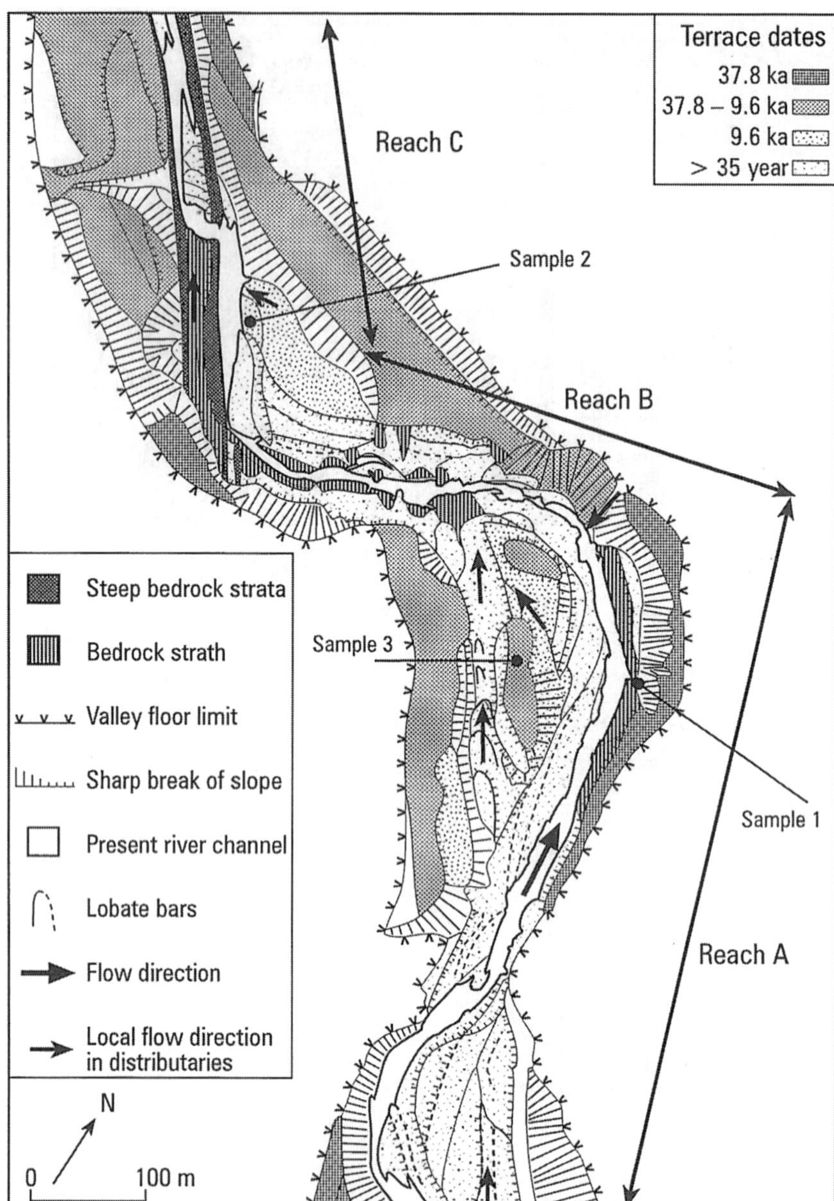

Figure 14.3   Geomorphological map of the Bergantes study site showing the spatial distribution of sedimentary units, and the suggested date of deposition of river terraces and bars. Dates of deposition are obtained from IRSL samples (error range is discussed in the text) and tree coring. Location of IRSL samples and the division of the study site into sub-reaches are also shown. Discrimination of the river terraces into levels representing aggradation phases is explained in the text and displayed in Figure 14.5

STUDY REACH DESCRIPTION AND SEDIMENTATION STYLE

The Bergantes study site is characterised by three markedly different sub-reaches termed A–C (Figure 14.3). Reach A is a 150–200 m wide aggrading 'alluvial reach' (Figure 14.4a) which begins immediately downstream of a bedrock constriction. The present low-flow channel exhibits features that are typical of many upland gravel-bed rivers with a subdued pool–riffle topography and a channel morphology with high width–depth ratio, which promotes braiding and extensive splay and gravel sheet deposition (Ferguson and Werritty, 1983). There is no clear differentiation between channel and floodplain (i.e. no abrupt bank edges or vegetation line) suggesting that the entire valley floor is active during moderate floods.

The valley floor sedimentology in reach A is dominated by very well-sorted gravel lobes (D_{50} c. 55 mm), which are typically 50–200 m long, 20–50 m wide and 0.3 m thick with sharp accretionary avalanche faces on their downstream margins. Lobe morphology is very similar to the low-relief bars described by Ashmore (1982) for gravel-bed braiding rivers, and represents the common form of mid-channel bar deposition in unconfined, transport-active gravel-bed rivers (e.g. Ferguson et al., 1992). Each gravel lobe consists of stacked gravel sheets, one to two grain diameters thick, which are often preserved as openworks at the lobe front. The gravel sheets probably owe their origin either to gravel overpassing (Carling, 1990) or movement *en masse* as a diffuse gravel sheet (Ashworth and Powell, 1992). Gravel sheet migration and superimposition has recently been described as a common textural feature of bartop sedimentation in ephemeral channels of southeast Spain (Conesa-García, 1995).

Towards the end of reach A there is a chute cutoff with a 25 m wide distributary that isolates an island on the inside of the floodplain before the main channel turns sharply west. The island crest is 10 m above the present channel bed and is an outcrop of resistant bedrock superimposed by 1–2 m of fine-grained flood deposits. The chute distributary is presently infilling with a series of 1–1.5 m high gravel lobes. The entrance to the chute is 1.7 m above the present channel bed and so requires a moderate flood to become active.

The right margin of reach A is bordered by a continuous, 15 m thick, Pleistocene river terrace (Figure 14.4a) deposited on a horizontal bedrock strath surface. Gully dissection of the Pleistocene gravels (D_{50} c. 30 mm) and undercutting during floods supplies sediment to the channel, but most of the valley floor sediments are derived from upstream through the constriction at the head of reach A.

The boundary between reaches A and B is abrupt and is a direct result of the exposure of bedrock on the channel bed and floodplain. In reach B, the bedrock dips westwards by 30–50° with the strike of the bedding planes running perpendicular to the main valley floor and river direction (Figure 14.4b). A difference in bedrock strength leads to the formation of a distinct 'step–pool' sequence (Whittaker and Jaeggi, 1982), with 6 m deep pools separated by 20 m long bedrock benches. The step–pool sequence traps small amounts of gravel in the deep scours but also promotes deposition of cobble splays at the channel margins (cf. Ferguson and Werritty, 1983). There is no local source of sediment in this reach and, in contrast to the aggrading alluvial reach A, reach B is a self-regulating, stable channel morphology which efficiently flushes through any sediment that enters the reach.

Reach B changes abruptly to a different channel morphology 240 m downstream at the entrance to the bedrock-confined reach C. The change in channel morphology is initiated

Figure 14.4   (a) Alluvial reach A looking upstream. Note the gravel splays at the channel margin and the 15 m high Pleistocene terrace. (b) Step–pool sequence in reach B. Flow is towards the camera and channel width is *c*. 20 m. (c) Bedrock-confined reach C. Flow is left to right and channel wetted width is *c*. 15 m. Note the 8 m high steeply dipping bedrock stratum that borders the channel

by the flow in reach B meeting a steeply dipping (60–70°) resistant bedrock stratum which runs parallel to the previous bedrock 'steps' and outcrops as a smooth 8 m high wall (Figure 14.4c). This stratum is 6–8 m wide and deflects the river through 90°, forcing it along a 15 m wide bedrock-confined channel (Figure 14.4c). The river manages to cut through this bedrock wall 180 m downstream but is immediately deflected back again through 90° as it contacts another parallel stratum of resistant rock. There is little exposed gravel in reach C and limited opportunity to develop an asymmetric thalweg or bars. The most common form of sedimentation in reach C is through the trapping of fine sand in bedrock niches up to 8 m above the channel bed and regions of slackwater.

Reach C has a 20 m wide distributary (Figure 14.4c) which is accessed at the head of the reach via an eroded gap in the bedrock wall. In order to bridge this gap, the local flow depth must exceed 3 m and the presence of fresh trash lines of vegetation in the distributary suggests that this can occur regularly. The opportunity for deposition high above the channel bed is enhanced by the structurally confined nature of reach C whereby any increase in discharge can only be accommodated by a local increase in depth until the adjacent bedrock strata are overtopped.

VALLEY FLOOR ALLUVIAL STRATIGRAPHY AND GEOCHRONOLOGY

Three sediment samples (see location in Figure 14.3) were taken from the principal Holocene and Pleistocene alluvial units of the Bergantes site for subsequent grain-size and IRSL dating analysis. Sample 1 was taken from a 12–14 m high Pleistocene terrace, at a point 3.8 m above the present channel bed on the east bank of reach A and was dated at 37.8 ± 4.2 ka. This is consistent with dates reported by Fuller et al. (1996) for river terraces with surfaces at similar heights located downstream at the Guadalope–Bergantes confluence and on the Guadalope at Castelserás (15 m and 17.5 m above the present river, respectively). Sample 2 was collected from the base of a 0.75 m thick silty sand unit capping a 7 m high fill terrace on the east bank of reach C, and was dated at 9.6 ± 1.6 ka. Early Holocene valley floor aggradation has also been recognised at the confluence between the Bergantes and the Guadalope rivers, dated at 9.0 ± 1.0 ka by Fuller et al. (1996). Sample 3 was taken 10 m above the present bed level in fine-grained flood deposits located on the island which separates the main channel from the distributary on the left bank of reach A. This was dated at 0.53 ± 0.08 ka and establishes a historic age of AD 1395–1545 for this unit.

In order to link the individual fluvial units of reaches A–C to the dating framework provided by the IRSL and distinguish between different periods of aggradation, the geomorphological map (Figure 14.3) was combined with the surveyed spot heights on each of the mapped units. Figure 14.5 shows the longitudinal profile of river terraces and alluvial units. Four levels have been defined and grouped around an IRSL date or a minimum age established by tree-ring dating. The spatial distribution of sedimentary units and their suggested date of deposition are also shown in Figure 14.3 .

The highest terrace level shown in Figure 14.5, 12–20 m above the low-flow water edge, is formed of trunk river and tributary alluvial fan material deposited between 33.6 and 42 ka (sample 1). A second level (Figure 14.5) between 9–10 m high, though at present undated, is bracketed between fluvial units ranging in age from 37.8 and 9.6 ka. It is capped, however, by fine-grained sediment dated at AD 1395–1545 (see sample 3) indicating that deposition during extreme flood events has occurred on this surface in historic

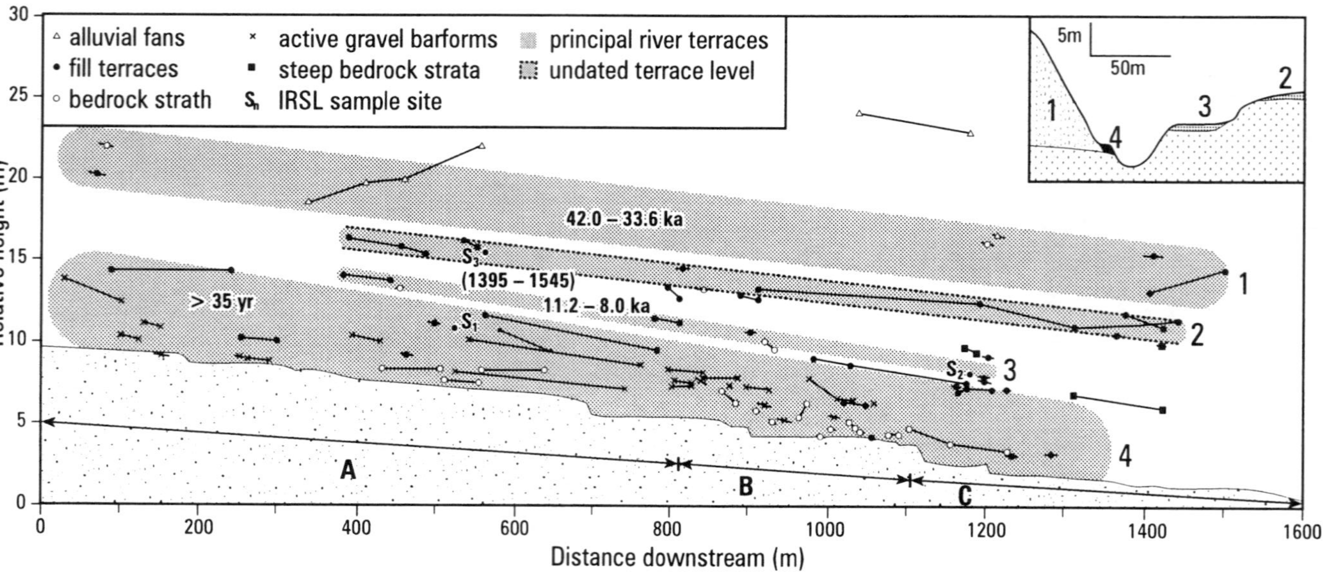

Figure 14.5 Longitudinal profile of the Bergantes study reach containing the heights of fill terraces, bedrock strath, barforms and steeply dipping bedrock strata above the low-flow water level. Four levels representing periods of aggradation have been defined by correlating alluvial units of similar sedimentology and height above the present channel. Each period of alluviation has been dated using IRSL and tree coring (see text and Figure 14.3). The surface of level 1 is 20 m above the present channel in some locations, and the position of level 2 is dotted because it is undated although it lies between deposits dated at 37.8 and 9.6 ka. Sediments dated at AD 1395–1545 may have been deposited on the surface of terrace level 2 during an exceptional large flood event in historic times (see text). The inset shows the relationship between the four terrace levels in a generalised valley cross-section

times. The third terrace level, 6–7 m high (Figure 14.5), indicates an aggradation episode that has been dated at 9.6 ± 1.6 ka (sample 2). Recent sedimentation (probably < 100 years) is represented by gravel lobes and bars between 3 and 5 m above the low-flow level, and form the fourth terrace level (Figure 14.5). The approximate age of this level has been defined by dendrochronology analysis of trees growing on the gravel units, which yield a minimum age of deposition of 35 years.

The interpretation of the alluvial sequence suggests either that the Early Holocene terrace (level 3; inset, Figure 14.5) could have been buried by younger deposits similar to those of historic age that cap level 2 (inset, Figure 14.5), and that incision followed forming the lower level 3 terrace surface, or that selective deposition of overbank fines in extreme flood events in historic times has led to the burial of the older sediments that constitute level 2 (inset, Figure 14.5). In either case, the presence of isolated patches of fine-grained sediment up to 10 m above the present channel indicates a selective depositional/erosional pattern in bedrock channels and confirms that sporadic flood events in recent historic times have resulted in overbank deposits being selectively preserved in zones probably related to areas of flood slackwater (Baker and Kochel, 1988).

HISTORICAL CHANGES IN CLIMATE, FLOODING AND SEDIMENTATION STYLE

References to flood events recorded in historic sources are presented in Table 14.1. This database only represents recorded historic flood events for the Bajo Aragón rivers, and the scarcity of information for the mediaeval and post-mediaeval period is compensated by more detailed information on floods collected for other southern tributaries of the Ebro (18 floods between AD 1250 and 1490, and eight events between AD 1490 and 1790) by Comisión Técnica de Inundaciones (1985). A paucity of large flood events for the whole of the Iberian Peninsula, with the exception of the Ebro basin and Mediterranean rivers (Turia and Júcar), between AD 1290 and 1400 has also been reported by Benito et al. (1996), indicating that this spatial pattern of flooding results either from typical atmospheric conditions of south/southwestern flow affecting the Mediterranean, or by 'cold pools' which are responsible for the major flood events affecting the Mediterranean rivers and the Ebro basin. The Bergantes catchment flooding pattern is directly influenced by these atmospheric conditions which result in sudden increases in discharge (Figure 14.2b shows peak discharges high above the mean discharge plotted in Figure 14.2a for station 31) and represents a rapid hydrograph rising limb. There is a good correlation between information provided by historic sources (Table 14.1) and gauging data (Figures 14.2a,b) as highlighted for the large flood events recorded in AD 1957, 1962, 1965, 1967 and 1977. The flashiness and high magnitude of floods in the Bergantes catchment have great repercussions not only for the mode and rate of sedimentation in this river but also for the whole Guadalope basin. This pattern resembles the behaviour recognised on ephemeral streams in southeast Spain by Conesa-García (1995), where it appears that it is the rare catastrophic flood events that are responsible for the overall changes in the sedimentology and morphology of the floodplain and valley floor.

The response of the valley floor evolution to a change in flood regime is also dependent on the degree of anthropogenic disturbance to the catchment which generally leads to reduced vegetation density, slope degradation and subsequent valley floor aggradation (Bull, 1991). Benavente et al. (1991) have described land-use changes in the Bajo Aragón

Table 14.1   Historical records of large floods in the Bajo Aragón region. Data are taken from Comisión Técnica de Inundaciones (1985). Archives usually document large floods that caused infrastructural damage or widespread overbank flooding. Historical records are more reliable in the 20th century

Year	Discharge (m^3s^{-1})	Recorded floods in the Bajo Aragón region
1311	*	Guadalope in Alcañiz
1801	*	Guadalope in Caspe
1863	*	Guadalope in Caspe
1869	*	Guadalope in Caspe
1878	200	Guadalopillo in Calanda
1879	*	Martín (Oliete, Albalate, Samper)
1885	109	Martín (Oliete, Albalate, Samper)
1887	*	Martín (Oliete, Albalate, Samper)
1889	230	Matarraña
1891	*	Guadalope in Caspe
1897	*	Martín (villages in its margins)
1897	36	Matarraña (Pena) in Beceite
1898	*	Martín (villages in its margins)
1898	*	Guadalope in Alcañiz
1900	156	Martín in Oliete
1903	30	Pena (Matarraña) in Valderrobres
1904	185	AguasVivas in Moneva
1904	500	AguasVivas after joining the Arroyo Moyuela
1905	10	AguasVivas in Moneva
1909	33	Pena (Matarraña)
1913	20	Pena (Matarraña)
1915	76	Martín in Hijar
1917	160	Guadalope in Alcañiz
1920	900	Guadalope in Alcañiz
1921	*	Guadalopillo (Alcorisa, Foz-Calanda, Calanda)
1922	*	Guadalopillo (Alcorisa, Foz-Calanda, Calanda)
1929	*	Martín in Hijar
1932	774	Bergantes in Zorita
1932	900	Guadalope in Alcañiz
1933	*	Bergantes in Aguaviva
1943	*	Matarraña in Valderrobres
1945	*	Guadalope in Alcañiz
1945	600	Martín in Hijar
1945	*	Seco in Oliete
1945	*	Matarraña in Maella
1946	250	Martín in Oliete
1947	*	Martín in Hijar
1951	*	Guadalope in Alcañiz
1957	*	Guadalope in Alcañiz
1957	*	Guadalope in Alcañiz
1957	*	Matarraña in Valderrobres and Beceite
1962	910	Bergantes in Zorita
1963	*	Martín in Hijar
1965	640	Guadalope in Alcañiz
1966	124	Pena (Matarraña) in Beceite
1967	*	Matarraña in Valdetorres
1967	1560	Bergantes in Zorita
1967	1019	Guadalope in Alcañiz
1969	137	Matarraña in Beceite
1977	655	Guadalope in Alcañiz

* No information available on the magnitude of the flood peak (records usually only include a qualitative description of the flood impact)

region experienced since the Early Neolithic when intense grazing and vegetation clearance increased the vulnerability of catchments to erosion. Work also undertaken in other regions of Spain (e.g. Alonso and Garzón, 1994) suggests that episodes of aggradation may have been encouraged by human disturbance in the Mediterranean areas since the Early to Middle Holocene.

Alluviation in the Bergantes River during mediaeval times (AD 1395–1545) is likely to have been influenced both by climatic change towards wetter conditions and abundant runoff that occurred in the western Mediterranean at the beginning of the Little Ice Age (Macklin et al., 1994), and by anthropogenic causes, namely deforestation and overgrazing, which have been observed in the region by Davis (1994) for the period 1.4–0.4 ka. This is corroborated by work undertaken by Palanques et al. (1990) and Guillén and Palanques (1992) which suggested that progradation of the Ebro delta could be linked to intense anthropogenic activity at the end of the Middle Ages.

CONCLUSIONS

Little attention has been paid to the style and rate of sedimentation in upland Mediterranean rivers which are strongly influenced by local tectonics and bedrock outcrop. This chapter has described the spatial pattern of alluviation in a structurally controlled reach of the unregulated Bergantes River. Three sedimentary environments have been distinguished that relate to differences in structural controls: (a) an alluvial reach pinned at both ends by a bedrock constriction, where gravel lobes are deposited and reworked by relatively frequent (five to ten year recurrence interval) floods; (b) a step–pool sequence with 6 m deep scours separated by 20 m long bedrock benches and occasional cobble splays at the channel margins; and (c) a straight reach, confined by bedrock, where sedimentation is restricted to trapping of fine sediment as fill terraces and in pockets created by the irregular weathering of the steeply dipping bedrock.

IRSL and tree-ring dating of fluvial units provide information on the nature of the sedimentary processes in this reach and demonstrate four distinct phases of valley floor sedimentation in the last 40 000 years. Significant aggradation occurred 37.8 ± 4.2 ka related to the formation of a Late Pleistocene terrace consisting of trunk river and tributary alluvial fan deposits. Additionally, an aggradation phase has been recognised during the Late Glacial–Early Holocene transition, 9.6 ± 1.6 ka. In historic times, AD 1395–1545, fine-grained sediments have been deposited during exceptional large flood events on the surface of a previous terrace which is believed to have been formed sometime between 37.8 and 9.6 ka. Relatively moderate floods in recent decades have been responsible for transporting coarse material forming a series of gravel sheets and lobes 3–5 m above the present channel. Changes in flooding, anthropogenic activity and climatic change towards drier conditions may help explain recent aggradation and channel changes.

ACKNOWLEDGEMENTS

This research was undertaken whilst Alma López-Avilés held a School of Geography PhD studentship at the University of Leeds. Additional funding was provided by Mr Amando and Mrs Juana López-Avilés and the Leeds 'Superslugs' research project. We are extremely grateful for help and advice offered by Mr Jose Antonio Benavente (School of Archaeology of Alcañiz), Ms Alejandra Almunia Martín (Natural Risks Department at the Civil Protection Unit of Zaragoza), and Mr Lorenzo Polanco and Mr Eloy Medrano (Confederación Hidrográfica del Ebro). Dr Ian Fuller kindly processed the IRSL samples. Two anonymous referees provided valuable and perceptive comments.

REFERENCES

Alonso, A. and Garzón, G., 1994. Quaternary evolution of a meandering gravel river in central Spain. *Terra Nova*, **6**, 465–475.

Ashmore, P.E., 1982. Laboratory modelling of gravel braided stream morphology. *Earth Surface Processes and Landforms*, **7**, 201–225.

Ashworth, P.J. and Powell, M.D., 1992. Discussion of Laronne, J.B. and Duncan, M.J., Bedload transport path and gravel bar formation. In P. Billi, R.D. Hey, C.R. Thorne, and P. Taconni (eds), *Dynamics of Gravel-Bed Rivers*, Wiley, Chichester, 200–202.

Aurell, M., Meléndez, A. and Meléndez, G., 1990. Caracterización de la secuencia Oxfordiense en el sector central de la Cordillera Ibérica. *Geogaceta*, **8**, 73–76.

Bailiff, I.K., 1992. Luminescence dating alluvial deposits. In S. Needham and M.G. Macklin (eds), *Alluvial Archaeology in Britain*, Oxbow, Oxford, 27–35.

Baker, V.R. and Kochel, R.C., 1988. Flood sedimentation in bedrock fluvial systems. In V.R. Baker, R.C. Kochel and P.C. Patton (eds), *Flood Geomorphology*, Wiley-Interscience, New York, 123–138.

Batalla, R.J. and Sala, M., 1995. Effective discharge for bedload transport in a subhumid Mediterranean sandy gravel-bed river (Arbúcies, north-east Spain). In E.J. Hickin (ed.), *River Geomorphology*, Wiley, Chichester, 93–103.

Benavente, J.A., Navarro-Cases, C., Ponz-Palacios, J.L. and Villanueva-Herrero, J.C., 1991. El poblamiento antiguo del área endorreica de Alcañiz (Teruel). *Boletín del Taller de Arqueología de Alcañiz*, **2**, 36–92.

Benito, G., Machado, M.J. and Pérez-González, A., 1996. Climate change and flood sensitivity in Spain. In J. Branson, A.G. Brown and K.J. Gregory (eds), *Global Continental Changes: the Context of Palaeohydrology*, Geological Society, London, Special Publication 115, 85–98.

Bull, W.B., 1988. Floods; degradation and aggradation. In V.R. Baker, R.C. Kochel and P.C. Patton (eds), *Flood Geomorphology*, Wiley-Interscience, New York, 157–166.

Bull, W.B., 1991. *Geomorphic Responses to Climate Change*, Oxford University Press, New York.

Carling, P.A., 1990. Particle over-passing on depth-limited gravel bars. *Sedimentology*, **37**, 345–355.

Carling, P.A. and Grodek, T., 1994. Indirect estimation of ungauged peak discharges in a bedrock channel with reference to design discharge selection. *Hydrological Processes*, **8**, 497–511.

Collier, R.E.Ll., Leeder, M.R. and Jackson, J.A., 1995. Quaternary drainage development, sediment fluxes and extensional tectonics in Greece. In J. Lewin, M.G. Macklin and J.C. Woodward (eds), *Mediterranean Quaternary River Environments*, Balkema, Rotterdam, 31–45.

Comisión Técnica de Inundaciones, 1985. *Estudio de inundaciones históricas: mapa de riesgos potenciales*, Comisión Nacional de Protección Civil, IV–VIII.

Conesa-García, C., 1995. Torrential flow frequency and morphological adjustments of ephemeral channels in south-east Spain. In E.J. Hickin (ed.), *River Geomorphology*, Wiley, Chichester, 169–192.

Davis, B.A.S., 1994. *Palaeolimnology and Holocene environmental change from endoreic lakes in the Ebro Basin, north-east Spain*, PhD thesis, University of Newcastle-Upon-Tyne.

Duller, G.A.T., 1996. Recent developments in luminescence dating of Quaternary sediments. *Progress in Physical Geography*, **20**(2), 127–145.

Ferguson, R.I. and Werritty, A., 1983. Bar development and channel changes in the gravelly river Feshie, Scotland. In C.D. Collinson and J. Lewin (eds), *Modern and Ancient Fluvial Systems*, International Association of Sedimentologists, Special Publication 6, Blackwell, Oxford, 181–193.

Ferguson, R.I., Ashmore, P.E., Ashworth, P.J., Paola, C., Powell, D.M. and Prestegaard, K.L., 1992.

Measurements in a braided river chute and lobe: 1. Flow pattern, sediment transport and channel change, *Water Resources Research*, **28**, 1877–1886.

Fuller, I.C., Macklin, M.G., Passmore, D.G., Brewer, P.A., Lewin, J. and Wintle, A.G., 1996. Geochronologies and environmental records of Quaternary fluvial sequences in the Guadalope basin, Northeast Spain, based on luminescence dating. In J. Branson, A.G. Brown and K.J. Gregory (eds), *Global Continental Changes: the Context of Palaeohydrology*, Geological Society, London, Special Publication 115, 99–120.

Fumanal, M.P. and Carmona, P., 1995. Paleosuelos Pleistocenos en algunos enclaves del País Valenciano. *El Cuaternario del País Valenciano*, 125–134.

Guillén, J. and Palanques, A., 1992. Sediment dynamics and hydrodynamics in the lower course of a river highly regulated by dams: the Ebro River. *Sedimentology*, **39**, 567–579.

Harvey, A.M. and Wells, S.G., 1987. Response of Quaternary fluvial systems to differential epeirogenic uplift: Aguas and Feos river systems, southeast Spain. *Geology*, **15**, 689–693.

Ibáñez, M.J., Pellicer, F. and Yetano, L.M., 1983. Rasgos geomorfológicos del contacto entre la Cordillera Ibérica y la Depresión del Ebro. *Geographicalia*, **18**, 3–19.

IGME, 1978. *Mapa Geológico de España: Peñarroya de Tastavins. Scale 1:50 000*, Instituto Geológico y Minero de España, 520/30–20.

IGME, 1981. *Mapa Geológico de España: Tortosa. Scale 1:200 000*, Instituto Geológico y Minero de España, 41.

Lewin, J., Macklin, M.G. and Woodward, J.C., 1995. *Mediterranean Quaternary River Environments*, Balkema, Rotterdam.

Macklin, M.G. and Passmore, D.G., 1995. Pleistocene environmental change in the Guadalope basin, Northeast Spain: Fluvial and archaeological records. In J. Lewin, M.G. Macklin and J.C. Woodward (eds), *Mediterranean Quaternary River Environments*, Balkema, Rotterdam, 103–113.

Macklin, M.G., Passmore, D.G., Stevenson, A.C., Davis, B.A. and Benavente, J.A., 1994. Responses of rivers and lakes to Holocene environmental change in the Alcañiz region, Teruel, north-east Spain. In A.C. Millington and K. Pye (eds), *Environmental Change in Drylands: Biogeographical and Geomorphological Perspectives*, Wiley, Chichester, 113–130.

Macklin, M.G., Lewin, J. and Woodward, J.C., 1995. Quaternary fluvial systems in the Mediterranean basin. In J. Lewin, M.G. Macklin and J.C. Woodward (eds), *Mediterranean Quaternary River Environments*, Balkema, Rotterdam, 1–29.

Maldonado, A., 1975. Sedimentation, stratigraphy, and development of the Ebro Delta, Spain. In M.L. Broussard (ed.), *Delta Models for Exploration*, Houston Geological Society, Houston, 311–338.

Mensua, S. and Ibáñez, M.J., 1973. La Depresión de Mas de las Matas: Una cubeta en el contacto entre la cuenca del Ebro y las montañas Ibéricas. In *J.M. Casas Torres. Homenaje a una Labor (25 años de docencia)*, Universidad de Zaragoza, Zaragoza, 191–213.

Ministerio de Agricultura, 1977. *Mapa de Cultivos y Aprovechamientos: Peñarroya de Tastavins (Teruel). Scale 1:50 000*, Ministerio de Agricultura, 520/30–20.

Nelson, C.H., 1990. Estimated post-Messinian sediment supply and sedimentation rates on the Ebro continental margin, Spain. *Marine Geology*, **95**, 395–418.

Palanques, A., Plana, F. and Maldonado, A., 1990. Recent influence of man on the Ebro margin sedimentation system, northwestern Mediterranean Sea. *Marine Geology*, **95**, 247–263.

Peña-Monné, J.L., Gutiérrez-Elorza, M., Ibáñez, M.J. *et al.*, 1984. *Geomorfología de la provincia de Teruel*, Instituto de Estudios Turolenses, Teruel.

Poesen, J.W.A. and Hooke, J.M., 1997. Erosion, flooding and channel management in Mediterranean environments of southern Europe. *Progress in Physical Geography*, **21**(2), 157–199.

Pueyo-Mur, J.J., 1978. La precipitación evaporítica actual en las lagunas saladas del área: Burjaraloz, Sástago, Caspe, Alcañiz y Calanda (provincias de Zaragoza y Teruel). *Revista del Instituto de Investigaciones Geológicas*, **33**, 5–56.

Rodríguez-Vidal, J., 1982. Distribución morfoclimática en la depresión media del Ebro: procesos dominantes y modelado actual. *Estudios geológicos*, **38**, 43–50.

Sala, M., 1982. Datos cuantitativos de los procesos geomórficos fluviales actuales en la cuenca de la Riera de Fuirosos (Montnegre, macizo litoral Catalán). *Cuadernos de Investigación Geográfica*, Colegio Universitario de La Rioja, **8**, 51–68.

Sánchez-Navarro, J.A., San Román-Saldaña, J.L., de Miguel-Cabeza, J.L. and Martínez-Gil, F.J., 1990. El drenaje subterráneo de la Cordillera Ibérica en la Depresión del Ebro: aspectos geológicos. *Geogaceta*, **8**, 115–118.

Simón-Gómez, J.L., 1990. Algunas reflexiones sobre los modelos tectónicos aplicados a la Cordillera Ibérica. *Geogaceta*, **8**, 123–129.

Smith, A.G. and Woodcock, N.H., 1982. Tectonic synthesis of the Alpine-Mediterranean region: a review. In *Alpine Mediterranean Geodynamics*, American Geophysical Union, Geodynamics Series 7, 15–38.

Thornes, J.B., 1987. The Palaeo-ecology of erosion. In J.M. Wagstaff (ed.), *Landscape and Culture: Geographical and Archaeological Perspectives*, Blackwell, Oxford, 37–55.

Whittaker, J.G. and Jaeggi, M.N.R., 1982. Origin of step-pool systems in mountain streams. *Proceedings of the American Society of Civil Engineers, Journal of the Hydraulics Division*, **108**, 758–773.

Wintle, A.G., 1993. Luminescence dating of aeolian sands: an overview. In K. Pye (ed.), *The Dynamics and Environmental Context of Aeolian Sedimentary Systems*, Geological Society, London, Special Publication 72, 49–58.

Wohl, E.E., 1992. Bedrock benches and boulder bars: Floods in the Burdekin Gorge of Australia. *Geological Society of America Bulletin*, **104**, 770–778.

Wohl, E.E., Greenbaum, N., Schick, P. and Baker, V.R., 1994. Controls on bedrock channel incision along Nahal Paran, Israel. *Earth Surface Processes and Landforms*, **19**, 1–13.

Woodward, J.C., 1995. Patterns of erosion and suspended sediment yield in Mediterranean river basins. In I.D.L. Foster, A.M. Gurnell and B.E. Webb (eds), *Sediment and Water Quality in River Catchments*, Wiley, Chichester, 365–389.

PART IV
Late Pleistocene Palaeohydrology

CHAPTER 15

New Approach to the Ice Age Paleohydrology of Northern Eurasia

MIKHAIL G. GROSSWALD

Institute of Geography, Russian Academy of Sciences, Moscow, Russia

INTRODUCTION

Continental paleohydrology is a new branch of the geosciences. It is aimed at applying the achievements of the hydrological sciences, in particular the models of hydrological cycles and water balance, to the scenarios of the past. It also concerns the introduction of various hydrological parameters and techniques into paleogeography (Starkel, 1995). This new branch comprises fluvial paleohydrology, paleolimnology, and paleoglaciology, three sub-branches that, until recently, operated separately. Also, paleoclimatology dealing with the models of evaporation and atmospheric precipitation is affiliated with continental paleohydrology.

One of the central targets of this new branch is to gain insights into continental paleohydrology of the ice ages. The great glaciations were times of global reorganization in the Earth's hydrological system; the glaciations caused profound changes in all the elements of the system – the glaciers, rivers, lakes, and the ocean – and they also brought about major changes in the means of replenishment and transportation of atmospheric moisture.

Of all the ice ages, these changes can be effectively reconstructed only for the last one, the Late Pleistocene glaciation, spanning the time interval of 125 to 10 ka. With small exceptions, all the glacial evidence coming from landforms, from ice-sheet and deep-sea drillings, from peatbogs, and from lacustrine and river-terrace stratigraphies, that is well enough preserved to be translatable into the language of paleohydrology, belongs to the last glaciation. For this reason, we must restrict paleohydrological analysis to the framework of the last great glaciation.

Paleohydrological changes appeared in the glaciated parts of continents in two major ways: first, by bringing about reorganization of drainage systems, and second, by altering their hydrological balance (Teller, 1995).

PROGLACIAL DRAINAGE SYSTEMS

Glaciation of the continental margins and mountain ranges

According to this author's concept, the largest ice sheets covered the high-latitude margins of Eurasia, in particular, to the polar continental shelves and coastal lowlands. Thus, marine ice sheets prevailed in Pleistocene Eurasia. Spreading centers of the ice sheets were located near the present-day shoreline, on the continental margin, and the ice flowed both outward, to the deep ocean, and inward, to the continent interiors. In other words, the ice moved up the major rivers catchments linked to the Arctic and Pacific Oceans (Hughes et al., 1977; Grosswald, 1983, 1997; Grosswald and Hughes 1995). However, unlike the case in North America, the ice sheets in Eurasia never, either during their retreat or maximum stage, reached the continental divide.

There were also extensive glacier complexes in the Eurasian mountains. So far, our knowledge of them is incomplete, patchy, and controversial. We lack reliable data on the amplitudes of equilibrium-line depression in the mountains, and the available estimates are often over-cautious, subjective, and diminutive. It is believed that an equilibrium-line depression equal to 500–600 m prevailed throughout the former USSR, while in some mountains, like the Pamirs and eastern Tien Shan, it was as small as 200–300 m.

However, a number of cases from the Caucasus, Pamirs, Tien Shan and Trans-Baikalian mountains demonstrated that those small equlibrium-line depressions are not true. As a rule, they originated from errors in determining the extents of past glaciation, in particular, where the Late Glacial and Holocene end moraines were mistaken for the outer limits of maximum glacier expansion. The results are quite different where the limits are accurately determined in the field and confirmed by interpretation of space images. For instance, in the northern Tien Shan, where we managed to accurately locate the 'maximum' Late Pleistocene ice margins, the equilibrium-line depression measured by the Hoefer method along five profiles turned out to be 1150–1200 m (Grosswald et al., 1994); in the East Sayan and Khamar Daban mountains (west of Lake Baikal) it amounted, according to Kuhle's observations, to 1400–1500 m (Grosswald and Kuhle, 1994). These results match the Pleistocene equilibrium-line depression for the Caucasus and Altai mountains, where Vardaniants (1938), based on field data and his own technique, estimated it as equal to 1200 m.

Numerical modeling based on the assumption of a 1000–1200 m equilibrium-line depression suggested that ice caps and Spitsbergen-style glacier complexes formed in the ice age Tien Shan, Pamirs, Transbaikalia and other mountains of Central Asia and Siberia, so that both the mountains and their piedmonts were completely ice-covered (Grosswald et al., 1994). The mountains of Eurasia form a semi-continuous system of ranges extending from the Pamirs and Tien Shan to Chukchi Peninsula, hence their ice cover comprised a virtually continuous chain of ice caps and extensive glacier complexes. Extending from the Pamirs to Chukchi Peninsula, the ice-cap chain connected the marine ice sheets of the Eurasian margins with the great glacier system of the Central Asian Highland. A particularly important part was played by the Chersko–Kolyman glacier complex: it became wedged between the East Siberian, Chukchian (Beringian), and Okhotsky marine ice sheets resulting in the formation of a huge composite superglacier. By its areal extent, it probably surpassed the Scandinavian and Barents–Karan ice sheets. Incidentally, the same conclusion comes from modeling experiments of Verbitsky and Oglesby (1992) and Marsiat (1994).

Based on relatively sparse evidence, the concept of a continuous chain of ice caps and glacier complexes in the mountains of ice age Eurasia should be considered a working hypothesis that must be tested by analysing additional field data, high-resolution space images, and new results of modeling.

Glaciation and its impact upon drainage systems

Growth of ice sheets and mountain glacier complexes led to profound changes in the topography of Eurasia. Broad and continuous flat-topped icy ranges, 2.3 to 3 km a.s.l. high, emerged along the Arctic and Pacific continental margins, now submerged; in the highlands, glaciers filled intermontane basins and transverse valleys and built up over the tops and slopes of the ranges. As a result, the chain of initially separate mountain massifs was changed into an unbroken barrier separating the catchments of Siberian rivers from those of the Amur River and the now closed basins of Central Asia.

These orographic changes were similar throughout the entire Quaternary ice ages, and all of them were favorable for river impoundment and lake formation. Ice sheets of the continental margins blocked the rivers flowing into the Arctic Ocean, the Bering and Okhotsk Seas. The closed basins formed by the glacial blockade were capable of containing enormous lakes. Such a setting probably persisted during the Early and Middle Pleistocene, as well as throughout the Weichselian, i.e. the Late Pleistocene glaciation. For this reason, the proglacial drainage systems of different ages had to be similar. Thus, I share the concept of repetitive development of the proglacial hydrological systems, which suggests that the systems copied themselves time after time throughout the Quaternary. As shown by Teller (1987, 1995), this concept applies well to the proglacial paleohydrology of North America. There are different opinions on this subject, though: according to Arkhipov et al. (1995), the Eurasian drainage systems evolved from glaciation to glaciation, changing their extent and structure, and flowing first into the Caspian Sea, the next time into the Black Sea, and the third time into the Mediterranean.

The mountain glacier complexes of Eurasia, as noted above, heightened and enhanced the mountain barrier, which diagonally crosses the continent. As today, it played the part of a drainage divide between the main catchments of Eurasia, but its role was enhanced. Among other consequences, it cut off the Selenga River catchment from Lake Baikal, and the Irtysh River catchment from the river's upper reaches. During glacial retreat, some openings and passages appeared in the ice barrier; the shrinking ice gave way to rivers and ice-dammed lakes; the lakes tended to periodically burst through their ice dams causing cataclysmic floods across the piedmonts (Baker et al., 1993; Baker, 1996; Grosswald and Rudoy, 1996).

A strong link between the morphology and size of the Eurasian paleohydrological basins with the geography of past ice sheets is demonstrated by Figure 15.1. Of the four sketches, only Figure 15.1d is based on my latest ice-sheet reconstruction, while the rest show how the idea of the basins evolved. According to Figure 15.1a, the proglacial hydrology was restricted to the *pradolinas* of Poland and northern Germany. According to Figure 15.1b, which took into account the occurrence of Barents–Karan ice sheet, the paleohydrology includes the lakes and spillways of Europe and West Siberia. In Figure 15.1c, showing the larger ice-sheet system (due to adding of the East Siberian ice sheet), the western catchment also becomes larger (to include the Lena–Vilyuy Lake), and a second, east-bound drainage system appeared. As it was, the geography of past ice sheets

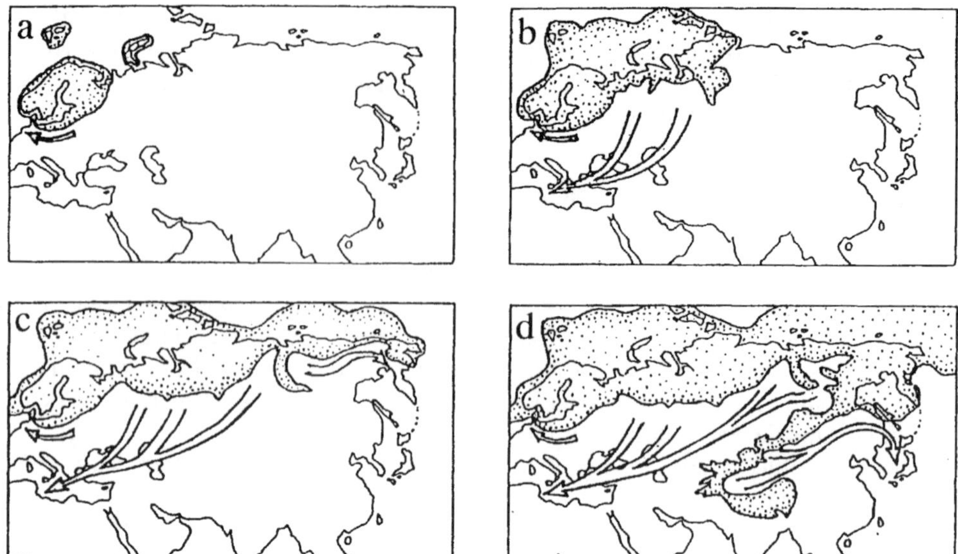

Figure 15.1   Changing concept of the Late Weichselian proglacial drainage systems in northern Eurasia. The arrows point out generalized directions of the past meltwater flow. (a) Model based on Gerasimov and Markov (1939); (b–d) models based on Grosswald (1983)

was uncovered step by step, feature by feature. And, concurrently, also step by step, the concept of proglacial paleohydrology was developing. As a rule, each new paleohydrological model was first postulated (based on land topography and ice-sheet outlines), and only afterwards substantiated by evidence.

Trans-Siberian drainage system

This system was a direct consequence of the north Eurasian ice sheets. During the last glacial maximum, it extended from Chersky range in the east to the Alps in the west, and drained a catchment area amounting, according to new estimates, to $23 \times 10^6 \, km^2$, thus being more than three times the size of the present-day Amazon basin. Major constituent elements of the system were Novo-Euksinian (Black), Khvalyn (Caspian), and Aral 'seas', as well as the Mansi, Yenisey, and Lena-Vilyuyan lakes. Their total areal extent was in excess of $3 \times 10^6 \, km^2$, while the runoff was focused first in the Black Sea, and then, via the Sea of Marmora, into the eastern Mediterranean Sea (Grosswald, 1983; Grosswald and Kotlyakov, 1989; Grosswald and Hughes, 1995).

My latest reconstruction (Figure 15.1d) is different from the previous one (Figure 15.1c) in its eastern part, as it does not display the second Siberian drainage system. This change in the reconstruction is due to our (Grosswald and Hughes, 1995) revision of glacial paleogeography of northeastern Siberia and Beringia. We came to realize (see above) that during the last glacial maximum the ice sheets of that region were too large, and the ice-free terrains too small for the development of the eastern system.

Reconstruction of the Trans-Siberian drainage system is substantiated by a wealth of geological data, such as the occurrence of Upper Pleistocene lacustrine deposits, *pradolinas*, spillways, lake terraces and shore lines. Radiocarbon ages exist for major paleolakes and spillways, in particular of Mansi Lake in west Siberia, Turgai Spillway, the Aral, Khvalyn, and Novo-Euksinian 'Seas' (Kvasov, 1979; Arkhipov et al., 1980; Grosswald, 1983; Goncharov, 1986). As for the system's linkage with the Mediterranean Sea, it is confirmed both by geomorphology of the connecting channels and by the fact of simultaneous (with glacial maximum) changes in the sea itself, i.e. by its cooling and diminishing salinity (Thiede, 1978; Thunnell, 1979; Grosswald and Kotlyakov, 1989).

A key role for documenting the Trans-Siberian meltwater drainage system belongs to the aforementioned spillways. The latter include large and well-known channels like the Mylvan, Keltman, Manych, Turgay, and Kas-Ket, and smaller overflow channels like the Upper Dnepran and Upper Volgan groups, or virtually unknown ones like the Vilyuy–Tunguskan. To this latter group belong the Tompo and East Khandyga channels (right-hand tributaries of the Aldan), which linked upper reaches of the Yana and Indigirka rivers with the Lena–Vilyuyan Lake.

Some channels, in particular the Keltman, Turgay and Kas-Ket spillways, were subjected to special studies (Goretsky, 1964; Ryabkov, 1975; Grosswald, 1983). The others (with small exceptions) retain their geomorphic features and can be examined on space images. Broad vistas are open to the latter type of research, which was recently demonstrated by Baker (1996). For example, his and his associates' examination of Manych Spillway area, Russia, resulted in establishing that 'it displays a suite of landforms characteristic of cataclysmic flooding: streamlined hills, elongate depressions, and scabland erosion. The channel width was found to be immense (35 km) which further indicated the great scale of paleoflow. Very large cross-sectional area of the paleochannel (approximately 10^6 m^2 at the 42 m Khvalyn level, yielding 28 m flow depth) suggested that at a flow velocity of 10 m s^{-1}, required for bedrock scabland erosion of this type, the estimated flood discharge amounted to 10^7 m^3 s^{-1}, which is comparable to that of ice-dammed lakes outbursts like those of glacial Lake Missoula' (Baker, 1996, p. 100).

In general, however, the present knowledge of the Eurasian spillways is inadequate. This is unfortunate since the geographical position and geomorphology of the spillways can provide invaluable evidence for the directions and regime of meltwater flow, as well as for location of past ice dams. I would emphasize that the spillway geography, like that in Figures 15.2 and 15.3, makes it quite clear that the north- and east-flowing rivers of northern Eurasia were glacially dammed and diverted.

The Trans-Siberian system did not persist for long in the way it is shown in Figure 15.1d. As soon as it matured, the levels of its lakes began to become lower due to the erosional deepening of the spillways. Other factors causing changes in the system included glacio-isostatic movements of the crust and oscillations of the ice margin. The ice-dammed lakes shifted northward to follow the retreating ice sheet, or retransgressed in response to ice readvances, and their outflow was frequently rerouted. Radial (relative to the ice sheets) meltwater flow turned marginal and then returned to radial. Intervals of 'normal' outflow alternated with those of cataclysmic outbursts. These were the features making proglacial paleohydrology highly complicated and subject to rapid changes. Such features may be best illustrated by the Late Glacial history of North American proglacial drainage systems (Prest, 1970; Teller, 1987).

For Eurasia, this type of 'time-slice-after-time-slice' reconstruction is a task for the future. So far, we only know that, during the last glacial maximum, the Siberian meltwater was discharged into the Mediterranean Sea; and that during the Late Glacial it was first

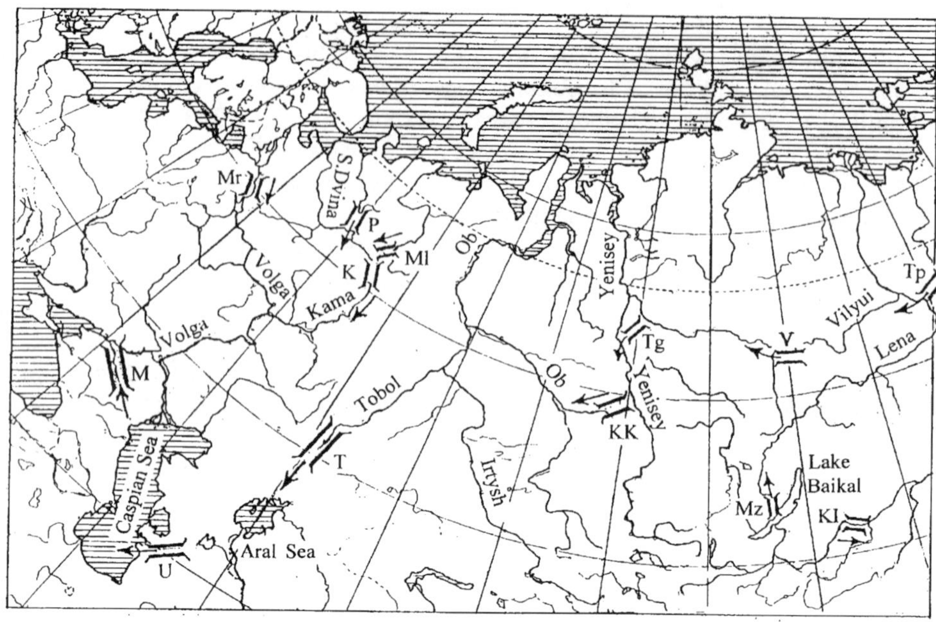

Figure 15.2　Major spillways in north-central Eurasia: evidence for occurrence of an ice dam on the Arctic margin of the continent. Letters refer to spillways: Mr, Maryinsky; P, Pegishdor; U, Uzboy; T, Turgay; KK, Kas-Ket; Tg, Tungusky; V, Vilyuysky; Tp, Tompo; Mz, Manzurka; KI, Khilok-Ingodan; M, Manych; K, Keltman; ML, Mylvan

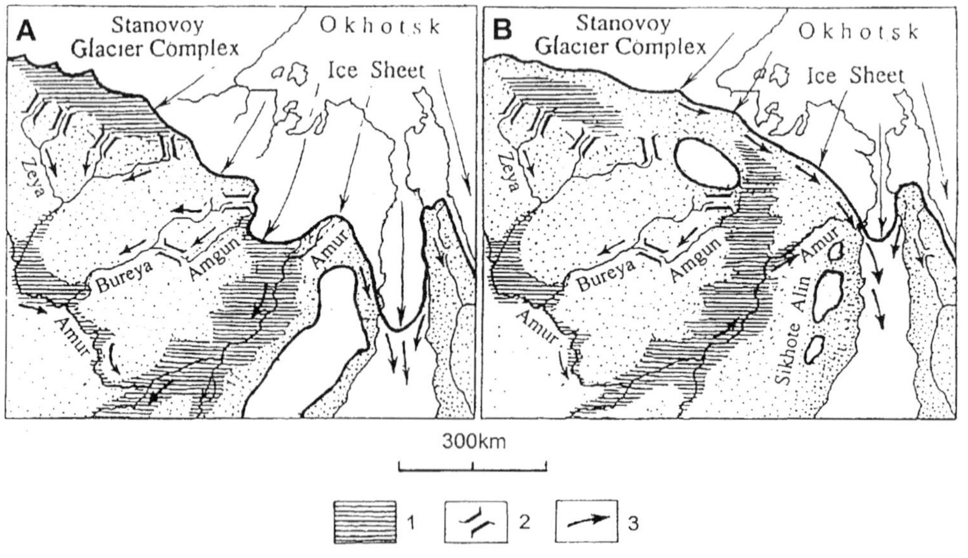

Figure 15.3　Evidence for reorganizations of the drainage network in catchment areas of the Amur and Uda Rivers, as caused by the damming effects of the past Okhotsk Sea ice sheet. Legend: 1, Late Quaternary lakes; 2, spillways; 3, hypothetical past water flow (after Grosswald and Hughes, 1998)

routed toward the North Sea, and then, at 13–11 ka, to the Barents Sea. Due to the latest collapses and surges of the Barents–Karan ice sheet, which occurred at about 10 and 8.5 ka (Grosswald and Krass, 1998), some returns to earlier scenarios of northern blockage were taking place. However, the ice dams created by surges were only short-lived, so that meltwater runoff kept alternating its routing toward the Barents Sea and the Baltic Ice Lake

Gobi–Amuran proglacial drainage system

The problem of this drainage system is posed here for the first time. To do this, we had to enhance and reconsider our knowledge of the last glaciation in the mountain areas of Siberia and Central Asia. Only now, when a new concept of that glaciation is developed, can we see that the Amur River used to be the axis of a huge glacial catchment. In fact, the Amur's entire flow was bordered by big mountain glacier complexes, so that its water balance had to reflect the strong influence of glaciers. In addition, there are certain grounds to speculate that the catchment could be strongly enlarged by annexing the (now closed) basins of Central Asia, in particular, the west Mongolian, south Trans-Baikalian, Gobian, Junggarian, and Tarim groups of basins.

The present-day climate of the Central Asian basins is arid, their landscapes belong to desert types, and their drainage is closed and does not reach the ocean. However, during the ice ages, the environment changed dramatically. The climate turned wetter, large glaciers formed in the mountains, and coniferous forests covered the ice-free slopes (Murzayev, 1966; Devyatkin et al., 1978). Here, the ice ages were coeval with the great-lake stages, and the presently dry basins became filled with large bodies of water (Murzayev, 1949; Kuznetsov and Murzayev, 1963; Selivanov, 1965; Kuznetsov, 1968; Murzayeva et al., 1982).

A single water basin with its surface at 1260 m a.s.l., which is 500 m higher than the present level of Lake Uvs Nur, formed in the Basin of Great Lakes of Mongolia during the glacial maximum (Devyatkin et al., 1978). Another huge lake with the surface at 850–900 m a.s.l. filled the basin of the Selenga River in south Trans-Baikalian area (Osadchiy, 1995). The lakes of the Gobi Desert became many times larger than they are today (Pachur et al., 1995), and an enormous freshwater 'sea' probably emerged where the present-day Lake Djalai Nur occurs (V.E. Murzayeva, oral commun.). Lacustrine deposits linked to moraines are reported from Junggaria (Selivanov, 1965). Also, convincing evidence for an extensive lake filling the Tarim Basin was uncovered by the Sino-German Kunlun Shan Taklamakan Expedition of 1986 (Jäkel and Zhenda, 1991).

The great extent of of the past Central Asian lakes and their synchroneity with glaciations is confirmed by modern research. However, here (as in arid southwestern North America) it is still unknown whether these lakes were integrated into a single drainage system, because the direct evidence for connecting paleovalleys is sparse, and that for paleolevels of the lakes is incomplete. As in the case of the Owens Lake basin of Sierra Nevada, California (Benson et al., 1996), it will take special techniques and considerable effort to find indications that the lakes in question did overflow their basins and to date such events. As for geological evidence for the overflow, such as fossil shorelines, especially the highest ones, they are poorly preserved, having been either destroyed by wind erosion or concealed under thick accumulations of eolian sand. High-stand episodes were short, and their traces could be easily obliterated. 'For this reason', points out Kuznetsov (1968, p. 79), 'the evidence of fossil lake formations, such as terraces and shorelines, is not sufficient for establishing maximum levels of the paleo-lakes'.

I expect that the quest for past river beds and high-level shorelines will benefit from interpretation of high-resolution space images of Central Asia, but it will not be before some time. As of today, I can only put forth a hypothesis based on the assumption of a humid ice-age climate of the region. Under humid conditions, the lakes had to expand and their levels rise, so that they would inevitably turn into a single fluvial system discharging into the Amur River. In particular, the ice-dammed lake of the Selenga basin would flow eastward, via the Khilok and Ingoda rivers into the Shilka, while the rest of the Central Asian basin lakes would also flow eastward and northeastward, along the fossil Trans-Gobian superriver into the Djalai Nur and farther east into the Argun River; the latter, together with the Shilka, make up the upper reaches of the Amur.

Again, it should be stressed that all these changes were directly or indirectly coupled with glaciation: indirectly, when lake transgressions were caused by meltwater inflow, and directly, when the mechanism of ice-damming was involved. The example of glacially controlled reorganizations in the river system of the Baikalian region can illustrate this mechanism's operation. As we demonstrated before (Grosswald and Kuhle, 1994), a floating ice shelf, which formed in Baikal during the early stages of glaciation, would plug the lake's exit into the Angara River, thus forcing the lake level to rise by 300–350 m. As a result, the lake water outflow was rerouted into the Lena River, via the Manzurka spillway, which was studied in detail by Logachev et al. (1964). This, as is clear now, was a transient setting, when the glaciers were still growing. On reaching their maximum extent, suggested by modeling experiments based on the assumption of a 1200 m equilibrium-line depression (Fastook and Grosswald, 1998), the ice would completely cover Lake Baikal and impound the Selenga River, thus cutting the Mongolian and Trans-Baikalian meltwater runoff from the Angara–Yenisey system. The composite paleolake of the Selenga and Upper Vitim basins would rise to 850–900 m a.s.l. (Osadchiy, 1995). Until now, geologists tended to associate the lake either with increase in rainfall, or with effects of tectonic faulting and vertical displacement of crustal blocks. Neither seems probable to me. A more realistic mechanism of this lake's formation is ice-damming. It was probably the ice damming that made the lake form, and caused its flow to be rerouted eastward, via the Khilok and Ingoda rivers into the Shilka (Figure 15.4).

In summer seasons the meltwater produced by ablation of Central Asian glaciers would overfill the Trans-Gobian superriver, and force it to flood the Gobi. The alluvial silt blanket deposited would be immediately, i.e. concurrently with its formation, eroded (deflated) and redeposited by winds. Thus, it was probably a major source of the Chinese loesses.

The areal extent of the Gobi–Amuran paleocatchment would amount to about $6.5–7 \times 10^6$ km^2, which, of course, was only the case during maximum stages of glaciation. This extent would definitely shrink in the course of deglaciation. Several reasons can account for that parallel shrinkage. First, the catchment diminished due to lowering or disappearance of ice dams, such as the Baikalian, so that some parts of the catchment would 'return' into the Trans-Siberian drainage system. Second, the catchment shrank when, during the early Late Glacial, the climate returned to its arid mode (Pachur et al., 1995), so that the major Central Asian basins reclosed and lost their fluvial links with the Amur.

The concept of extensive glaciation in Central Asian mountains may also help in resolving the Lake Balkhash–Alakul puzzle, i.e. to shed light on the origin of the giant furrow containing the lakes. The hypothetical ice-dammed lake of Junggar Basin would

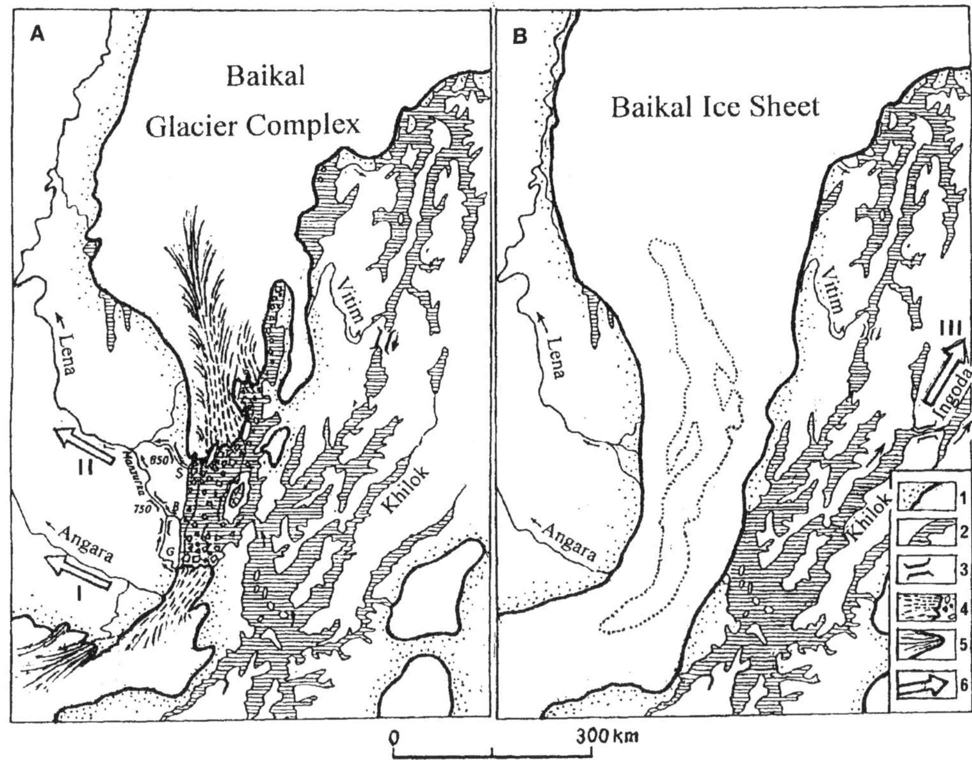

Figure 15.4   Reorganizations of proglacial drainage in the Baikalian region in conjunction with early, transient stages (A) and maximum extent (B) of last glaciation. The Baikalian outlets alternate between the Angara (I), Lena (II), and Amur (III) rivers. (A) After Grosswald and Kuhle (1994); (B) after Osadchiy (1995) for the paleolakes, and Fastook and Grosswald (1998) for the ice dome. Legend: 1, ice sheet and glaciers; 2, proglacial lakes; 3, spillways; 4, floating ice shelf; 5, largest outlet glaciers; 6, main direction of water discharge

burst through its weakening dam, thus the runoff from Junggaria would go, via the Junggarian 'Gate', into eastern Kazakhstan, taking on the form of cataclysmic floodings. In this hypothetical context, the furrow, along with its suite of geomorphological features, would be regarded as a product of scabland erosion.

Another paleolake, the Lena–Vilyuyan, would discharge meltwater through a spillway in the Vilyuy's upper reaches (now concealed under recent sediments and the Vilyuyan Water Reservoir), and further, via the Tunguska River, into the Yenisey. On entering the Yenisei Valley, the meltwater discharged, perhaps cataclysmically, along the margin of the south-facing ice lobe of the valley (Goncharov, 1986) and eroded a suite of giant furrows and related forms resembling scabland.

Of special interest is the evidence for late Pleistocene reorganizations of drainage network in catchments of the rivers flowing into the Sea of Okhotsk (Figure 15.3). There, in valleys of the Amur, Amgun, Togur, and Uda rivers, the deposits of paleolakes occur, while mountain ranges and uplands of the region are on a regular basis breached by canyons and through-valleys. There is evidence for past diversion of the Middle Amur flow southwestward, to the Yellow Sea along the Sungary and Lyaokhe valleys, as well as for deflection of the Uda River toward the Lower Amur (along a chain of paleovalleys and lakes extending parallel to the sea's shoreline). Also, clear geomorphological evidence attests to former diversion of the Lower Amur to the southeast, toward the Tatar Strait (Nikolskaya, 1969).

Until now, these drainage network reorganizations were considered to be linked to recent neotectonic processes. In particular, the paleolakes and river diversions are associated with the impact of uplifting fault blocks, and mountain range breaching with crustal faulting and antecedent river incision (Lebedev, 1995). However, this idea of tectonic control seems (at least to me) inconsistent with the hydrological system displayed in Figure 15.3. The changes in the system would have been inevitably random had they been tectonically controlled, whereas the figure offers a continuous, coherent spatial record of hydrographic changes which fits well into the setting produced by damming effects of the past Okhotsk Sea ice sheet. The margin of the past Okhotan ice sheet, having advanced for 300–350 km southwestward of the present shoreline, would ensure such consequences as the formation of paleolakes in the Uda and Middle Amur river valleys; water overflow across the Tukuringra–Dzhagdy range; deflection of the Amgun River flow into the Bureya River; and a complete turnover of the Lower and Middle Amur flow, resulting in its discharge into the Yellow Sea (along the Sungary–Lyaokhe valley) and/or the Sea of Japan (up the Ussury valley).

As glacial retreat progressed, the lake-and-river geography experienced step-by-step changes. When the ice margin got closer to the shoreline, the geographical setting become favorable for marginal meltwater runoff. A few marginal waterways then provided shortcuts between the Uda River mouth and the Lower Amur. The water from the Amur would first keep draining up the Sungary and Ussury valleys, but later it would be rerouted southeastward, into the Tatar Strait. This stage is documented by conspicuous through-valleys, cutting across the northern Sikhote Alin range. It was along them that the marginal meltwater discharge of the system reached the sea.

Hence, the meltwater discharge of the Gobi–Amuran drainage system alternated between the Yellow Sea and the Sea of Japan. The Sea of Japan, which became isolated from the Pacific due to sea-level lowering and closure of the Tsushima and Soya Straits, was turned into a closed 'pool' topped by a lid of fresh water, implying an enhancement of sea-ice coverage in winters and general cooling of the regional climate (Pletnev, 1985; Oba et al., 1991).

ON THE WATER BALANCE OF THE SYSTEMS

The hydrological balance of the Eurasian paleodrainage systems is virtually unknown. The available data on their constituents are sparse and controversial. To the best of my knowledge, there are no quantitative data on the paleoprecipitation, temperature, evaporation, and runoff discharges for the whole system catchments, while the available data for some parts of the catchments are haphazard and mismatching. As for the qualitative estimates at hand, they vary strongly and often contradict each other.

Trans-Siberian drainage system

It is broadly believed that the ice age climate of Siberia was very dry. For this reason, in a number of reconstructions, the system or its considerable portions are described as closed (e.g. Kvasov, 1979; Astakhov, 1991). The rivers are inferred as having reduced flow, and the levels of the paleo-Caspian 'Sea' and other internal water bodies are interpreted as low (e.g. Kislov and Surkova, 1996). A considerable aridization of the Late and Early Weichselian climate was also implicit in reconstructions by Astakhov (1991) and Arkhipov et al. (1995). Amplitudes of lowering of temperature and evaporation tend to be underestimated. For example, Kislov and Surkova (1996) based their model on the assumption that the cooling and drop in evaporation in the Caspian region equaled 1–4 °C and zero, respectively.

It should be remembered that quite a few geologists in Russia are still rejecting the idea of extensive glaciation of the Arctic. With these 'minimalistic' or 'diluvialistic' concepts in mind, they never acknowledge, let alone evaluate, the parts played in the continent's water balance by meltwater inflow and glacially controlled reorganizations of drainage networks. As for the reasons behind the known, indisputable regime and level variations of internal 'seas', they are mostly interpreted as resulting from mere changes in precipitation–evaporation balance (Kalinin et al., 1966).

Different implications come from paleobotanical and geomorphological evidence. Based on them, the glacial climate of Siberia and the Russian Plain was relatively humid, and some lowering in rainfall was more than compensated for by the drop in evaporation, which may have been up to 50% at the 7–8 °C cooling (Brackenridge, 1978). For example, I based my estimates on the assumption that discharges of the Siberian paleorivers did not differ much from their present discharges, so that their southward deflection would bring about very rapid, geologically instantaneous freshening and a rise in levels of the Aral and Caspian seas (Grosswald, 1983).

Gobi–Amuran drainage system

The ice age climate and paleohydrology of this system is as inadequately known as those of the Trans-Siberian system, though a considerable headstart was taken in Central Asia creditable to the works by Murzayev, Kuznetsov, Devyatkin, and others. According to their findings, confirmed by more recent explorations, the climate of the Central Asian basins, now arid, turned much moister during the past cold intervals. In this change to more mesic mode, the linkage of the Central Asian and Arctic climates becomes palpable, particularly the influence of the Eurasian ice sheet upon the environments of the entire continent. According to modern concepts of paleoclimatology (those of Flohn, Kutzbach, and others), the high-latitude ice sheets would create a circumpolar zone of high atmospheric pressure, which, in turn, would deflect the tracks of Atlantic cyclones by about 20° southward, bringing about increases in snowfall and glacierization of the south Siberian and Central Asian mountains. The mean cooling of Central Asia was large enough (by 8–9 °C) to cause a two-fold diminishing in evaporation, whereas the absolute temperatures were still relatively high, and the climate was not polar, but merely moderately cool (Pachur et al., 1995). These factors, relatively high precipitation rates and temperatures entailing intensive snow and ice melt in summers, are suggestive of water-rich rivers and high-level lakes (Kuznetsov, 1968; Nikolskaya, 1969). It is just that kind of paleoenvironment which is implied by the evidence of palynology, fossil lake shorelines, paleontology, and geochemistry of the basin sediments.

BRIEF DISCUSSION AND CONCLUSIONS

A sketch map (Figure 15.5) offers a rough outline of Eurasian ice sheets coupled with proglacial drainage systems. This reconstruction is tentative; it needs to be verified and substantiated by new evidence. The paleogeography displayed in Figure 15.5 is a working hypothesis; its basic premise is as follows.

The spatial pattern formed by ice sheets and mountain glaciers of northern Eurasia was that of a giant 'Z', of which the upper horizontal was the Arctic marine ice sheets, the lower horizontal was the problematic Tibetan (Qinghai–Xizang) ice sheet, and the connecting diagonal was the glacier complexes of the Siberian and Central Asian mountains. Two major glacier features made their first appearance on this map: the largest composite ice sheet in Pleistocene Eurasia covering land and shelves of the Siberian northeast, and the continuous diagonal ice-and-rock barrier of the Central Asian and Siberian mountains.

The geography of the ice sheets and glaciers was the main factor determining the

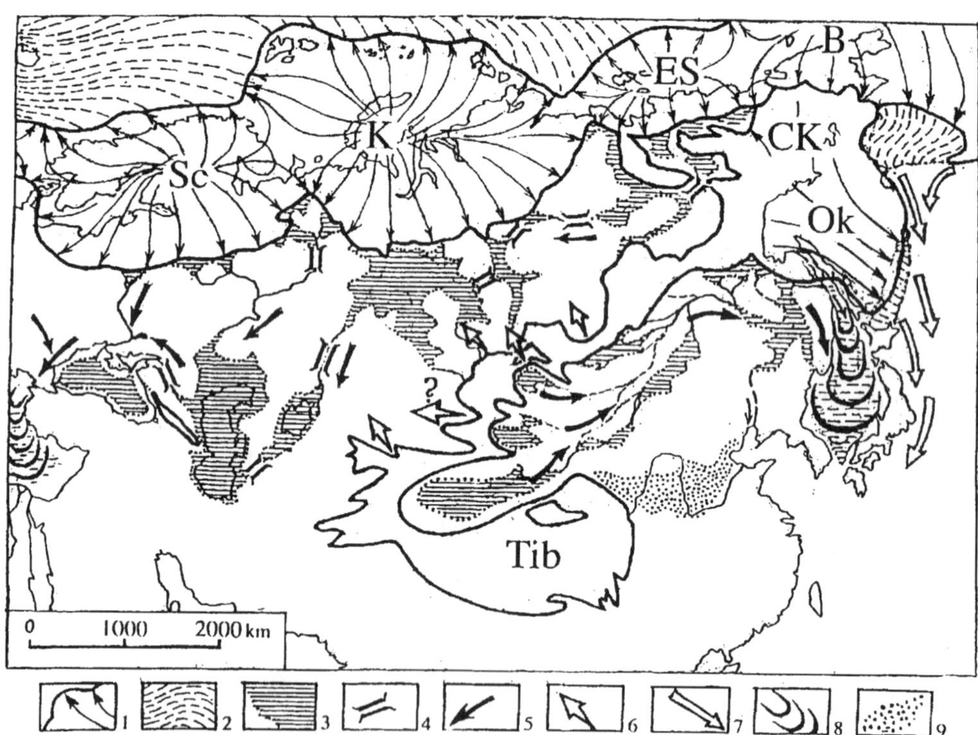

Figure 15.5   Last glaciation and proglacial drainage systems in Northern Eurasia: a tentative reconstruction. Legend: 1, ice sheets and ice flowlines; 2, floating ice shelves and their flowlines; 3, proglacial lakes; 4, major spillways; 5, general directions of meltwater flow in drainage systems; 6, flooding caused by outbursts of ice-dammed lakes; 7, drifting iceberg armadas; 8, freshwater plumes upon the seas; 9, loess-covered terrain. Ice sheets and glacier complexes: Sc, Scandinavian; K, Kara center of the Barents–Karan; ES, East Siberian; B, Chukchi centre of the Beringian; CK, Chersko–Kolyman; Ok, Okhotan; Tib, Tibetan

general plan of the continent's proglacial drainage systems. The Trans-Siberian system, having an areal extent of $23 \times 10^6 \, \mathrm{km}^2$, was situated in high latitudes; its flow was directed to the west, into the Atlantic and its seas. The second drainage system, the Gobi–Amuran, with an extent of nearly $7 \times 10^6 \, \mathrm{km}^2$, occurred south and east of the first, all in temperate latitudes. Its water was flowing eastward, into the Pacific. The mechanisms of ice damming and glacially controlled runoff diversions were involved in reorganizations of the Eurasian paleodrainage on a scale unmatched elsewhere.

The hydrological balance of the drainage systems must be a target for future research. As of today, we can only tentatively suggest, based on geomorphological and paleobotanical evidence, that the systems were relatively water-rich and contained large flow-through lakes. An important role in their balance was played by meltwater inflow, which, among other things, was confirmed by analysis of M. Kunakhovich (unpublished data). Based on the data on areal extent, heights, and equilibrium line altitudes from Figure 15.5 and topographic maps, and allowing for a mean temperature drop of $7\,^{\circ}\mathrm{C}$, Kunakhovich came up with the following meltwater inflows: into the first system, about $1400 \, \mathrm{km}^3$ per year; into the second system, in excess of $700 \, \mathrm{km}^3$ per year.

As for the values for the rest of the balance items, they can be based only on guesswork. One new approach to the problem may be through developing a special numerical model, which would take into account both changes in the circulation of the atmosphere and the distribution of precipitation under specific conditions of ice age Eurasia.

The discharge of the Eurasian meltwater into the Mediterranean and Japan seas brought about sizable changes in the sea temperature, salinity, horizontal and vertical circulation, as well as in the character of biotas and types of sedimentation. Studies of these changes are normally within the realm of oceanography, and they are virtually unknown to 'land-based' paleogeographers. Nevertheless, knowledge of the changes and their linkage to the facts of continental paleohydrology would enrich both the sciences.

The reconstructed paleodrainage systems and their changes can provide new insights into unsolved problems of other Earth and life sciences, such as geomorphology, Quaternary geology, and biogeography. For instance, the proglacial lakes offered refugia in which northern freshwater fish might manage to survive the ice ages. Lakes along with the spillways and paleorivers probably played the role of major avenues permitting the northern salmon, trout, Siberian smelt and some other ichthyospecies, as well as the Baikalian crustaceans such as *Palassea quadrispinosa* and polar seal to migrate to more temperate habitats (Kudersky, 1987; Segerstråle, 1982). Stream erosion, especially scabland-type erosion associated with cataclysmic outbursts of ice-dammed lakes, accounts for specific assemblages of landforms which, until recently, remained inexplicable. Large bodies of water that formed in intermontane basins of Central Asia strongly changed the regional climate, while their evaporation enhanced the snowfalls in glaciated mountains, and the alluvium of the outlets probably was a major source of Asian loess. As a corollary, it may be pointed out that the available data on ice age paleohydrology of northern Eurasia seem to strengthen the case for a continuous Eurasian ice sheet.

ACKNOWLEDGEMENTS

This research was supported by the Institute of Geography, Russian Academy of Sciences. It constitutes a part of our (with Terence J. Hughes, University of Maine, USA) joint project 'EISMAP – Eurasian Ice Sheet Mapping and Prediction'. The paper was presented at the Symposium on Continental Palaeohydrology in Toledo, Spain, in September 1996. I extend my thanks to the organizers and the Logovaz-International Science Foundation for making my participation possible.

REFERENCES

Arkhipov, S.A., Astakhov, V.I., Volkov, I.A., Volkova, V.S. and Panychev, V.A., 1980. *Paleogeografiya Zapadno-Sibirskoi Ravniny v Epokhu Pozdne-Zyryanskogo Lednikovogo Maksimuma* [*Paleogeography of the West Siberian Plain during the Late Zyryan Glacial Maximum*], Nauka, Novosibirsk (in Russian).

Arkhipov, S.A., Ehlers, J., Johnson, R.G. and Wright, H., 1995. Glacial drainage towards the Mediterranean during the Middle and Late Pleistocene. *Boreas*, **24**(3), 196–206.

Astakhov, V.I., 1991. The fluvial history of West Siberia. In L. Starkel, K.J. Gregory and J.B. Thornes (eds), *Temperate Palaeohydrology*, Wiley, Chichester, 381–392.

Baker, V.R., 1996. Megafloods and glaciation. In P. Martini (ed.), *Late Glacial and Postglacial Environmental Changes*, Oxford University Press, New York, 98–108.

Baker, V.R., Benito, G. and Rudoy, A.N., 1993. Paleohydrology of Late Pleistocene superflooding, Altay Mountains, Siberia. *Science*, **259**, 348–350.

Benson, L.V., Burdett, J.W., Kashgarian, M., Lund, S.P., Phillips, F.M. and Rye, R.O., 1996. Climatic and hydrologic oscillations in the Owens Lake Basin and adjacent Sierra Nevada, California. *Science*, **274**, 746–749.

Brackenridge, R., 1978. Evidence for a cold, dry full-glacial climate in the American Southwest. *Quaternary Research*, **9**(1), 22–40.

Devyatkin, E.V., Malayeva, E.M., Murayeva, V.E. and Shelkoplyas, V.N., 1978. Pluvial Pleistocene lakes in the Basin of Great Lakes, West Mongolia. *Izvestiya Akademii Nauk, Seriya geografichskaya*, **5**, 37–51 (in Russian).

Fastook, J.L. and Grosswald, M.G., 1998. Quaternary glaciation of Lake Baikal and adjacent highlands: Modeling experiments. *IPPCCE Newsletter*, **11**, 35–45.

Gerasimov, I.P. and Markov, K.K., 1939. *Chetvertichnaya Geologiya* [*Quaternary Geology*]. Uchpedgiz, Moscow (in Russian).

Goncharov, S.V., 1986. Limit of the last ice sheet in the middle reaches of the Yenisei River: position and age. *Doklady Akademii Nauk SSSR*, **290**(6), 1436–1439.

Grosswald, M.G., 1983. *Pokrovnye Ledniki Kontinentalnykh Shelfov* [*Ice Sheets of the Continental Shelves*], Nauka, Moscow (in Russian).

Grosswald, M.G., 1997. Late Weichselian ice sheets in Arctic and Pacific Siberia. *Quaternary International*, **45/46**, 3–18.

Grosswald, M.G. and Hughes, T.J., 1995. Paleoglaciology's grand unsolved problem. *Journal of Glaciology*, **41**(138), 313–332.

Grosswald, M.G. and Hughes, T.J., 1998. Evidence for Quaternary glaciation of the Sea of Okhotsk. *IPPCCE Newsletter*, **11**, 3–25.

Grosswald, M.G. and Kotlyakov, V.M., 1989. Great proglacial drainage system of Northern Eurasia, and its role in the inter-regional correlations. In *Chetvertichnyi Period. Paleogeogragiya i Litologiya*, Shtiintsa, Kishinev, 5–13 (in Russian).

Grosswald, M.G. and Krass, M.S., 1998. Deglaciation of the Barents-Kara shelf: role of gravitational collapses and surges. *Data of Glaciological Studies*, Moscow, 85, 71–84.

Grosswald, M.G. and Kuhle, M., 1994. Impact of glaciations on Lake Baikal. *IPPCCE Newsletter*, **8**, 48–60.

Grosswald, M.G. and Rudoy, A.N., 1996. Quaternary ice-dammed lakes in the Siberian Mountains. *Izvestiya Akademii Nauk, Seriya Geografichskaya*, **6**, 112–126 (in Russian).

Grosswald, M.G., Kuhle, M. and Fastook, J., 1994. Würm glaciation of Lake Issyk–Kul area, Tian Shan Mts.: a case study in glacial history of Central Asia. *GeoJournal*, **33**(2/3), 273–310.

Hughes, T.J., Denton, G.H. and Grosswald, M.G., 1977. Was there a late-Würm Arctic ice sheet? *Nature*, **266**, 596–602.

Jäkel, D. and Zhenda, Z., 1991. Reports on the 1986 Sino-German Kunlun Shan Taklimakan Expedition. *Die Erde*, Erg.-H., **6**, 1–202.

Kalinin, G.P., Markov, K.K. and Suyetova, I.A., 1966. Variations in levels of the earth's water basins during the recent geological past. *Okeanologiya*, **6**(5), 737–746.

Kislov, A.V. and Surkova, G.V., 1996. Variations of the apparent evaporation from the Kaspian Sea and changes in the sea's level during the Holocene and Late Pleistocene. *Vestnik Moskovskogo Universiteta, Geografiya*, **2**, 75–83.

Kudersky, L.A., 1987. The ways the northern elements of ichthyophauna were formed in the Northern European USSR. In *Problemy Teorii i Praktiki Rybokhozyaistvennoi Nauki [Theoretical and Practical Problems in the Science of Ichthyology and Fishery]*, Promrybvod, Leningrad, 102–121 (in Russian).

Kuznetsov, N.T., 1968. *Vody Tsentralnoy Azii [Waters of Central Asia]*, Nauka, Moscow (in Russian).

Kuznetsov, N.T. and Murzayev, E.M., 1963. Lake-stages in the Quaternary history of Central Asia. In *Ozera Poluaridnoy Zony [Lakes of the Semi-Arid Zone]*, Izdatelstvo Akademii Nauk SSSR, Moscow (in Russian).

Kvasov, D.D., 1979. *The Late Quaternary History of Large Lakes and Inland Seas of Eastern Europe*, Suomalainen Tiedeakatemia, Helsinki.

Lebedev, S.A., 1995. On the influence of tectonic thresholds upon river-network reorganizations within the Lower-Amur basins. *Geomorfologiya*, **1**, 47–51 (in Russian).

Logachev, N.A., Lomonosova, T.K. and Klimanova, V.M., 1964. *Kainozoiskiye Otlozheniya Irkutskogo Amfiteatra [Cenozoic Deposots of the Irkutsk Amphitheater]*. Nauka, Moscow (in Russian).

Marsiat, I., 1994. Simulation of the Northern Hemisphere continental ice sheets over the last glacial–interglacial cycle: experiments with a latitude–longitude vertically integrated ice-sheet model coupled to a zonally averaged climate model. *Palaeoclimates*, **1**(1), 59–98.

Murzayev, E.M., 1949. On the paleogeography of the Northern Gobi. In *Trudy Mongolskoy Ekspeditsii [Proceedings of the Mongolian Expedition]* **38**, 41–62 (in Russian).

Murzayev, E.M., 1966. *Priroda Sintsyana i Formirovaniye pustyn Tsentralnoy Azii [The Nature of Xinjiang, and formation of the Central Asian Deserts]*. Nauka, Moscow (in Russian).

Murzayeva, V.E., Konopliova, V.I., Devyatkin, E.V. and Serebrianny, L.R., 1982. Pluvial environments of the Late Pleistocene in the arid zones of Asia and Africa. *11th INQUA Congress, Abstracts of Papers* 1, 185–186 (in Russian).

Nikolskaya, V.V., 1969. Geomorphology and paleogeography. In V.V. Nikolskaya, and A.S. Khomentovsky (eds), *Yuzhnaya Chast Dalnego Vostoka [Southern Part of Far Eastern Russia]*, Nauka, Moscow, 40–66 (in Russian).

Oba, T., Kato, M., Kitazato, H. *et al.*, 1991. Paleoenvironmental changes in the Japan Sea during the last 85,000 years. *Paleoceanography*, **6**(4), 499–518.

Osadchiy, S.S., 1995. Evidence for a maximum transgression of Lake Baikal. *Geografiya i Prirodnye Resursy*, **1**, 179–189 (in Russian).

Pachur, H.-J., Wünnemann, B. and Zhang, H., 1995. Lake evolution in the Tengger Desert, Northwestern China, during the last 40,000 years. *Quaternary Research*, **44**(2), 171–180.

Pletnev, S.P., 1985. *Stratigrafiya donnykh otlozheniy i paleogeografoya Yaponskogo morya v pozdnechetvertichnoye vremya (po planktonnym foraminiferam) [Deep-sea stratigraphy and paleogeography of the Sea of Japan during the Late Quaternary (as derived from plankton forams)]*. Academy of Sciences Center, Vladivostok (in Russian).

Prest, V.K., 1970. Quaternary geology. In R.J.W. Douglas (ed.), *Geology and Economic Minerals of Canada*, Geological Survey of Canada, Economic Geology Report 1, 675–764.

Ryabkov, N.V., 1975. River systems reorganization in the northeastern Russian Plain in the context of former glaciations of the area. In *Problemy Perestroyki i Perekhvata Rechnykh Dolin [Problems of Reorganization of River Valleys]*, Nauka, Moscow, 59–70 (in Russian).

Segerstråle, S., 1982. The immigration of glacial relicts into Northern Europe in the light of recent geological research. *Fennia*, **160**(2), 303–312.

Selivanov, E.I., 1965. *Geomorfologiya Dzhungariyi [Geomorpology of Junggaria]*. Nedra, Moscow (in Russian).

Starkel, L., 1995. Introduction to global palaeohydrological changes. In K.J. Gregory, L. Starkel and V.R. Baker (eds), *Global Continental Palaeohydrology*, Wiley, Chichester, 1–20.

Teller, J.T., 1987. Proglacial lakes and the southern margin of the Laurntide Ice Sheet. In W.F.

Ruddiman and H.E. Wright Jr (eds), *North America and Adjacent Oceans during the Last Deglaciation*, Geological Society of America, Boulder, Colorado, 39–69.

Teller, J.T., 1995. The impact of large ice sheets on continental palaeohydrology. In K.J. Gregory, L. Starkel and V.R. Baker (eds), *Global Continental Palaeohydrology*, Wiley, Chichester, 109–129.

Thiede, J., 1978. A glacial Mediterranean. *Nature*, **276**, 680–683.

Thunell, R.C., 1979. Eastern Mediterranean Sea during the last glacial maximum: an 18,000-year B.P. reconstruction. *Quaternary Research*, **11**(3), 353–372.

Vardaniants, L.A., 1938. On the past glaciation of the Altai and Caucasus. *Izvestiya Gosudarstvennogo geograficheskogo obschestva*, **70**(3), 386–404.

Verbitsky, M.Y. and Oglesby, R.J., 1992. The effect of atmospheric carbon dioxide concentration on continental glaciation of the Northern Hemisphere. *Journal of Geophysical Research*, **97**(D5), 5895–5909.

CHAPTER 16

Mountain Ice-dammed Lakes of Southern Siberia and their Influence on the Development and Regime of the Intracontinental Runoff Systems of North Asia in the Late Pleistocene

ALEXEI RUDOY

Geography Department, Tomsk State Pedagogical University, Siberia, Russia

INTRODUCTION

For several decades the catastrophic failures of Pleistocene pre-glacial North American lakes Spokane and Missoula (e.g. Bretz, 1923; Pardee, 1942) have been considered unique and characteristic only of the Channeled Scabland on the Columbia basalt plateau and adjoining regions. In monographs and textbooks the plateau, with its complex of geomorphological traces of catastrophic lake failures, has been suggested as a 'special case' of catastrophic transformations against the background of 'normal' evolution of the Earth's surface.

At the beginning of the 1980s the first evidence of systematic outburst floods from giant Pleistocene ice-dammed water bodies of intermontane basins was found in Central Asia, in the Altai mountains (Figure 16.1), in the valleys of the Chuya, the Katun, the Chulishman and the Bashkaus rivers (Rudoy, 1981, 1984; Butvilovskij, 1985). There also appeared studies of Quaternary glaciohydrology of north Mongolian basins (Grosswald and Rudoy, 1996; Grosswald, 1987) and of the depression of Issyk-Kul lake (Grosswald et al., 1994). Calculated and analytical data also showed the possibility of immense Late Pleistocene glacial floods in Zabaikalie and Pribaikalie (Rudoy, 1987; Grosswald and Kuhle, 1994).

The Altai discovery of systematic superfloods invalidates the supposed uniqueness of the cataclysmic outbursts from the widely known Missoula Lake. In turn, this discovery encouraged the author to look at the recent history of ancient glaciers from new, palaeoglaciohydrological positions. It also made the studies of glacial cataclysmic superfloods a powerful geologic factor, the role of which in the creation of modern landscapes has not been practically considered before at all.

Palaeohydrology and Environmental Change. Edited by G. Benito, V. R. Baker and K. J. Gregory
© 1998 John Wiley & Sons Ltd.

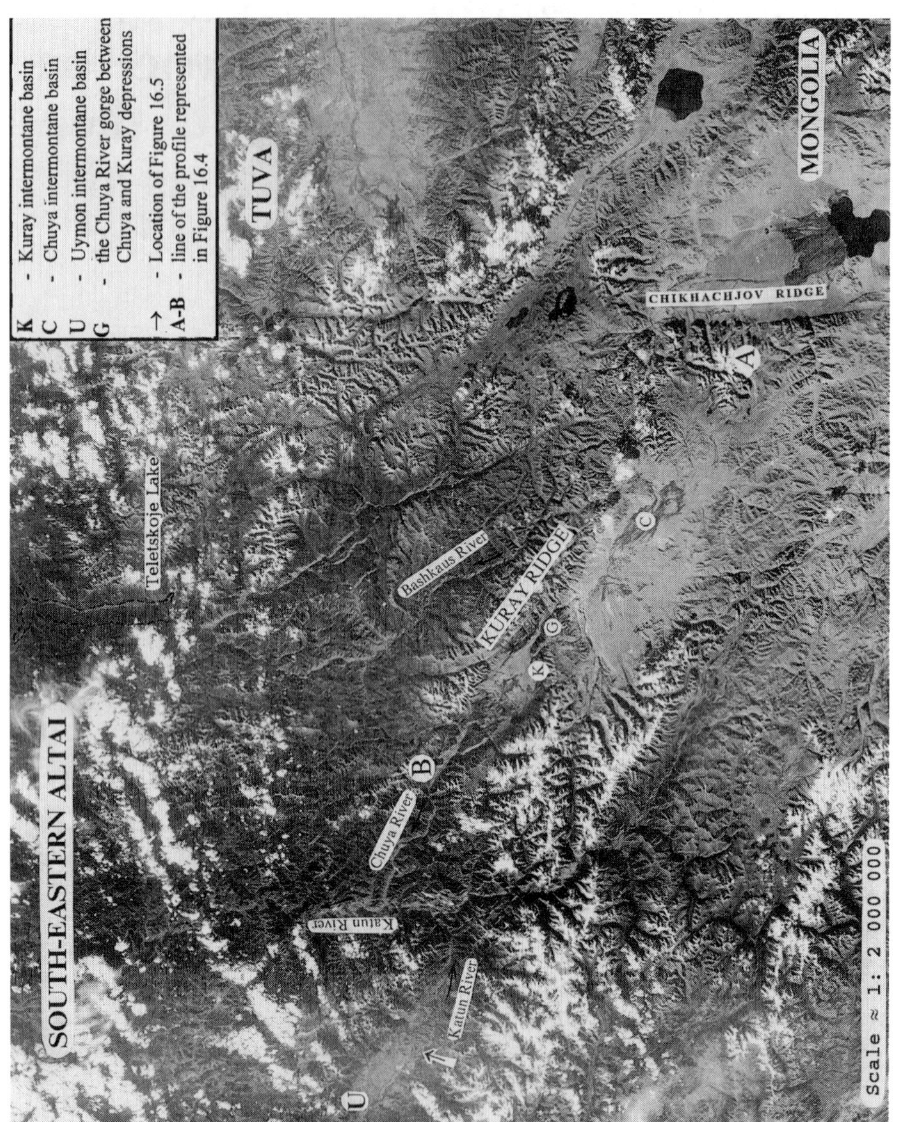

Figure 16.1   Location of the Altai mountains in south Siberia.

FUNDAMENTALS OF THE THEORY OF DILUVIAL MORPHOLITHOGENESIS

The main peculiarity of the regime of ice-dammed lakes is their cataclysmic emptying (jökulhlaups). All jökulhlaups are very rapid and rarely take longer than 10–15 days. Periods of maximum discharges for outburst flooding usually take less than 10% of this time. The magnitudes of water discharges grow sharply and become immense during short intervals at the culmination of the hydrograph peaks. Today, the number of modern ice-dammed lakes amounts to thousands (in the Alps, the Pamirs, the Caucasus, Tien Shan, Kara-Korum, Alaska, Patagonian Andes, Scandinavia, Iceland and elsewhere). Their outbursts lead to large floods, which can result in tragic and possibly irreversible national demographic consequences because of the great number of human victims.

The Pleistocene glaciers of dry land and continental shelves were accompanied by the rise and growth of giant near-glacier water reservoirs whose dimensions were many times greater than modern ones. Discharges of outburst superfloods from these often exceeded 1×10^6 m^3 s^{-1}. Water velocities were dozens of metres per second, and depths of such floods were more than hundreds of metres (Table 16.1). The initial land surface underwent immense transformations during very short time intervals as a result of the geological work of these glacial outburst superfloods.

The widespread occurrence of ice-dammed lakes of different types during glacial periods, their systematic failures caused by unstable ice dams, and the immense consequences of these failures, all led the author to identify a new complex of exogenetic processes – diluvial. Diluvial processes of relief formation are processes of Earth surface transformation by cataclysmic water streams from outbursting ice-dammed lakes. Therefore, these floods are called diluvial (Rudoy, 1987).

Where glaciers dam river valleys and impound lakes, the lakes break the ice dams, and wash away or destroy the glacier traces in the main runoff valleys out of lake basins. Glacial processes are more active at high hypsometric levels, while diluvial processes are active at lower levels, where in most cases their influence exceeds that of the glaciers which gave rise to these processes. This is the essence of the diluvial theory.

The author's concept of the cause–consequence connections of glacial and diluvial processes in the Altai mountains has been tested in all great river valleys of southern Siberia during the last decade. Presently it can be extrapolated theoretically to all regions of the Earth having a similar palaeoglaciohydrological situation for a given time. The theory of diluvial morpholithogenesis affords a hypothesis to consider in regions where the morphostructural appearance is similar to the Altai, and where Quaternary glaciation was able to block the river outflow from intermontane basins. The mountains and uplands of the Baikal region belong to this category. Here, the whole mountainous diluvial morpholithocomplex can be found. Moreover, judging by the size of dammed reservoirs, it is much more grandiose than that in the Altai–Sayany mountain district. On the other hand, the diluvial morpholithogenesis concept lets us solve another problem: to first assume the presence of large pre-glacial lakes at the upper reaches of rivers containing evidence of diluvial floods within these valleys, and consequently to reconstruct the dimensions of the glaciers. Geological ages of diluvial accumulative deposits will be similar both to the age of corresponding ice-dammed lakes and to the age of glaciers blocking them. Such direct and reversed extrapolations are true for lowland regions as well.

Table 16.1   Hydrological characteristics of the Quaternary ice-dammed lakes

Lake or lake system	Area (10^3 km^2)	Hw (m)	Vmax, (km^3)	Hj (m)	Discharge (10^5 m^{3}s^{-1})	Source
Stolper, North Germany		5			0.037	Piotrovskij (1994)
Kan Lakes, Altai	0.26(?)		19(?)		0.10	Rudoy et al. (1989)
Porcupine, Alaska					1.34	Thorson (1989)
Ulagan Lakes, Altai	0.12	300	14		10–20	Butvilovskij (1985)
Abay, Altai	0.32(?)	300	23(?)	230	0.14	Rudoy et al. (1989)
Uymon Lakes, Altai	0.12(?)	200	200	217	1.9	Rudoy et al. (1989)
Yaloman Lakes, Altai	0.017–0.04	760	120	830	2.0	Rudoy et al. (1989) Rudoy (1995b)
Jassater Lakes, Altai	0.6	270	100	300(?)	2.0	Rudoy et al. (1989) Rudoy (1995b)
Darkhat Lakes, Mongolia	2.6	200(?)	250	430	4.0	Grosswald (1987)
Missoula, North America	7.5	635	2514		170	Pardee (1942); O'Connor and Baker (1992)
Chuya-Kuray Lakes, Altai	12	900	3500	> 1000	180	Baker et al. (1993); Rudoy (1981, 1984, 1995a); Rudoy and Baker (1993)

For comparison, the mean annual runoff of the Amazon is of 7000 km^3 (about 15% of the world annual runoff). Hw, depth of the lake near the dam; Vmax, lake volume; Hj, ice dam thickness calculated according to the formula by Nye (1976) and geomorphologically reconstructed

Climatic conditions at definite time intervals (glacial periods, epochs, ice phases and oscillations) conditioned glacier formation and existence. Glaciers dammed river flows that led to the formation of lakes. Ice dams failed when lake depressions became overfilled with meltwater and the release of energy occurred. The quantity of energy was determined by the water mass, valley declivity, geometry and roughness of the outflow channel.

The regime of ablation and rate of intermontane depression filling with meltwater led to a periodicity of about a century for outbursts from giant ice-dammed lakes during the last deglaciation. The repetition of cataclysmic outbursts from the largest pre-glacial basins of the central and southeastern Altai is also geological (Rudoy, 1995a).

Cataclysmic outbursts of basinal ice-dammed lakes of Central Asia produced water streams with discharges of over $1 \times 10^6\,\mathrm{m}^3\,\mathrm{s}^{-1}$. Maximum water discharges reached about $18 \times 10^6\,\mathrm{m}^3\,\mathrm{s}^{-1}$ when the Chuya and Kuray ice-dammed lakes, the greatest in the Altai mountains, failed (Baker et al., 1993). Momentary flood velocities would exceed $40\,\mathrm{m}\,\mathrm{s}^{-1}$, while the depths of a superflood would reach 400 m. These were the greatest known freshwater floods on the Earth.

The high discharges and velocities of cataclysmic glacial superfloods provided their ability to generate enormous amounts of work (Benito, 1997). Systematic failures of giant ice-dammed lakes led to immense transformations of the initial land surface. The latter included the development of deep gorges, channels of water outflow and erosional forms, and accumulation of thick units of loose sediments, including ramparts and terraces, giant current ripples and diluvial berms.

Zones of energetic local vortices and still broader areas of backwater would develop at the expansions of the main runoff valleys (or in the intermontane basins), as well as at the curvatures or bends and other alternations in the plan configuration of river channels (Figure 16.2).

Destructive and accumulative forms developed in such a way as to comprise paragenetic morpholithological associations in preglacial and glacial zones, which interacted with the repeated influence of catastrophic floods from failing ice-dammed lakes (scablands). The process of scabland development was named 'diluvial', and the constructive and destructive work of glacial superfloods performs the essence of the diluvial morpholithogenesis (Rudoy and Baker, 1993).

Ice-dammed lakes are one of the most important components of Quaternary continental nival–glacial systems at the local, regional and, probably, global levels. The rise of such lakes and their existence are conditioned by glaciers. The lake regime is determined mostly by the glacier regime. The margins of Eurasian and North American ice sheets impounded immense amounts of freshwater. As a result, there appeared vast ice-dammed freshwater seas on the pre-glacial landscape (Bretz, 1923; Baker and Bunker, 1985; Arkhipov et al., 1995; Grosswald and Hughes, 1995). In the mountains, meltwater concentrated within huge intermontane basins blocked by glaciers. Ice-dammed lakes in the mountains and on the plains were connected via a network of diluvial runoff channels.

The area of ice-dammed Lake Mansi in eastern Siberia was over $600 \times 10^3\,\mathrm{km}^2$, while the total area of all ice-dammed seas of the lowlands and tablelands of northern Asia was not less than $3 \times 10^6\,\mathrm{km}^2$ (Rudoy, 1995b). Areas of basinal ice-dammed lakes were much less. However, owing to the deeply dissected relief their depths were hundreds of metres. The volume of basinal ice-dammed reservoirs in the Altai exceeded $7.3 \times 10^3\,\mathrm{km}^3$, and their total area was more than $27\,000\,\mathrm{km}^2$ (Rudoy, 1995a). Water streams, with volumes

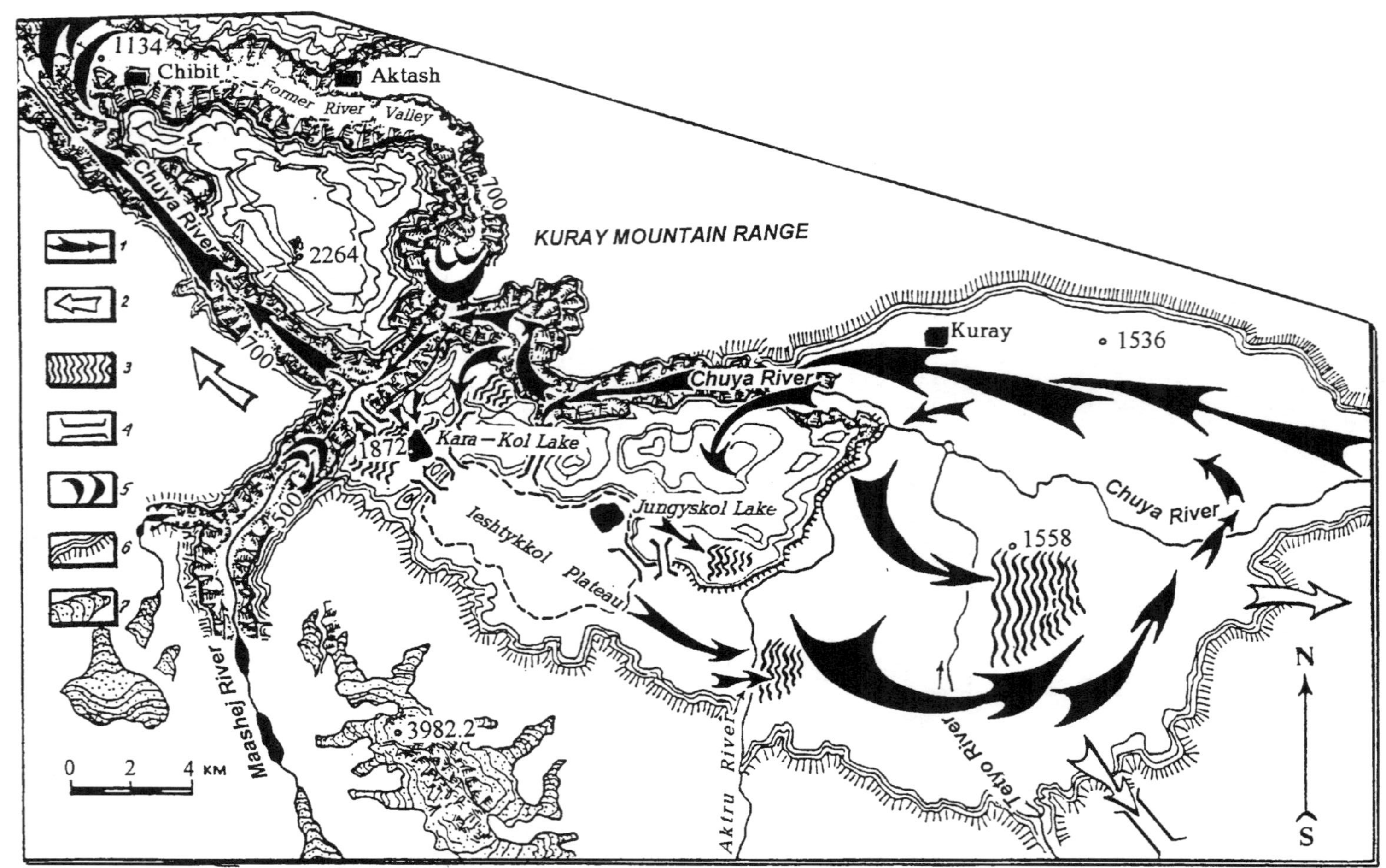

1134
Chibit
Aktash
Former River Valley
Chuya River
700
KURAY MOUNTAIN RANGE
700
2264
Kuray
1536
Chuya River
1
2
3
4
5
6
7
700
500
1872
Kara–Kol Lake
Chuya River
Jungyskol Lake
Ieshtykkol Plateau
1558
Maashej River
3982.2
Aktru River
Tetyo River
N
S
0   2   4 KM

Figure 16.2   Palaeohydrologic sketch of Kuray intermontane basin in the Altai mountains, about 11.5 ka (Rudoy et al., 1989; Rudoy, 1995a,b). Legend: 1, direction of diluvial floods from field evidence; 2, inferred direction of diluvial floods; 3, field of giant current ripples; 4, spillways and gorges crossed by the outbursts; 5, end moraines; 6, boundary of intermontane basins; 7, modern glaciers. The palaeoflood direction has been reconstructed according to the orientation of the fields of diluvial dunes, spillways and gorges. A flood peak discharge of 18×10^6 m^3 s^{-1} was calculated in a site between the villages of Chibit and Kuray. Maximum flood velocities were about 45 m s^{-1}, flood depth during peak discharge exceeded 400 m, and stream power reached 10^6 W m^{-2} (Baker et al., 1993)

of thousands of cubic kilometres, flowed catastrophically to the water surface of the ice-dammed freshwater basins of eastern and western Siberia.

Simultaneous outflow of ice-dammed lakes in the Altai only caused a 12 m increase in the level of Mansi Lake, but this was 4 m more than the lake needed to flow over Turgay spillway (see Chapter 15). Cataclysmic emptyings of the ice-dammed lakes of Darkhat and Khubsugul could not but cause the same effect in the Yenisei Basin. As a result, Kas-Ket spillway came into action, and it set to work the whole grandiose runoff system of ice-dammed freshwater seas of northern Asia (see Chapter 15).

A periodicity of about a century for outbursts from the greatest basinal ice-dammed lakes in southern Siberia led to a similar periodicity of outflows of surface waters several metres deep out of pre-glacial lowland basins into the Atlantic basin. Moreover, the runoff system of Siberian ice-dammed seas itself is inferred to have been extremely dynamic (Arkhipov et al. 1995; Grosswald and Hughes, 1995). Thus, we may infer that the modern hydrographic network of lowland Asia which bore Quaternary glaciation is conditioned mainly by climatic, not tectonic, processes. Morphological associations of lowland scablands must have developed. Lowland scablands, very different from those of the mountains, possess a set of special features. The latter allow us to differentiate their relief from typical erosional–denudational landscapes and from typical accumulative fluvial plains of the morpholithogenesis concept. In this respect, vast territories of Putorana Plateau, Tungus trapps plateau and many other regions of middle and eastern Siberia may have been modified by cataclysmic flooding. Here, the palaeoglaciohyd-rological situation of the last glacial period must have been very aggressive, developing scabland similar to the Channeled Scabland of the Columbia Plateau.

The load of large pre-glacial lakes upon the crust is another problem of Quaternary glaciohydrology. On the one hand, the load of lake water several hundred metres deep could not itself be very influential for any noticeable sagging of local parts of the continental crust. On the other hand, this load appeared and disappeared periodically and very rapidly owing to the filling and cataclysmic emptying of the lakes. The latter made some continental locations rather unstable ('diluvial tremble') and caused inevi-table local earthquakes, fissures and faults within the depressions, plug rockfalls and landslides. From the scientific point of view, it is noteworthy that this periodic 'limnoisos-tasy' may serve as another example of the endogenous component of the mor-pholithogenesis.

Seismic instability of large ice-dammed basins is conditioned by periodical and sharp fluctuations of the load on lake-depression bottoms. It could also have served as a stimulus to ice movements which resulted in new damming of ice outflow, lake depress-ions being again filled with water.

DAM AND OUTFLOW MECHANISMS OF QUATERNARY ICE-DAMMED LAKES IN MOUNTAIN AREAS

Development of ice-dammed lakes

Proving the recurrence of cataclysmic outbursts of giant Quaternary ice-dammed lakes (Baker and Bunker, 1985; Rudoy, 1984, 1988a), modern research focuses mainly on the study of the last stages of pre-glacial lake evolution. Early catastrophies remain almost invisible to Quaternary glaciohydrology, thus we know little of the mechanisms of repeated dammings of lake depressions, their causes, and the dynamic type of the glaciers which would dam intermontane basins.

Most modern ice-dammed lakes which have had jökulhlaups are dammed by pulsating glaciers. Every repeated filling of the basins with meltwater is preceded by a new shift of the dam glacier. If surges do not occur then no lake appears, i.e. there are no outbursts and no diluvial floods occur either. In such a case neither diluvial sediments nor the relief would appear.

Modern analogies show that all ice-dammed Pleistocene lakes of south Siberia (their total area was not less than $100 \times 10^3 \, km^2$ with a total water volume of over $60 \times 10^3 \, km^3$) were dammed by pulsating glaciers. The number of cataclysmic lake outbursts which have been proved geologically or analytically equals the number of shifts of the glaciers which would dam every individual glacier basin (if those emptyings were realized by means of outbursts, sweeping down ice dams). Within the Altai mountains there existed several dozen of large ice-dammed lakes during the last ice age (25–14 ka). Their areas were around $100 \, km^2$ each, and at least five of them had areas of over $1000 \, km^2$ with water volumes of over $300 \, km^3$ (Chuya-Kuray, Uymon, Jassater lake systems and Teletskoje, Bertek and Tarkhat lakes) (Figure 16.3). Those lakes occupied intermontane basins of various morphological types. They are nearly evenly spread over the whole territory of the Altai mountains. This means that pulsation glaciers which might dam those basins were equally characteristic of all altitudinal climatic zones of the Altai (at least during the late ice age), even in cases of lakes that emerged and outburst only once. The pulsating glacier dimensions were one to two orders greater than modern ones, since the dimensions of the lakes dammed by them exceeded the modern ones considerably. The thickness of ice dams was inversely proportional to the number of their outbursts. Consequently, the diluvial flood discharges were directly dependent on the dam thickness.

Repeated outbursts of big pre-glacial plain lakes in North America and North Eurasia may also testify to great surges of outlet glaciers along the periphery of the Quaternary ice covers which would block river valleys. This fact partly confirms previously made conclusions about surges which used to be characteristic of both mountain glaciers and ice sheets.

Ice-dammed lakes and ledoyoms

Many intermontane depressions were formerly fully occupied by glaciers from the surrounding mountains during glaciation maxima and would turn into giant ledoyoms. The Russian term 'ledoyom' means an 'ice body' by analogy with a 'water body'. These ledoyoms would begin to function as independent glaciation centres. Those centres fed huge valley glaciers within river valleys that came out of the ledoyoms (Chuya, Kuray, Ulagan, Julukul and many other intermontane depressions of south Siberia). Calculations of the melt glacier runoff volume at the maximum and the post-maximum of the last glacial have shown that it was much greater than the contemporary runoff. Within the Chuya intermontane depression this volume was to be $8.8 \, km^3$ per year, i.e. it was over 30 times more than the contemporary volume. Thus, the transgressions of the pre-glacial lakes and those of the glacier were synchronized (Rudoy et al., 1989). This means that under a glacier culmination, the intermontane basins must have already been occupied by dammed lakes. The thermoluminescence and ^{14}C dating of lake-glacier varved 'clays'

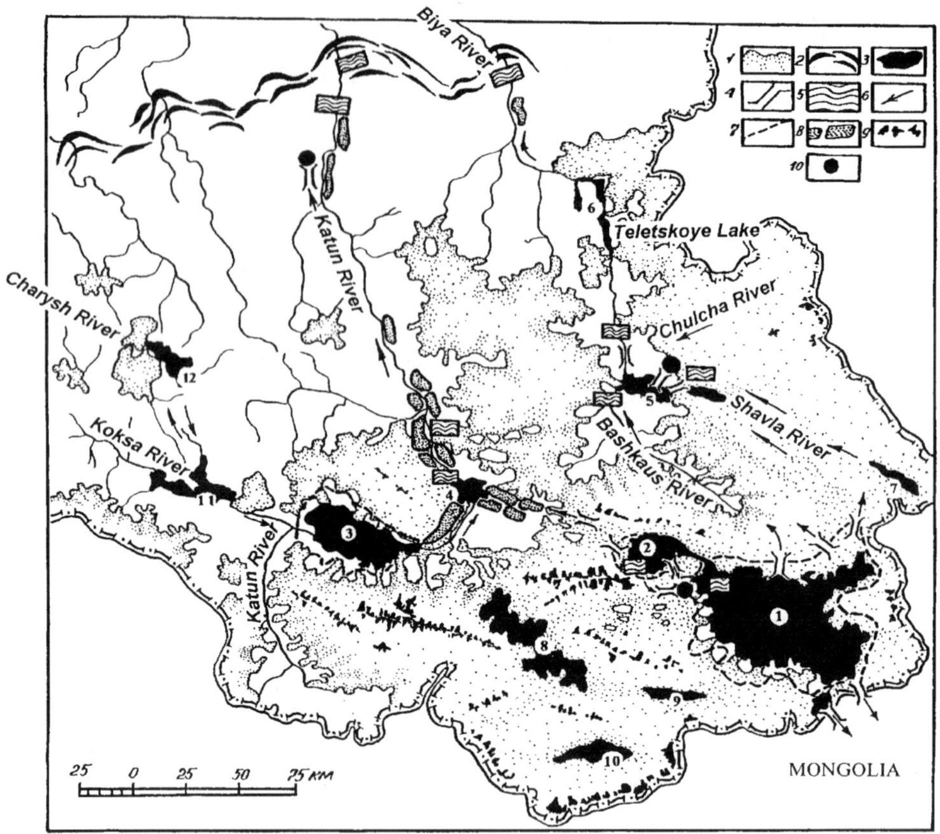

Figure 16.3   Palaeoglaciohydrologic scheme of the Altai, about 14 ka. Legend: 1, boundaries of ice complexes; 2, probable limit of ice spread at the stage of the last glacial maximum; 3, ice-dammed lake; 4, spillway; 5, giant current ripples; 6, direction of diluvial floods; 7, maximum lake boundaries; 8, diluvial terraces and ramparts; 9, modern glaciers; 10, dry waterfalls. The names of the ice-dammed lakes are: 1, Chuya Lake; 2, Kuray Lake; 3, Uymon Lakes; 4, Yaloman Lake; 5, Ulagan Lakes; 6, Teletskoje Lake; 7, Julukul Lake; 8, Jassater Lake; 9, Tarkhat Lakes; 10, Bertek Lake; 11, Abay Lake; 12, Kan Lakes. Lake dimensions are reconstructed according to lake terraces, deposits and glacier marks. Maximum dimensions of ice-dammed lakes are reconstructed according to the spillways locations, outburst valleys and diluvial ramparts. Boundaries of glacial complexes are drawn according to the position of the snow-line which, at 18–20 ka and 14 ka, was 1200 and 800 m, respectively, lower than the present

confirmed the calculations given above: 32–25 ka for the varved clays of the Chagan–Usun River. Along this river there existed a long and narrow bay of the Chagan ice-dammed lake during the interglacial of the Late Würm Ice Age (Rudoy and Kirianova, 1994).

The interrelations of glaciers with ice-dammed lakes had several possible forms in south Siberia, depending on the basin topography, the snow-line depression in the surrounding mountains, and the glaciation energy:

(1) A ledoyom only (without a lake) – the so-called classic ledoyom.
(2) A water body and a ledoyom together (Figure 16.4). This model deals with the case when mountain valley glaciers (Figure 16.4(1), 16.4(3)) went down from the mountains into lake depressions and joined to form 'shelf' glaciers on the surface of a large ice-dammed lake, thereby covering the ice-dammed lakes. There may have been 'catch lakes' (Figure 16.4(2)). When a 'catch lake' outflows water, the glacier ice drops onto the bottom of the lake and leaves certain geological forms of 'dead' ice. It also projects eskers and kames, which have previously developed on the surface of the 'shelf' glacier, onto the lake bottom (Figure 16.5; Rudoy, 1990). The eskers and kames are not laid on the ground moraine, as typically occurs, but on the lake-glacier sediments.
(3) Ledoyoms of 'aufeis' type. This model involves cases when the equilibrium limit-line of the glacier goes lower than the lake surface (Figure 16.6). At the same time there appear complicated forms, which consist of a thick lense of meltwater which is armoured by lake ice, aufeis and glacier ice, and by snow-firn sequences, too. Surfaces of such ice-dammed lakes, consequently, are involved in the zone of glacier accumulation and they turn into independent ice centres with subradial ice outlets. Possible analogies of such an evolution mechanism of pre-glacial lakes and glaciation are thick water lenses under 3–4 km thick units of ice cover at the location of Dome B and Dome C and Vostok Station in Eastern Antarctica (Rudoy, 1995b).
(4) An ice-dammed lake only.

Under different extensions of the glacier at different time periods, one and the same basin underwent different sequences of the lake-glacier events.

Ice-dammed lake outbursts

The outbursts of ice-dammed lakes involve a wide range of possible mechanisms, from slow leakage and thermo-erosion to cataclysmic breakages and outbursts of ice. In the main runoff valleys out of some basins there are moraine fragments of the glaciers that used to dam the lakes. These moraines are associated with the outburst sites at the upper levels of the runoff channels along the outflow routes from depressions. We have studied such forms at the site between Chuya and Kuray basins (Figure 16.4, site G), below the Kuray basin on the slopes of the site of Barotal, in the Katun River valley below the site of Sok-Yaryk and at some other locations. It is very likely that, at the stages of late deglaciation, the outflows of Chuya, Kuray and Uymon ice-dammed lakes in the direction of main valleys would mostly occur via intra- and near-glacial channels and via under-glacial spillways. The lakes would not reach this maximum volume because of the main outflow decrease and diminished thickness of the dams. The dams were never completely destroyed at those stages as, for example, with the failure of Strandline Lake in Alaska in September 1982 (Sturm and Benson, 1985). This lake had a volume of 7×10^8 m^3, and the water outflow velocities were estimated by the authors of the paper as equalling 14 m s^{-1}. After the cataclysmic failure of the lake, which took only 5 h, the runoff channels remained open for about a year longer, after which they closed again. Mathews (1973) describes a cataclysmic outburst of Summit Lake in December 1965. The lake outflowed via an intraglacial tunnel of a regular shape with maximum diameter of 13 m and length of nearly 13 km, the water discharge being 3200 m^3 s^{-1}. After a

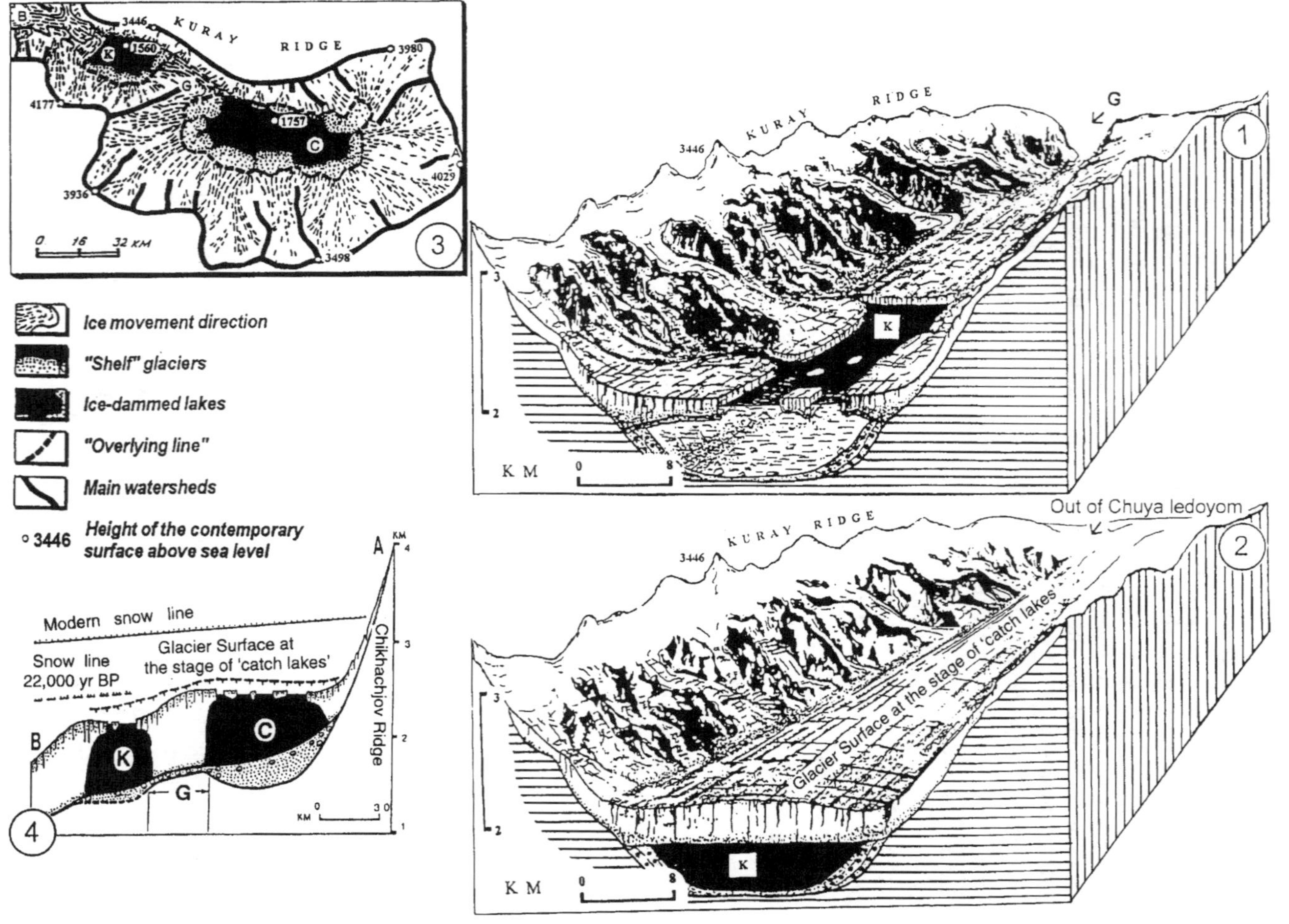
KURAY RIDGE
3446
1560
K
3980
4177
G
1757
C
A
4029
3936
3498
0 16 32 KM
3
Ice movement direction
"Shelf" glaciers
Ice-dammed lakes
"Overlying line"
Main watersheds
3446 Height of the contemporary
surface above sea level
KURAY RIDGE
G
3446
K
3
2
K M 0 8
1
Out of Chuya ledoyom
KURAY RIDGE
3446
Glacier Surface at the stage of 'catch lakes'
3
2
K M 0 8
K
2
Modern snow line
Glacier Surface at
the stage of 'catch lakes'
Snow line
22,000 yr BP
A KM 4
3
B
K
C
2
G
0 3 0
KM 1
Chikhachjov Ridge
4

Figure 16.4 Stages of the 'catch lake' development within intermontane basins of south Siberia about 25–22 ka. K and C, Kuray and Chuya ice-dammed lakes; G, Chuya river gorge between Chuya and Kuray depressions. (1) Transgressions of the ice-dammed lakes within the intermontane basins were synchronous with the ice extension. With a snow-line depression at 800 m, the mountain glaciers reached the level of the ice-dammed lakes within the mountain basins covering the surface (25–22 ka). (2) When the 'shelf' glaciers joined, they flow on the ice-dammed lake surface giving rise to 'catch lakes' (22–20 ka). As the snow-line lowered about 300 m (20–18 ka), independent glacier centres in the intermontane depressions developed. (3) Geomorphological sketch of the Kuray and Chuya ice-dammed lakes. (4) Profile across the Chuya and Kuray basins (see (3) and Figure 16.1 for location)

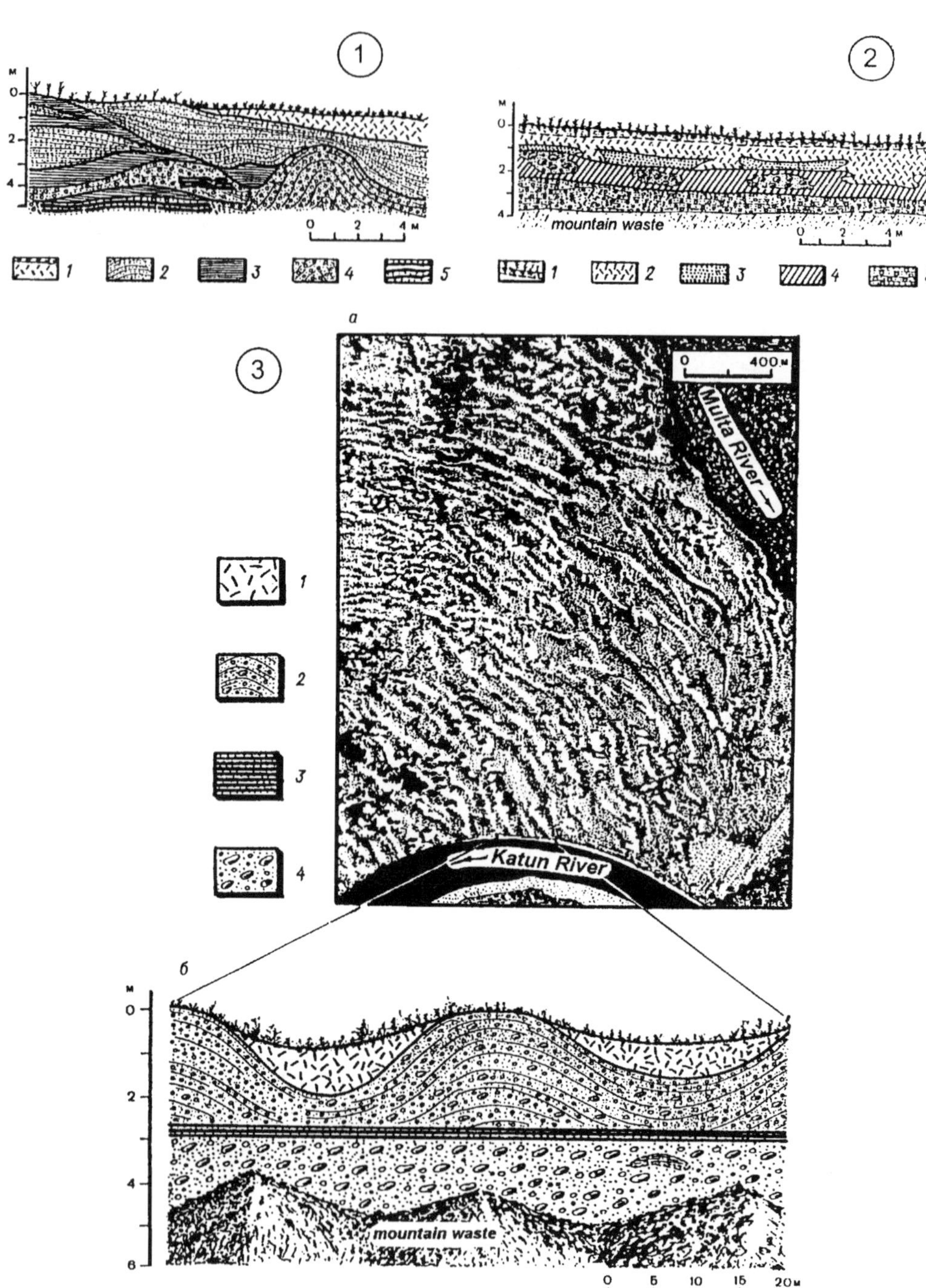
1
2
3
а
0   400 м
Multa River
1
2
3
4
Katun River
б
mountain waste
0   5   10   15   20 м
mountain waste

cataclysmic outburst of Abdukagor Lake in the Pamirs in 1973, immediately below it there remained some fragments of both Medvezhij (Bear) Glacier, which had dammed the lake and front moraines of the Glacier of the Russian Geography Society, which was situated 5 km down the valley. But still further down the stream of the Vanch River, the valley turned out to be washed by glacial floods along practically its full stretch (Rudoy, 1995a).

It is very likely that the mechanism of underglacial outflows of lakes would prevail at the glacier culminations, though the outflows themselves would occur much more rarely. They might be similar to those described by Piotrovskij (1994) for the Late Pleistocene Stolper Lake in northern Germany in the North Sea, and by Brennard and Shaw (1994) for the Late Quaternary glaciers of southern Lake Ontario.

PROBABLE CHRONOLOGY AND DISCUSSION

Observations for modern mountain glacial basins show that alternations in ice runoff volumes are closely connected with the glacier area (e.g. Dikikh, 1993; Panov, 1993). Generally, the total area of glaciation increases with the lowering of the firn-line and the expansion of ice accumulation areas. Consequently, the quantity of icemelt outflow increases. During ice ages glacial discharges grew considerably as compared to initial or modern discharges. This would happen for especially great depressions of the snow-line (in the Altai and Sayan they were over 1200 m). The icemelt outflow increase for these conditions is indicated by the calculations of Rudoy et al. (1989). Therefore, intermontane basins having narrow and deep runoff channels would be rapidly impacted by meltwater concentration within the basins as a response to ice damming. It follows that there were always pre-glacial lakes in the mountains and on the plains when glaciers grew large enough to block river valleys. This important theoretical premise allows the establishment of a chronology of lake-glacial events despite the limited set of reliable absolute dates. This was accomplished by proceeding from the speed of filling of intermontane basins with meltwater and from considering the hydraulics of diluvial floods (Rudoy and Baker, 1993).

The absolute ages of the lake outbursts are uncertain, as are the duration of the diluvial form development and the period of multigenetic sedimentation. The outburst occurred over minutes, hours and days, but diluvial forms and sediments required at least many tens or hundreds of years to develop. This is why the period of accumulation for the sediments of several floods may engender some error in defining the absolute age. The dating of the sediments that belong to the intermediate events between floods is similarly problematic.

Basin filling with meltwater at least up to the levels interpreted as lake terraces (2200 m) took about 100 years for the Chuya basin and about 30 years for the Kuray basin. Further

---

Figure 16.5 Marginal ice formations in Uymon intermontane basins in the Central Altai (see Figure 16.1 for location). (1) esker exposures in the central part of the basin. Legend: 1, brown loess; 2, coarse and fine-grained grey sand; 3, brown sand layers, occasionally in lenses; 4, gravel and pebbly gravel beds, the bedding has a 'dome-shaped' appearance; 5, coarse-grained sands horizontally bedded (lake origin). (2) Esker profile at Nizhniy Uymon. Legend: 1, sod; 2, brown loess; 3, dark-grey coarse-grained sand; 4, loess-type loamy sand with moraine interlayers and lenses; 5, clearly bedded lake gravel and pebbly gravel. (3) Marginal eskers of the rim of the glacier tongue which were developed at the contact of the glacier and the ice-dammed lake. Legend: 1, brown loess; 2, gravel and pebble gravel with scarce boulders, bed planes follow the topography; 3, dark brown loamy soil; 4, moraine

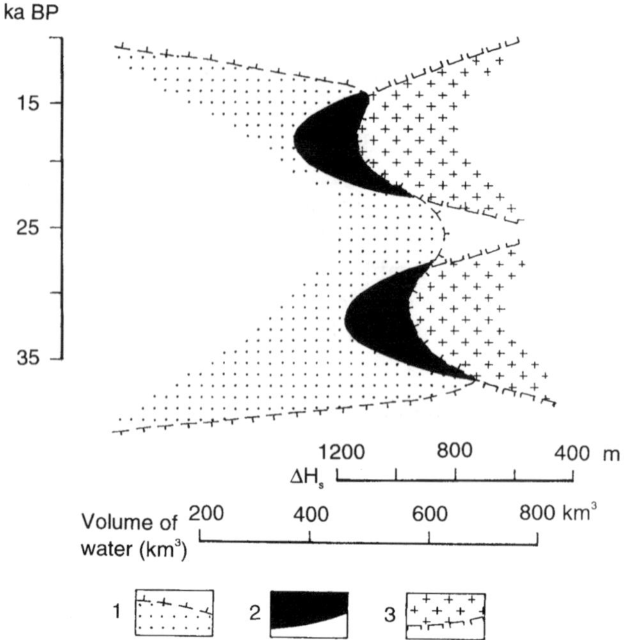

Figure 16.6   Diagram representing the formation mechanism of an 'aufeis'-type ledoyom within ice-dammed lakes of the intermontane basins of south Siberia (the Chuya intermontane basin was taken as an example). Legend: 1, water-body stage; 2, 'aufeis'-type ledoyom ('aufeis'-type ice body) stage; 3, snow-line depression ΔHs

filling of the basins with water ceased because the lakes emptied and/or because the snow-line lowered to the level of water bodies and below.

Maximum lake transgressions, i.e. the maximum storage of meltwater, was synchronous with the glacier maxima. However, at that time the ice dams were so thick that it is very unlikely the lakes could generate the outbursts with periodicity of about a century, mentioned above. Regular water outflows from ice-dammed lakes must have occurred during early and late glacial epochs when ice dams were unstable. At the Quaternary glacial maxima the snow-line depression would cause the transformation of the lakes into 'aufeis' ledoyoms. At these times the lakes were buried under lake ice, aufeis and glacier ice, and they could not break up for thousands of years.

Reconstructions of the Kuray and Chuya ice-dammed lakes are based on their preserved lake terraces. It is obvious that the ancient shorelines cannot establish the greatest water storage in the basins, since maximum water volumes coincided with maximum glacier dimensions, including periods of 'aufeis' ice bodies. At these stages foothill and 'shelf' glaciers from the valley tributaries to the basins terminated at the shoreline of the water bodies. These were lakes within ice walls which left no trace when their walls melted during glacial degradation.

We can speculate as to the number of 'super-outbursts' of the Chuya–Kuray lake system which had discharges in excess of 1×10^6 m³ s⁻¹. Such maximum discharges occurred for cases of total water outflow from the lakes. Examples of modern pre-glacial lakes show that complete emptying of pre-glacial lakes is less common than partial water outflow for the various mechanisms of ice-dam deformation (e.g. Post and Mayo, 1971; Nye, 1976). Thus, the most likely scenario is that of rare, individual (possibly, a single) phenomenal jökulhlaups with discharges of over 15–$20 \times 10^6\,\mathrm{m}^3\,\mathrm{s}^{-1}$, against a background of systematic outbursts of ice-dammed lakes, with a periodicity of about a century, and discharges less than $1 \times 10^6\,\mathrm{m}^3\,\mathrm{s}^{-1}$.

The period of lake-depression filling was estimated as 100–130 years for the Late Pleistocene ice-dammed Lake Darkhat in north Mongolia (Grosswald, 1987). These calculations were based on the correlation of precipitation and evaporation with annual runoff in the basin of the Darkhat depression at the Würm glacial maximum. The time of filling of 100–130 years gives a periodicity of about a century to the repeated cataclysmic emptying of Lake Darkhat into the Yenisei basin. We can probably consider such a frequency of cataclysmic emptying a characteristic feature of large ice-dammed lakes of Central Asia, if we assume similar climatic conditions. The task now is to find some trustworthy geological proof of this hypothesis. Such proof might be provided by diluvium in the flood terraces of the mountains. Geological exposures of the diluvial piedmont cone in the southwest or central Siberia might yield evidence of repetitious flood sedimentary layers reflecting this periodicity.

Bearing in mind the synchronism of diluvial processes with initial and final ice stages, one should also comprehend that the scabland structure observed nowadays is nevertheless conditioned mainly by the work of the last superfloods from the last ice-dammed lakes broken up during the degradation period of the last glaciation.

There are several trustworthy radiocarbon dates from various parts of the Altai. These show that the last phenomenal cataclysmic failure of the Chuya–Kuray system of ice-dammed lakes left giant current ripples in the Kuray and Yaloman basins, and at the site of Platovo–Podgornoje, in the Altai foothills. These were emplaced not later than 13 ka, based on two facts. The age of vegetation remnants from lake loamy soils in frost mounds of the plateau of Yeshtikkol in the western part of the Kuray basin is $10\,845 \pm 80$ [14]C yr BP (CO AH-2346), according to V. A. Panitshev. Thus, there was no water in that part of the Kuray basin later than approximately 11 ka. Marlstones and vegetation remnants within the sediments of the North Altai yield dates of $13\,890 \pm 200$ and $12\,750 \pm 65$ [14]C yr BP (CO AH-779). Maloletko (1980) associates these dates with the time of snow-line depression and the carrying of erratic boulders over the foothills. Thus, Maloletko concludes that the cataclysmic flooding took place in the Katun River valley at 13 ka. After that geological date, the lakes would degrade simultaneously with the degradation of the glaciers which fed them. This does not, of course, exclude their outbursts, but the hydraulics of those failures could not be especially large.

Rudoy discovered vegetation remnants in the exposure of a frost mound (pingo) in the central part of Chuya intermontane basin. V. A. Panitshev and L. A. Orlova dated these at 3810 ± 105 ^{14}C yr BP (CO AH-2146). There are reasons to think that the frost mounds themselves developed even later, about 2140 years ago (Rudoy, 1988b). Consequently, there was no lake in the basin by that time. This evidence contributes to the conclusion that ice-dammed lakes of the intermontane basins in Siberia disappeared completely later than 5 ka. The once vast water bodies split into a number of small lakes, the remants of which are preserved today.

CONCLUSIONS

1. Ice-dammed lakes in the mountains and on plains developed whenever glaciers grew large enough to block river valleys.
2. Discovery of the systematic cataclysmic glacial superfloods which formed the Central Asian scabland modifies the 60-year-old idea of uniqueness for failures of glacial Lake Missoula in North America. Numerous outbursts of ice-dammed lakes in intermontane basins of southern Siberia and western North America mean that diluvial processes can now be viewed as normal exogenous processes of relief formation which produce geomorphic work under certain oroclimatic conditions.
3. The diluvial morpholithogenesis theory is based on the proposition of a connection between glacial and diluvial processes. The latter can occur under palaeoglaciohydrological situations similar to those studied in any region of the Earth.
4. Diluvial processes of relief formation are among the most powerful known erosional Pleistocene processes. Diluvial processes generated by glaciers transformed the glacial and pre-glacial landscapes of many regions and created characteristic plains and mountain scablands.
5. The 1990s is the decade of natural disaster reduction (UNESCO). In this regard, it is desirable to consider the role of short-term, but powerful and systematically repeated, processes in estimating the degree of natural risk. Such processes may create forms that persist for thousands of years. The consequences of these processes are the more tragic the more unexpected their recurrence. Studies of past recurrences will remove this unexpected quality.

ACKNOWLEDGEMENTS

I thank Marina Kirianova for participating in all recent high-mountain Altai expeditions and for active discussion. I am grateful to Elena Hubert, who translated materials into English and shared the hardships of joint expeditions of 1993–1995 to the mountains of Siberia. I also thank the editors who helped improve the clarity of the text. This research was supported by the Russian Foundation of Fundamental Research, Grant No. 9705-65878.

REFERENCES

Arkhipov, S.A., Ehlers, J., Johnson, G. and Wright, H.E., 1995. Glacial drainage towards the Mediterranean during the Middle and Late Pleistocene. *Boreas*, **24**, 196–206.
Baker, V.R. and Bunker, R.C., 1985. Cataclysmic Late Pleistocene flooding from Glacial Lake Missoula: A Review. *Quaternary Science Reviews*, **4**, 1–41.
Baker, V.R., Benito, G. and Rudoy, A., 1993. Palaeohydrology of Late Pleistocene superflooding, Altay Mountains, Siberia. *Science*, **259**, 348–350.
Benito, G. 1997. Energy expenditure and geomorphic work of the cataclysmic Missoula flooding in the Columbia River gorge, USA. *Earth Surface Processes and Landforms*, **22**, 457–472.
Brennard, T.A. and Shaw, J., 1994. Tunnel channels and associated landforms, south-central Ontario: their implications for ice-sheet hydrology. *Canadian Journal of Earth Sciences*, **31**(3), 505–521.
Bretz, J.H., 1923. The Channeled Scabland of the Columbia Plateau. *Journal of Geology*, **31**, 617–649.
Butvilovskij, V.V., 1985. Katastrophicheskie sbrosi lednikivo-podprudnikh osyor Yugo-Vostoch-

nogo Altaia i ikh sledi v reliefe [Cataclysmic failures of ice-dammed lakes in the South-East Altai and their traces in the relief]. *Geomorphologia*, **2**, 65–74 (in Russian).

Dikikh, A.N., 1993. Lednikovij stok rek Tian-Shania i yego rol v formirovanii obshtshego stoka [Glacial outflow of the rivers in the Tian-Shan and its role in the general outflow formation]. *Data of Glacial Research Academy of Science – USSR*, **77**, 41–50 (in Russian).

Grosswald, M.G., 1987. Poslednee oledenenie Saiano-Tuvinskogo nagoria: morphologia, intensivnost pitania, podprudnyie osyora [Late glacial of SayaniTuva highland: morphology, tensity of alimentation, ice-dammed lakes]. In V.M. Kotliakov and M.G. Grosswald (eds), *Vzaimodeistviye oledeneniya s atmospheroi i okeanom [Interaction of the glaciation with the atmosphere and the ocean]*. Moscow, Academy of Science USSR, 152–170 (in Russian).

Grosswald, M.G. and Hughes, T.J., 1995. Paleoglaciology's grand unsolved problem. *Journal of Glaciology*, **41**(138), 313–332.

Grosswald, M.G. and Kuhle, M., 1994. Impact of glaciations on Lake Baikal. *Newsletter*, **8**, 48–60.

Grosswald, M.G. and Rudoy, A.N., 1996. Chetvertichniye lednikovapodprudniye osyora v gorakh sibiri [Quaternary Ice-Damned Lakes in Siberian Mountains]. *Proceedings of Russian Academy of Science. Seriya Geografichskaya*, **6**, 112–126 (in Russian).

Grosswald, M.G., Kuhle, M. and Fastook, J.L., 1994. Würm glaciations of Lake Issyk-Kul Area, Tian-Shan Mts.: A case study in glacial history of Central Asia. *Geo Journal*, **33** (2–3), 273–310.

Maloletko, A.M., 1980. O proishozhdenii Maiminskogo vala [About the origin of Maima terrace (Altay)]. *Questions of Geography of Siberia, Tomsk*, **13**, 89–92 (in Russian).

Mathews, W.H., 1973. Record of two jokulhlaups. *Symposium on the Hydrology of Glaciers*, Cambridge, 7–13 Sept. 1969, 99–110.

Nye, J.F., 1976. Water flow in glaciers: jokulhlaups, tunnels and veins. *Journal of Glaciology*, **17**, 181–207.

O'Connor, J.E. and Baker, V.R., 1992. Magnitudes and implications of peak discharges from glacial Lake Missoula. *Geological Society of America Bulletin*, **104**, 267–279.

Panov, V.D., 1993. *Evolutsia sovremennogo oledeneniya Kavkaza* [Evolution of the modern glaciation of the Caucasus], Hydrometeoizdat, St Petersburg (in Russian).

Pardee, J.T., 1942. Unusual currents in glacial Lake Missoula, Montana. *Geological Society of America Bulletin*, **53**, 1569–1600.

Piotrovskij, J.A., 1994. Tunnel-valley formation in north-west Germany geology, mechanisms of formation and subglacial bed conditions for the Bornhoven tunnel valley. *Sedimentary Geology*, **89**, 107–141.

Post, A. and Mayo, L.R., 1971. Glacier dammed lakes and outburst floods in Alaska. *Hydrological Investigations, Atlas HA-455*, US Geological Survey.

Rudoy, A.N., 1981. Kistorii prilednikovykh osyor Chuyskuy kotloviny, Gorniy Altai [To the history of pre-glacial lakes of Chuya basin, Altai Mountains]. *Data of Glacial Research, Academy of Science USSR*, **41**, 213–218 (in Russian).

Rudoy, A.N., 1984. Gigantskaia riab tetseniya – dokazatelstvo katastroficheskikh proryvov gliatsialnykh osyor Gornogo Altaia [Giant current ripples: proof of catastrophic breaches of glacial lakes in the Altay Mountain]. In *Contemporary Geomorphological Processes in Altay, Bysk*, 60–64 (in Russian).

Rudoy, A.N., 1987. Reliefoobrazuyushtshaia rol chetvertitsnykh lednikovo-podprudnykh osyor mezhgornykh kotlovin [The relief-formating role of Quaternary ice-dammed lakes in intermountain basins]. In N A. Logatchov (ed.), *Protsessy formirovaniya -reliefa Sibiri [Processes of the relief formation in Siberia]*, Nauka, Novosibirsk (in Russian).

Rudoy, A.N., 1988a. Rezhim lednikovo-podprudnykh osyor mezhgornykh kotlovin Yuzhnoi Sibiri [The regime of the intermontane ice-dammed lakes of South Siberia], *Data of Glacial Research, Academy of Science USSR*, **61**, 36–44 (in Russian).

Rudoy, A.N., 1988b. O vozhrastie tebelerov i vremeni okonchatelnogo ischeznovenia lednikovopodprudnykh osyor na Altaie [About pingo (tebelers) age and the time of Pleistocene ice-dammed lakes final disapperance in the Altay]. *All-Union Geographical Society Procedings*, **120** (4), 344–348 (in Russian).

Rudoy, A.N., 1990. Ledoyomy i lednikovo-podprudnyie osyora Altay v Pleistocene [Ledoyoms and ice-dammed lakes of the Altai in the Pleistocene]. *All-Union Geographical Society Proceedings*, **122**(1), 43–52 (in Russian).

Rudoy, A.N., 1995a. Geomorphologicheskiy effect i gidravlika pozhdnepleistocenovikh jokul-hlaupov lednikovo-podprudnikh osyor Altaia. [Geomorphological effect and hydraulics of the Late Pleistocene jokulhlaups of ice-dammed lakes in the Altai]. *Geomorphologia*, **4**, 61–76 (in Russian).

Rudoy, A.N., 1995b. *Chetvertichnaia glaciogidrologia gor Tsentralnoi Asii [Quaternary glaciohydrology of Central Asia Mountains]*, Doctorial thesis, Institute of Geography of Russian Academy of Science, Moscow (in Russian).

Rudoy, A.N. and Baker, V.R., 1993. Sedimentary effects of cataclysmic Late Pleistocene glacial outburst flooding, Altay Mountains, Siberia. *Sedimentary Geology*, **85**, 53–62.

Rudoy, A.N. and Kirianova, M.R., 1994. Ozyorno-lednikovaya podprudnaya formatsia i chetvertichnaya paleogeographia Altaia [Lacustrine and glacial dam formation and Quaternary palaeogeography in the Altai]. *Russia Geographical Society Proceedings*, **126**(6), 62–71 (in Russian).

Rudoy, A.N., Galakhov, V.P. and Danilin, A.L., 1989. Rekonstruktsiya lednikovogo stoka verhney Chuyi i pitanie lednikovo-podprudnykh osyor v pozdnem pleistocene [Reconstruction of glacial drainage of the Upper Chuya and the feeding of ice-dammed lakes in the Late Pleistocene]. *All-Union Geographical Society Proceedings*, **121**(3), 236–244 (in Russian).

Sturm, M. and Benson, C.A., 1985. History of jokulhlaups from Strandline Lake, Alaska, USA. *Journal of Glaciology*, **31**, 272–280.

Thorson, R.M., 1989. Late Quaternary paleofloods along the Porcupine River, Alaska: Implications for regional correlation. *US Geological Survey Circular*, **1026**, 51–54.

CHAPTER 17

Simulation of the Caspian Sea Level Changes During the Last 20 000 Years

A.V. KISLOV

Faculty of Geography, Moscow State University, Russia

AND G.V. SURKOVA

Institute of Geography, Russian Academy of Sciences, Moscow, Russia

INTRODUCTION

The most direct evidence for palaeohydrological changes comes from reconstruction of lake-level changes. The lake area (and its level) can be considered to tend to equilibrium with the changing climate and therefore to be an indicator of past climate. However, often there is a lack of explicit synchronization between global forcing factors and lake-level changes. Large lakes filter out short-term variability and respond to longer-term variations in the hydrological cycle. The means by which large lakes quasi-periodically fluctuate remains largely a mystery. Therefore the history of large lakes has been studied for decades, yet much remains unexplained.

The goal of this chapter is to evaluate whether hydrological anomalies in the Caspian Sea area and its catchment caused by climate change are sufficient to explain Caspian Sea level changes during the past 20 ka.

The Caspian Sea is a vast inland lake. It is situated in the western part of Central Asia. The sea is fed by several rivers, the greatest contribution (more than 80% from common volume of the runoff) being produced by the Volga River. The catchment area is spread out over most of the Russian Plain.

Figure 17.1 shows the variation of Caspian Sea level for different time scales (during part of the Quaternary, Holocene, the last 2 ka and current changes, respectively; the present-day Caspian Sea level altitude is -28 m). In this chapter we concentrate on the Late Quaternary period. We describe early attempts to simulate the connections between global climate variations and regional-scale water budget changes which were reflected by Caspian Sea level changes.

To solve this problem we use the results of a series of computer modelling experiments which have been carried out using a limited area model (LAM) nested in the climate general circulation model (GCM). The LAM has been applied over the Caspian Sea and its catchment.

Palaeohydrology and Environmental Change. Edited by G. Benito, V. R. Baker and K. J. Gregory
© 1998 John Wiley & Sons Ltd.

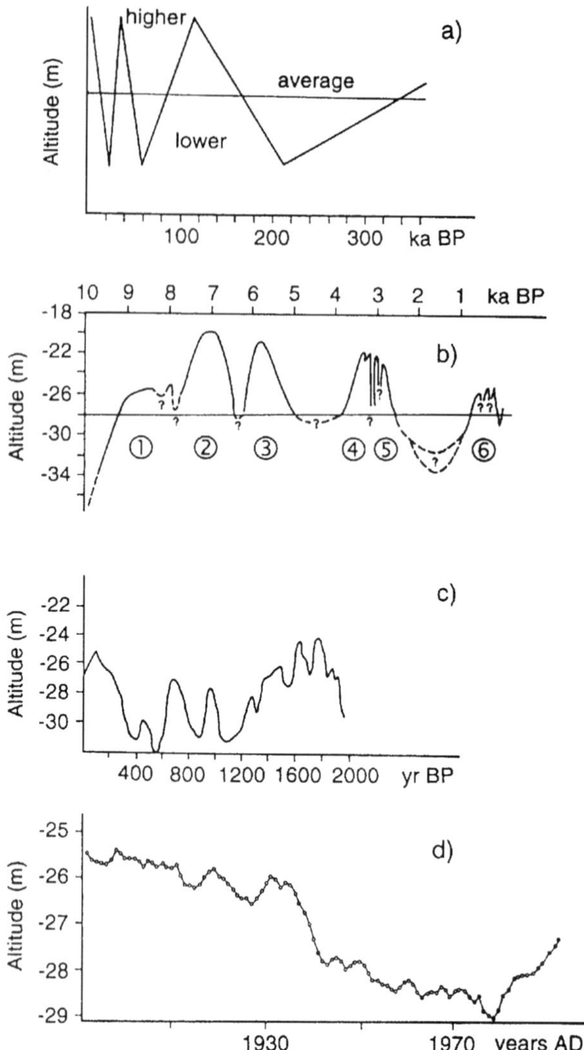

Figure 17.1   Caspian Sea level variation at different time scales. (a) Schematic Caspian Sea level
changes during last 350 ka. (b) Holocene Caspian Sea level changes; numerals in circles mark time
of so-called New Caspian Transgression stages (after Richagov, 1993). (c) Caspian Sea level changes
during the last 2000 years (after Varuschenko et al., 1987). (d) Modern Caspian Sea level changes. (b)
Reproduced by permission of Professor G.I. Richagov (Moscow State University). (c) Reproduced
by permission of Professor R.K. Klige (Moscow State University)

It is assumed that global climate is an approximate equilibrium which is uniquely
determined by the state of the external factors and boundary conditions. The steady-state
response is simulated by the model. This approach gives a 'snapshot' of a specific time in
the past. It is expected that the Caspian Sea surface area tends towards an equilibrium size
such that runoff (Y) is balanced by the excess of evaporation (E) over precipitation (P)
from the lake surface. In this situation the equilibrium modelling experiments with

climate model are therefore relevant to solve our hydrological task. The steady-state equation of the annual water budget for a closed lake is:

$$ef = YF \qquad (17.1)$$

where F is the area of the catchment, f is the area of the lake surface, $e = E - P$, and Y is the runoff from the catchment into the lake.

Variation of the lake area relative to the present may be expressed in the form:

$$\Delta f = F\Delta Y/e_0 + Y\Delta F/e_0 - f\Delta e/e_0 \qquad (17.2)$$

where e_0 represents the present-day condition.

Lake-level change (Δh) is described in the following form:

$$\Delta h = \phi(\Delta f) \qquad (17.3)$$

where $\phi(\Delta f)$ is the well-known empirical function which is determined by the morphology of the lake (Richagov, 1993). The symbol Δ in Equations 17.2 and 17.3 can be interpreted as the deviation from present-day values. In this situation $e_0 = E_0 - P_0$ is available from model control run for the present-day conditions and $e = E - P$ is determined from model simulation of the 'snapshot' of the past which is of interest to us. This approach allows us to evaluate the contribution of the different factors connected with variation of e, F and Y to change of the level as:

$$\Delta h = (\Delta h)_Y + (\Delta h)_F + (\Delta h)_e \qquad (17.4)$$

Assuming, based on reconstruction, that there was no change in the catchment configuration during the last 20 ka, we constrain $(\Delta h)_F = 0$. And, finally:

$$\Delta h = (\Delta h)_Y + (\Delta h)_e \qquad (17.5)$$

In the next section we describe our global and regional model as well as the method by which they are coupled. Finally, the simulation results together with observed data are presented.

MODEL

The computer modelling experiments were carried out using the simplified general circulation model (SGCM) and a LAM nested in the SGCM. These models are based on the Moscow State University (MSU10x15L3) climate model (Kislov, 1991). The LAM consists of the planetary boundary layer and surface including two layers of the vegetation and three soil layers (so-called 'LAMBLS'). The SGCM and LAMBLS has allowed investigation of the climate sensitivity to changes of the insolation distribution at the top of the atmosphere and to changes of the surface characteristics in the past. We will describe a state of stable equilibrium obtained from the time integration of the climate model.

The finite difference mesh of the SGCM has a spacing between grid points of $10°$ latitude and $15°$ longitude. It has three levels in the vertical direction. The SGCM has a daily time step. Our nesting techniques are used for the planetary boundary layer over the limited area only. It touches upon the three-dimensional exchange of heat and water, the calculation of precipitation, solar and terrestrial radiation etc., with the simple nested modification of the wind speed. The LAMBLS has a one hour time step. The finite difference mesh of the LAMBLS has a spacing between grid points of 2° latitude and 2° longitude. The LAMBLS horizontal grid point spacing may be chosen freely; it is limited only by the lack of detailed information about prescribed parameters of the surface. The interactive exchange of information between the SGCM and LAMBLS exists on each time-integration step of the SGCM.

Standard validation of the SGCM and LAMBLS has been carried out for different regions (subtropical desert, steppe, boreal forest and tundra). Compared to the driving SGCM, the SGCM and LAMBLS produces more realistic regional details of surface climate resulting from landscape inhomogeneities.

SIMULATION RESULTS

We present here results from some experiments for different epochs, each differing in its CO_2 concentration, ice-sheet distribution, sea-surface temperature, together with results from our simulation of present-day climate. The orbital parameters for the present day and various past simulations were calculated based on Milankovitch theory (Berger, 1994) in accordance with the data used by Kutzbach and Guetter (1986). We must stress that the radiocarbon time scale using for the estimation of the age of the events and solar absolute (calendar) time scale are not equivalent. They begin to diverge by more than 500 years beyond 5000 solar years ago and are greater in more ancient times (Stuiver and Reimer, 1993), but due to gradual changes of insolation at the top of the atmosphere the difference in radiation flux is practically insignificant during the last 10 ka. At a more ancient time they are not significant compared to the strong variation of the boundary conditions.

We begin the description of the modelling results from the warm culmination of the Holocene at 6 and 9 ka (the well-known Atlantic and Boreal Optimum). During these events there were strong orbitally induced changes in solar energy received by the Earth. Thus, at 6 ka northern hemisphere summer insolation was 8% greater and winter insolation was 8% less than today. Other external forcing was invariant with time (besides small changes in continental ice-sheet configuration: the Labrador Peninsula and centre of Canada were covered by the remains of the Laurentide ice sheet). The description of the boundary conditions and external forcing are given by Kislov (1993a).

Simulation studies of the change in climate during 6 and 9 ka performed by global climate models with sea-surface temperature either fixed (Kutzbach and Guetter, 1986) or calculated by simplified ocean models (Mitchell et al., 1988) provide some support for the astronomical theory, and demonstrate in general an agreement between modelled and observed data, but fail to obtain actual climatic conditions over some arid and semiarid regions (e.g. North Africa, Central Asia etc.). Earth-orbital radiative forcing does not seem sufficient to be able to explain these regional-scale discrepancies. Because no principal change (regarding present-day level) in other external forcing factors was found, it can be postulated that changes in landscapes (principally in vegetation cover) are required in order to explain satisfactorily the observed change in climate of arid and semiarid zones. To test these hypotheses, some equilibrium experiments with different

GCMs were carried out with the state of vegetation either prescribed or calculated by simplified models (Kislov, 1993a, 1994a,b; Clausen and Gayler, 1995; Harrison et al., 1995; Jolly et al., 1995; Noblet et al., 1995). A good agreement between simulated and observed data was found.

These points were confirmed by simulated data over the catchment of the Caspian Sea based on SGCM and LAMBLS modelling experiments. The first run was carried out for perpetual northern summer 6 ka BP using the change of the insolation flux at the top of the atmosphere only. A comparison of modelled data (Figure 17.2) with palaeodata (Velichko and Klimanov, 1990) reveals that palaeotemperature patterns closely resemble simulated data in middle latitudes, but positive simulated anomalies over the south and southeast parts of the region are at variance with reconstructed negative anomalies.

To improve this situation a second experiment was carried out in which the fixed properties of the vegetation and soil were modified based on well-documented reconstructions of its changes which were collected by Hastenrath (1985), Kadomura (1986), Zubakov (1986) and Varuschenko and Tarasov (1992). Landscape modification produced a pattern (see Figure 17.3b) of decreased surface temperature over the southern continental part of the region where the heat lost due to enhanced evapotranspiration (see Figure 17.3c) is more important than the increase of the surface radiation flux (see Figure 17.3a). A better agreement with palaeodata confirms the conclusion that climate–vegetation feedback is realized in arid and semiarid zones and substantially modifies the response of the climate system to orbital forcing.

The results of this experiment with SGCM and LAMBLS are used as a basis for investigation of the water budget change over the Caspian Sea basin. At first we focus on the variation of the excess $e = E - P$ over the sea surface and we will estimate its contribution to Caspian Sea level changes. At 6 ka the simulated summer average over the Caspian Sea surface (Δe) was obtained as 0.24 mm/day and $\Delta e/e_0 = 0.05$. Taking into account the seasonal variation of orbitally induced anomalies of insolation at the top of the atmosphere, our result is considered to represent a maximum value for the summer

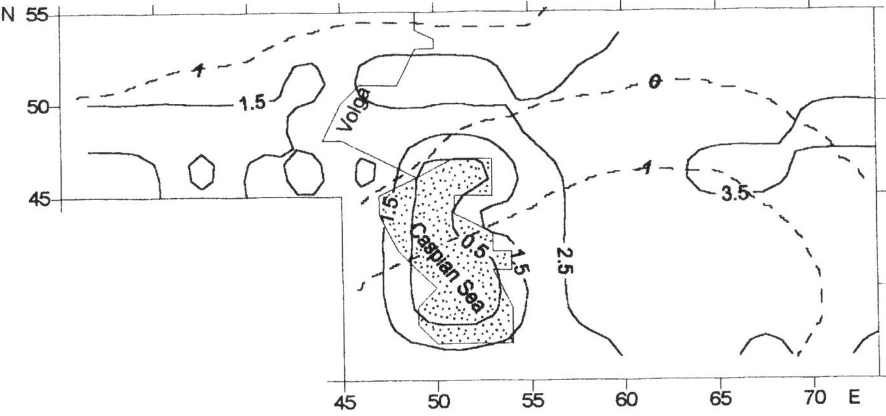

Figure 17.2   Northern summer surface air temperature (K) anomalies '6 ka minus present-day climate' as simulated by SGCM and LAMBLS (solid lines) and summer surface air temperature anomalies from palaeoreconstruction of the Atlantic Optimum of the Holocene (dashed lines)

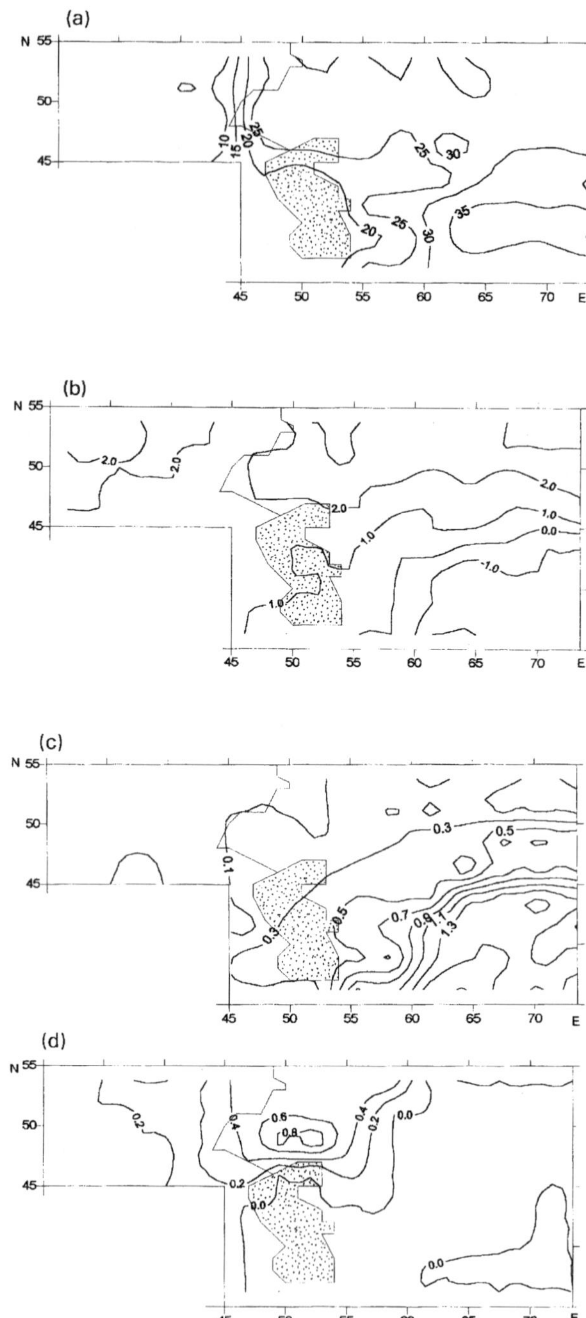

Figure 17.3 Deviations from present-day values for summer conditions over the Caspian Sea region during the Atlantic Optimum, simulated by SGCM and LAMBLS. (a) Surface radiation budget (Wm^{-2}). (b) Surface air temperature (K). (c) Rate of evaporation (mm/day). (d) Rate of precipitation (mm/day)

half-year. During winter the temperature decreases respond to the influence of the negative anomalies of insolation flux at the top of the atmosphere. There are weak differences in P and E between 6 ka and the present-day control run. These anomalies are statistically insignificant and we constrain them to zero. Taking these facts into account, we can estimate the annual averaged value $\Delta e/e_0$ as 0.025. Using Equations 17.2 and 17.3 we obtain $(\Delta h)_e = -0.5$ m.

Now we will evaluate the contribution of changes of runoff to the lake-level changes (see Equation 17.5). Runoff depends on climatic factors (P and E over the catchment) and hydrological characteristics of the catchment. A robust quantitative method to calculate runoff within GCM remains largely unavailable. Annual averaged runoff is considered to be the difference between precipitation and evapotranspiration over the catchment surface. Therefore the first step is to calculate the excess $P - E$ over the catchment area and to estimate its variation in the past. Based on the simulation results it was found that at 6 ka within the Volga River catchment the summer half-year averaged changes of $P - E$ were approximately 0.2 mm/day relative to present. During the cold season there were no statistically significant changes of $P - E$ over the catchment. We can express annual averaged $\Delta(P - E) = 0.1$ mm/day. This value is considered to be an approximation of the change of runoff (ΔY). Using Equations 17.2 and 17.3 we have $(\Delta h)_Y = 2.5$ m.

Our estimated value of $\Delta(P - E)$ was compared to simulated $\Delta(P - E)$ over the same territory which was considered by GCM experiments within the Palaeoclimate Modelling Intercomparison Project (PMIP). Based on the results of various GCMs (CCSR, ECHAM, GENESIS, LMCELMD, MRI, UGAMP, GFDL, CCM) we have found that over the Volga River catchment, against a background of the 6 ka annual averaged value, $\Delta(P - E)$ was found to be zero (PMIP workshop, 1995a). This result is contrary to our result. This deviation can be explained by the fact that we used a model with a more realistic description of the energy and water exchange between surface and atmosphere due to high spatial resolution. Replacing our results in Equation 17.5 we have the cumulative effect: $\Delta h = 2$ m.

At 9 ka the simulated hydrological fields over the Caspian Sea and its catchment are mainly similar to the simulated values at 6 ka . We have $(\Delta h)_e = -0.8$ m, $(\Delta h)_Y = 2.5$ m. Once again we have approximately $\Delta h = 2$ m.

Now consider the recent reconstructions. Velichko et al. (1987) noted that at 6 ka the annual averaged precipitation over the Volga River catchment was 0.2 mm/day more than at present. Guiot et al. (1993) showed that over this region anomalies of the excess precipitation minus potential evapotranspiration (PET) were more than 1 mm/day at 6 ka and they were mainly determined by anomalies of the precipitation. At 9 ka, ΔP over the Volga River catchment was found to be 0.7 mm/day and again $\Delta P = \Delta(P - PET)$. We expect, assuming that $\Delta(P - PET)$ approximates $\Delta(P - E)$, that $\Delta Y = (1 - 0.7)$ mm/day. Let us reflect that based on geomorphological data at 6 ka , $\Delta h = 7$ m and at 9 ka $\Delta h = 2.5$ m (see Figure 17.1).

It is recognized that reconstructed scenarios do not provide a particularly strong test of the absolute magnitude of both simulated or observed Caspian Sea level changes. However, they allow us to conclude that at 6 ka and 9 ka the region (including the Volga River catchment) was shown as wet and there were positive anomalies of runoff from the catchment into the Caspian Sea. The agreement between simulated and observed data allows us to conclude that SGCM and LAMBLS is able to simulate hydrological values with a useful level of accuracy.

The hydrological values over the Caspian Sea basin have been simulated using SGCM and LAMBLS for the Last Glacial Maximum (LGM) (20–21 ka) as well. During this period there were large changes relative to the present day in continental ice sheet configuration, major changes in optical properties of the atmosphere due to decrease of the greenhouse gases, and a significant global decrease of the sea-surface temperature. There were practically no changes in the insolation at the top of the atmosphere. A description of the boundary conditions and external forcing was given by Kislov (1994b).

Analysis of the geological and geomorphological records within the Caspian Sea basin showed that during the LGM sea level was decreased (absolute altitude then was $h = -64$ m (Varuschenko et al., 1987), remembering that the present-day value is -28 m) and the surface area of the sea was decreased. This was the so-called 'Enotauev-skya' regression of the sea level. We took this into account in modelling. 'New' land surface (former sea bottom) appeared along the sea shore, and landscapes there have been described according to palaeodata in the terms needed to run the model.

It was found that over the Caspian Sea surface $\Delta e = 0$ and consequently $(\Delta h)_e = 0$. The annual and catchment-averaged excess $\Delta(P - E)$ was found to be zero. Based on results of the PMIP experiments (PMIP workshop, 1995a, b) $\Delta(P - E) = 0$ (only one GCM has demonstrated $\Delta(P - E) = 1$ mm/day; others have represented $\Delta(P - E) = 0$). We can conclude that against a background of the LGM, the simulated hydrological response over the Russian Plain and western part of Central Asia was quite small. Therefore for this period changes of described $\Delta(P - E)$ values do not explain the observed Caspian Sea level decrease.

Simulated Caspian Sea level changes from various experiments (at 1, 2, 3, 4, 5, 6, 7, 8, 9, 11, 12, 20 and 21 ka) are summarized in Figure 17.4 as well as a reconstructed curve of the sea level (see Figure 17.1). It can be seen that through the Holocene the time variations of the modelled anomalies of the Caspian Sea level position are gradual and they gradually decayed after 6 ka to the present. This is an explicit quasi-linear reaction to the externally imposed radiative variations, which are complicated by the feedback between climate and vegetation.

The fluctuation Allerød–Younger Dryas–Boreal Optimum can be found in Figure 17.4. The boundary conditions and orbital forcing during the Allerød are similar to 9 ka (besides the globally decreased sea-surface temperature and changes in continental ice-sheet position). During the Younger Dryas an unusual level has been simulated as a reaction to the variation of the insolation, changes of the ice-sheet configuration, changes of CO_2 and strong cold anomalies of the temperature within the northern Atlantic Ocean (Kislov, 1993b). For the LGM the simulated anomaly is unrealistically small relative to the reconstructed data.

As seen from Figure 17.4, there was a lack of synchronization between global forcing factors and regional climate and hydrological changes. The shorter-term variations cannot be explained by changes in insolation, CO_2, sea-surface temperature, ice-sheet configuration or vegetation distribution.

CONCLUSION

A simplified GCM with an embedded limited area model was used to investigate the climate and hydrological regime sensitivity to the changes of insolation flux at the top of the atmosphere, variation of the CO_2 concentration and modification of the surface

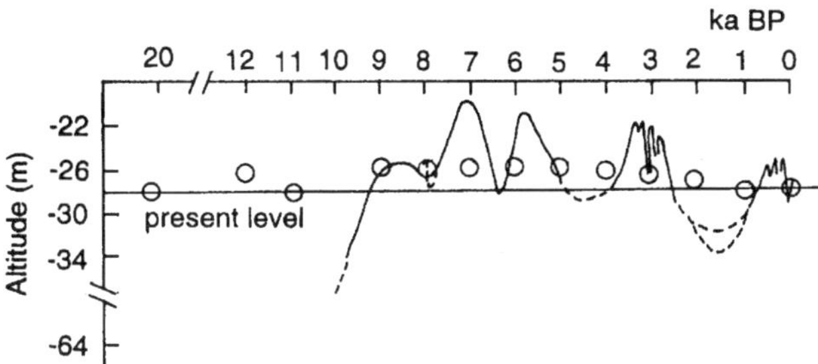

Figure 17.4   Simulated Caspian Sea level changes over the last 20 ka BP (circles) compared with data of geomorphological reconstruction (line; Richagov, 1993)

characteristics during the last 20 ka. It was shown that there was a lack of explicit synchronization between global forcing factors and regional climate and hydrological changes. Hydrological anomalies over the Caspian Sea area and its catchment simulated by climate modelling are not sufficient to explain Caspian Sea level changes during the past 20 ka, especially during the Late Pleistocene Ice Age.

To explain Caspian Sea level changes during the past 20 ka we must either implicate changes in some other non-climatological factors (tectonically induced variation of the basin configuration, intermittent interaction between Caspian Sea and Aral Sea, sun activity influence, etc.), or recognize that abrupt level variations occur due to inherently non-linear interactions in the system 'large lake–catchment' which are influenced by atmospheric forcing.

Discrepancies between the observation and simulations point to uncertainties in our understanding of what the past climate and hydrological variations actually were, uncertainties in the specification of the boundary conditions, or to inadequacies of the model itself.

REFERENCES

Berger, A., 1994. Astronomical theory of paleoclimates. In C.F. Boutron (ed.), *Topics in Atmospheric and Intersteller Physics and Chemistry*, Les Ulis, 411–452.

Clausen, M. and Gayler, V., 1995. The green Sahara at 6 ka BP: results from an interactive atmosphere biom model. *Global Analysis, Interpretation, and Modelling: First Science Conference, IGBP*, 25–29 September 1995, Garmish-Partenkirchen, Germany.

Guiot, J., Harrison, S.P. and Prentice, I.C., 1993. Reconstruction of Holocene precipitation patterns in Europe using pollen and lake-level data. *Quaternary Research*, **40**, 139–149.

Harrison, S.P., Kutzbach, J.E., Prentice, I.C., Behling, P.J. and Sykes, M.T., 1995. The response of Northern Hemisphere extratropical climate and vegetation to orbitally induced changes in insolation during the last interglaciation. *Quaternary Research*, **43**, 174–184.

Hastenrath, S., 1985. *Climate and Circulation of the Tropics*, Reidel, Dordrecht.

Jolly, D., Laaril, F., Harrison, S.P., Damnati, B. and Bonnefille, R., 1995. Intercomparison of simulations of the monsoon climate and vegetation of Africa at 6000 yr BP using pollen and lake data. *Global Analysis, Interpretation and Modelling: First Science Conference, IGBP*, 25–29 September, Garmish-Partenkirchen, Germany.

Kadomura, H., 1986. *Late glacial–early Holocene environmental changes in tropical Africa: a comparative analysis with deglaciation history*, Tokyo Meteorological University, Geographical Reports 21, 1–21.

Kislov, A.V., 1991. Three-dimensional model of atmospheric circulation with complete description of physical processes and simplified dynamics. *Izvestiya Russian Academy of Sciences, Atmospheric and Oceanic Physics*, **27**, 353–361.

Kislov, A.V. 1993a. A simulation of the climate of Holocene's optimum. *Izvestiya Russian Academy of Sciences, Atmospheric and Oceanic Physics*, **29**, 173–181.

Kislov, A.V., 1993b. Study of the genesis of the cold events during postglaciation. *Izvestiya Russian Academy of Sciences, Atmospheric and Oceanic Physics*, **29**, 182–186.

Kislov, A.V., 1994a. Study of forming factors in the warm climates of Holocene based on simplified general circulation model simulations, *Izvestiya Russian Academy of Sciences, Atmospheric and Oceanic Physics*, **30**, 443–450.

Kislov, A.V., 1994b. Study of the genesis of the global climate fluctuations during postglaciation. *Izvestiya Russian Academy of Sciences, Atmospheric and Oceanic Physics*, **30**, 601–607.

Kutzbach, J.E. and Guetter P.J., 1986. The influence of changing orbital parameters and surface boundary conditions on climate simulations for the past 18000 years. *Journal of Atmospheric Sciences*, **43**, 1726–1759.

Mitchell, J., Grahame, N.S. and Needhame, K.J., 1988. Climate simulation for 9000 years before present: seasonal variation and effect of the Laurentide ice sheet. *Journal of Geophysical Research*, **93**, 8283–8303.

Noblet, de N., Foley, J., Claussen, M. and Prentice, I.C., 1995. Palaeostates of the vegetation–atmosphere system: model experiments on orbital forcing and vegetation feedback. *Global Analysis, Interpretation and Modelling: First Science Conference, IGBP*, 25–29 September, Garmish-Partenkirchen, Germany.

PMIP workshop, 1995a. *Atlases for the 6 ka simulations*, Collonges-la Rouge, 1–6 October.

PMIP workshop, 1995b. *Atlases for the 21 ka simulations*, Collonges-la Rouge, 1–6 October.

Richagov, G.I., 1993. Sea-level changes of the Caspian Sea during last 10 000 years. *Moscow State University Reports*, Series 5, **2**, 38–49.

Stuiver, M. and Reimer, P.J., 1993. Extended ^{14}C data base and revised CALIB 3.0 ^{14}C Age Calibration Program. *Radiocarbon*, **35**, 215–230.

Varuschenko, S.I. and Tarasov, P.E., 1992. Landscape and climate conditions of arid regions of the North Hemisphere during Boreal warm culmination, *Water Resources*, **1**, 47–50.

Varuschenko, S.I., Varushenko, A.N. and Klige, R.K., 1987. *Change in the conditions of the Caspian Sea and lakes without drainage during palaeotime*, Nauka, Moscow.

Velichko, A.A. and Klimanov, V.A., 1990. Climate conditions over North Hemisphere 5–6 ka BP. *Izvestiya AN, Geography*, **5**, 38–52.

Velichko, A.A., Klimanov, V.A. and Belyaev, V.A., 1987. Caspian Sea and Volga river at 5.5 and 125 ka BP. *Priroda*, **3**, 15–23.

Zubakov, V.A, 1986. *Global Climatic Changes during the Pleistocene*, Gidrometeoizdat, Leningrad.

PART V
Tropical Zone Palaeohydrology

CHAPTER 18

Late Quaternary Landscape Instability in the Humid and Sub-humid Tropics

MICHAEL F. THOMAS

Department of Environmental Science, University of Stirling, UK

INTRODUCTION

Evidence for large-scale, and broadly synchronous, changes in hydrology across much of the humid and sub-humid tropics during the last 50 000 years was summarised in two recent papers (Thomas and Thorp, 1995, 1996) within which some indication was given of possible hillslope processes during each climatic phase. Most new publications tend to strengthen that argument in main outline, even if boundary dates and intervals require continuous revision. In particular, more evidence has emerged of significant environmental changes during the late Quaternary in central and eastern Brazil and in SE Asia, two regions about which considerable doubt has been cast in the past. This study takes up this question and argues that many landscapes in the tropics appear to have passed through one or more periods of marked instability.

Certain perceptions still linger in the appreciation of tropical palaeoclimate and palaeohydrology. Estimates of the 18 000 yr BP sea-surface temperatures (SSTs) for the tropical oceans remain persistently within 2°C of present-day temperatures, and bias discussions of the on-land evidence which suggests much greater temperature changes, in the range $4 \pm 2°C$ across much of tropical America and Africa (see Thomas, 1994, for summary). While lake-level evidence indicates major changes to the hydrological cycle between the glacial and post-glacial climates world-wide, a heavy emphasis has been placed on the dry tropics and particularly those zones bordering the Sahara desert, and the sub-tropical dry areas of southern Africa and Australia.

The problem with many global models of climate change is that they do not allow for the nature of ocean circulation or the dynamics of ocean–atmosphere interaction at a regional scale. Two recent papers illustrate this issue. In 1996, Linsley noted that the oxygen isotope record from the Sulu Sea, located in the 'warm pool' of the western Pacific, 'indicates that the Sulu Sea experienced no significant change in SSTs between glacial and interglacial times'. However, Thunell and Miao (1996) have derived a totally different picture from a neighbouring site (GGC-9) in the South China Sea which is subject to shallow water flows from the north. Here they estimate a reduction in glacial winter SSTs of 7°C, and of 1.5°C in summer, and they consider that a 7°C warming took place between 13 000 and 10 000 yr BP. It seems probable that similar trends would have been apparent off the Brazilian and the African coasts during this period. In fact, the depression of mean temperatures on land throughout the tropics by 5–8 °C at the Last Glacial Maximum (LGM) has been widely proposed (Thomas, 1994; Verstappen, 1994). It is also generally agreed that this cooling was accompanied by reductions of rainfall by 25–60% (Peters and Tetzlaff, 1990). Although the dryness of central and western Amazonia has previously been challenged (Colinvaux, 1987, 1991; Bush et al., 1990), recent studies appear to support the reality of this condition (Ledru, 1993; Iriondo and Latrubesse, 1994; Latrubesse and Ramonell, 1996; Van der Hammen and Absy, 1994). The diversity of relief in SE Asia has made generalisation very difficult, but Verstappen (1994) has attested to the widespread evidence for major climate and vegetation changes in that area.

A recent general circulation model (GCM) developed by Dong et al. (1996) offers versions of the climate and hydrology for the tropics of northern Africa and southern Asia at 115 ka, 21 ka and 6 ka. Major changes in parameters between 21 ka and 6 ka indicate a regional averaged precipitation over northern Africa (N of 10°N) for the LGM of − 40.62% compared to present day, and + 23.21% for the Holocene climatic optimum. Equivalent figures for southern Asia were − 64.44% and + 17.93% respectively. Comparable changes in soil moisture were predicted and, perhaps most interesting of all, the wetter climates of the Holocene were shown to have included higher peak daily rainfalls and soil moisture levels during the wet season. Such swings of 64–82% in mean rainfall between the driest and wettest periods of the last 21 ka, when combined with soil moisture changes, would provide the underlying rationale for major changes in hillslope and valley rocesses.

Rhind (1995) has also challenged the view that climate warming and greater evaporation lead to increased cloudiness and reduced rainfall, arguing from satellite studies that increased evaporation will lead to a thinning of low cloud, greater sunshine amounts, the formation of larger water droplets and increased precipitation. There is also some agreement that increased storminess would accompany such climate warming, particularly as it affected (or will in future affect) cyclonic areas of the tropics.

These and other considerations give added weight to the interpretation of Late Pleistocene and early Holocene stream sedimentation as responses to the switch from dry conditions to much wetter climates between c. 12 700 and c. 9000 yr BP, and that it is possible to infer that this change was accompanied by very high river discharges not only in large river systems such as the Nile and Amazon which have mountainous headwaters, but also in lowland tropical rivers, both large (such as the Niger) and small.

Catchment conditions leading to these high discharges must have involved major changes to slope hydrology and slope processes, and it is the purpose of this study to enquire into the landforms and sediments present in erosional catchments and steep lands. Aspects of such a study concern the occurrence of palaeolandslides and associated flow deposits, and also the existence and present status of alluvial fan deposits and related colluvia. However, reports on these features from tropical locations have been sparse and dated sequences are few.

PALAEOLANDSLIDES IN WARM TEMPERATE ENVIRONMENTS

Evidence has accumulated from temperate areas to suggest that clustering of mass movement events has occurred in mid-latitude west-coast climates, including the USA and Spain. Recent work from the Cantabrian region of Spain by González Díez et al.

(1996) has established the importance of a climatically driven rhythm of landsliding in that area, for which they propose five groups of landslides:

- after 500 yr BP
- 5000 – 500 yr BP
- 11 000 – 7000 yr BP
- 41 000 – 11 000 yr BP
- pre – 41 000 yr BP

They noted that the greatest frequency of landsliding was shortly after deglaciation and also after the climatic optimum of the mid-Holocene. Reneau et al. (1986) have also shown a clustering of dates for colluvium derived from hillslope hollows in California between 15 000 and 9000 yr BP, and both these studies demonstrate that, although landslides occur as discrete events in space and time, they respond to changes in the frequency of high magnitude rainfall events as expressions of changing climatic conditions. Mass movements that took place after deglaciation in high latitudes almost certainly involved factors other than precipitation changes.

PALAEOLANDSLIDES AND COLLUVIATION IN THE TROPICS

Very few dated landslides have been documented within the tropics, but indications of late Quaternary instability are significant. Shroder in 1976 published a study of mass movements on the Nyika Plateau of Malawi (around 2000 m a.s.l., precipitation up to 2000 mm per year; Figure 18.1 shows this and other major locations referred to in the text), and derived several radiocarbon dates, mainly from sediments exposed by a slope failure at Chelinda, involving a bog burst, in 1960. He noted repeated failure at this site which was probably a hillslope hollow, and derived dates:

- peat formation 4160 – 4900 yr BP
- second failure just prior to 5000 yr BP showing evidence of human interference and fires
- charcoal layer 9780 ± 140 yr BP
- basal date 10 170 ± 145 yr BP at − 3.3 m below first major movement

Shroder mapped 235 landslips of which 200 were on the eastern Nyika Plateau and this was attributed to higher precipitation (correlation coefficient 0.79) and to a thicker weathered mantle. A later study of valley sedimentation on the Nyika Plateau by Meadows (1983, 1985) indicated an influx of sediment shortly after 12 000 yr BP, continuing until *c.* 5000 yr BP, and preceded by a period of vigorous stream erosion.

In SE Brazil, Modenesi (1988) has examined large palaeolandslides on the Campos de Jordão, a part of the Serra de Mantequeira west of Rio de Janeiro (Figure 18.1C). Similar to the Nyika Plateau, this area is a 2000 m a.s.l. block mountain and has a mean annual temperature of 14 °C, with precipitation rising to 2800 mm per year. Modenesi described 'amphitheatres' 80–100 m deep and 100–500 m wide, all over the summit plateau or 'altos campos'. At least three generations of amphitheatre were recognised, with successive accumulation of earthflows or mudflows on lower slopes. Debris-flow lobes had accumulated over peat deposits on older colluvium forming gently convex 'lombas' revealing

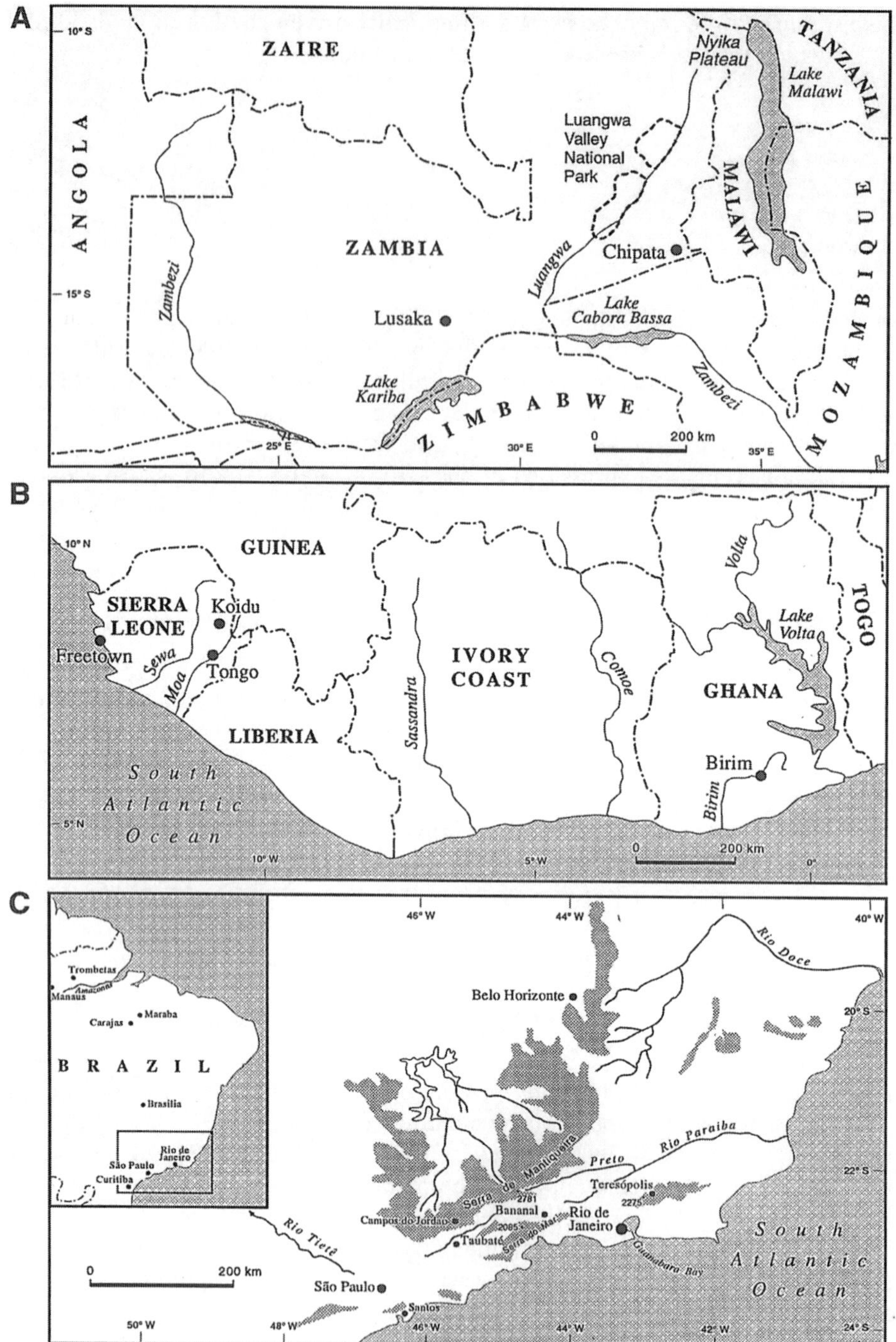

Figure 18.1   Locations referred to in the text

conspicuous bedding and inverted or disordered weathering sequences. She concluded that there was an initial mobilisation of huge masses of saprolite in a series of slump landslides, followed subsequently by shallow slides, rock falls and flows from back scarps. Modenesi's (1988) conclusion that, 'the evolution of slopes during the Quaternary is one of the less well known aspects of tropical geomorphology', remains true today.

LANDSLIDES FROM LOW ALTITUDE (< 1500 m) TROPICAL SITES

The landslide studies referred to above come from relatively high altitude sites (> 2000 m). However, widespread colluvium derived from mass movements in thick saprolite is found in the Bananal area, west of Rio de Janeiro, where associated fluvio-lacustrine sediments contain organic horizons dating to 9500–10 000 yr BP (De Moura et al., 1993), while in Minas Gerais State, intervals between fan deposition have been dated from organic horizons to 9500 and 8500 yr BP (Servant et al., 1989). A single date of 14 000 yr BP was offered by Pflug (1969) for fan deposition in the nearby Rio Doce valley.

The wider significance of colluvial deposits in humid tropical landscapes is emphasised by Ferreira and Monteiro (1985), who estimated that 50% of São Paulo State was underlain by colluvium, while many other studies have noted the importance of colluvial deposits in the humid tropics (see Thomas, 1994). However, studies of palaeolandslide activity from lowland humid tropical areas are confined to a few major occurrences, and many ancient landslides are concealed below the forest cover and have often been overlooked.

The fact that landslide activity is a regular and major part of the process system within the present-day climates of most tropical and sub-tropical rainforest areas must also be acknowledged. Their occurrence responds to high magnitude rainfall events and to periods of prolonged precipitation. Considerable documentation has been offered for areas such as Hong Kong, Japan, Puerto Rico and SE Brazil, and these reports emphasise the role of frontal troughs and cyclonic storms (hurricanes, typhoons) in generating frequent, shallow translational slides and flows (mainly debris slides and debris flows). Deep-seated landslides, on the other hand, may have prolonged histories and many recorded cases have been in active seismic areas such as New Guinea and Japan.

It is not possible to generalise for all tropical areas the threshold rainfalls likely to trigger landslides. Such figures depend on a number of local factors such as slope, nature and thickness of regolith, and rock type and fissility. Nevertheless, it is clear that 24 h rainfalls of c. 75 to > 100 mm are commonly indicated, together with antecedent rainfalls of > 250 mm, measured over varying periods from 24 h to 15 days. Peak rainfall intensities only affect the incidence of shallow debris flows, and these become frequent when intensities approach 100 mm h^{-1}. Although not applicable directly to other areas, the model offered by Lumb (1975) for Hong Kong remains a simple and useful guide.

Active landsliding is comparatively rare in the more seasonal tropical climates which give rise to deciduous woodlands and savanna vegetation formations and are associated with typical rainfalls of 800–1200 mm per year. Although high rainfall intensities occur within such climates and can lead to debris flows, they tend to occur within smaller storms, and periods of prolonged high rainfall are few. Where widespread landslide scars and debris are found in these environments, it is logical to look for evidence of climate change.

PALAEOFANS AND LANDSLIDES IN E ZAMBIA

One such area currently under investigation is in central Africa, where palaeolandslides have been investigated on part of the central African plateau in E Zambia, close to the Malawi border at latitude 15°S (Figure 18.1A). The general altitude is 1000–1200 m a.s.l., but steep residual hills rise to 1400–1600 m. The rainfall average is 1100 mm per year, and the natural vegetation is miombo woodland. The local landscape is dominated by a series of steep-sided, elongate hills with a relief of 200–400 m, while the intervening lowlands are made up of extensive, gently sloping land surfaces (slope pediments, glacis, rampas) that lead into broad, valley floors known as 'dambos' (see Thomas and Goudie, 1985). The basement rocks of the area are dominantly granulites which contain quartzite beds that exhibit near-vertical bedding. The granulites are known to be weathered to > 30 m depth and the saprolite is dominantly ferrallitic in composition. It is likely, therefore, that the regolith has been forming since pre-Quaternary times.

The rainfall regime at Chipata is typical of the seasonal tropics: the rainy season lasts approximately six months and totals average 1100 mm. The wettest year in a 50 year rainfall record was 1978 with 1600 mm and one other year exceeded 1400 mm. But when 24 h totals are plotted against antecedent (15 day) rainfall as used by Lumb (1975), all years fall into the category of minor risk of landslides. No active features have been seen and no local records of landslides are known. Yet all the major hills, and many minor ones, associated with the geological features as described, are dominated by landslide scars and debris accumulations.

These landslides are mostly of the slump-earthflow type: deep-seated failures showing some rotation. Many are morphologically complex but can be up to 0.5–1 km wide and 1–1.5 km long, often ending in quartzite boulder-dominated ramparts or lobes. They have been generated on 26–28° slopes and have produced headwall scarps maintaining slopes of at least 33–38°, often exceeding 100 m in height. Even these slopes appear stable under the prevailing climatic regime. The failures occurred within the weathered granulite, and quartzite rubble has been carried down from the upper slopes on and within the clay matrix. Even where the debris is apparently clast-supported there are clay skins up to 1 cm thick, and it would appear that a fluidised quartzite rubble has accumulated very rapidly as flow deposits up to 1.5 km from the landslide scars. Large masses of weathered granulite have also been brought down virtually intact. It is inferred from these features that most of these landslides have either been short-lived, rapid events or have at least been through a phase of rapid movement under hydrological conditions very different from those of today.

Where rivers exit from small catchments in the hills, they have formed alluvial fans which occasionally contain very coarse deposits of boulders. Everywhere the fans, which have slopes exceeding 10° near their apexes and decline to 3–5° towards their distal ends, have been dissected by the current drainage, and the deposits are often exposed as gravel terraces containing cobbles and sometimes boulders. The larger catchments around Chipata are only 10–20 km² in area, yet they have generated fans containing boulders up to 1 m in diameter. Such torrential deposits are also encountered along the margins of the Luangwa River, a major left-bank tributary of the Zambezi, which flows in a fault trough some 80 km to the west. Here, streams exiting from the Chindeni Hills have previously formed coarse fans that dip below the floodplain deposits of the contemporary river.

These fan sediments can be seen to be closely associated with the landslide debris at certain locations, and there is a possibility that the two processes were broadly contemporaneous. In one important section (Chipata Quarry) a very coarse boulder deposit is overlain by landslide debris in a manner that suggests a very short time gap between the two. In all cases, large quantities of water would have been required to trigger the processes leading to these deposits.

There is also evidence for widespread colluviation in this area, either associated with or taking place subsequent to fan formation. The sandy colluvium is up to 8 m thick across many of the broad glaçis-type slopes, and in places box-shaped gullies expose sections in which discontinuous cobble and boulder layers are found within the predominantly sandy sediment. The conditions for the formation of such sediments probably involved intense rainfalls on a poorly vegetated landscape, but at present no dates are available to relate these events to the periods of fan formation. In other sections shallow gullies, or palaeostream channels, containing coarse sediments at the base and infilled with sandy colluvium are seen and these were probably a part of the palaeodrainage system.

This landscape has clearly been formed by a process regime more energetic than today and there is an urgent need to date the features identified. Much has been written of the dry climates of the late glacial period in central and southern Africa, though far more is known of the sub-tropical south and of the high plateaus nearer the Equator, in Kenya and Tanzania, than for the vast deciduous woodland areas of central Africa. A 30–60 + % reduction in rainfall, as indicated by Peters and Tetzlaff (1990) and in the modelling exercise by Dong et al. (1996), at or near the LGM, could have reduced the rainfall of the Chipata area to 500–600 mm per year at the LGM. Under these conditions, acacia-dominated dry woodland and grassland landscapes were probably dominant, and the deep saprolite over the granulites would have dried out to considerable depths. What is not clear is the amount of additional moisture that would have been necessary to produce the high energy features of this landscape.

The widespread flooding and fluvial activity which occurred during the Pleistocene–Holocene transition (reviewed in Thomas and Thorp, 1995; Table 18.1) was clearly a response to increased rainfalls, but it was also a function of the status of the soil and plant cover and their response to enhanced rainfall amount and intensity. Nevertheless, much wetter conditions developed after c. 12 700 yr BP, possibly receding during the Younger Dryas, and then recovering towards a peak 10 500–post-9000 yr BP. It was not until this later date that a full recovery of vegetation was achieved, at least in many tropical rainforest areas. It seems likely that wetter conditions may have prevailed until c. 6000/5000 yr BP. Relict plunge pool deposits recently dated in N Australia (Nott, and Price, 1994; Nott et al., 1996) suggest greatly enhanced rainfalls prior to 20 ka and after 11 ka, lasting until 5 ka, with flood peaks twice to five times the magnitude of contemporary events. A summary of palaeohydrological conditions in the tropics is given in Table 18.1.

Deforestation today does not appear to lead immediately either to rapid sheet erosion or to landsliding, and the present climate does not contain rainfall patterns likely to lead to widespread deep-seated mass movement. The inference, therefore, from the existence of the landforms and sediments in the Chipata area must be that the landscape has passed through a crisis or crises in which enhanced rainfall may have taken place under conditions of poor vegetation recovery following the last Termination. Unfortunately we do not have dates for these features at present and cannot, therefore, press the argument far beyond the general principle.

Table 18.1 Timing of some late Pleistocene and early Holocene geomorphic events in the tropics

Location – features (Reference)	Years BP (^{14}C)
Dry phase fans	
Jos Plateau, Nigeria – fans, basal alluvium (Zeese, 1991)	18 000
Taubaté basin, Brazil – stone line (Riccomini et al. 1989)	18 000
Bamenda Highlands, West Cameroon – slope gravels (Tamura, 1986)	c. 20 000
Rio Doce, Brazil – valley aggradation (Pflug, 1969)	14 000
Wet phase indicators	
Nyika Plateau, Malawi – landslide basal deposits (Shroder, 1976)	10 170
Rio Paraíba do Sul, Brazil – fluvio-lacustrine sediments (De Moura et al., 1993)	10 000–9500
Minas Gerais, Brazil – organic horizons in fan deposits (Servant et al. 1989)	9500–8500
Ok Menga, Papua New Guinea – upper colluvium (Murray and Olsen, 1990)	c. 8000
High energy fluvial events	
Blue and White Nile systems – peak floods and sediment (Williams, 1980)	
(B & W)	c. 7000
(B only)	7500
(W only)	8400
(B & W)	12 500–11 000
Nyika Plateau – erosion/sedimentation (Meadows, 1985)	
sediments	< 12 000– > 5000
valley erosion	> 12 000
Koidu, Sierra Leone / Birim, Ghana – erosion/sedimentation phases (Thomas and Thorp, 1980, Thomas et al., 1985; Hall et al., 1985)	
	< 3000 – ongoing
	< 5000–4200
	< 8000–7000
bedrock channels	10 200–9500
bedrock channels	2700–11 500
N Territory, Australia – relict plunge pool deposits (Nott and Price, 1994; Nott et al., 1996)	
	11 000–4000
	30 000–20 000

LANDSLIDING IN THE WET TROPICS: THE CASE OF SIERRA LEONE

Observation and mapping of landslides in Sierra Leone, particularly in the mountainous Freetown Peninsula (Figure 18.1B), where monsoon rainfall regime exceeds 3000 mm per year, adds further to the general argument. This climate produces active landslides during years of exceptional rainfall, but these events are probably infrequent and unable to explain fully the distribution of old slides. Two major slides near Freetown were documented by the press and are dated to August 1945. The rainfall data for this period show clearly that the slides occurred after the wettest five day period since records began in 1876; 1121 mm rain fell between 5 and 9 August 1945 and 401 mm on 10 August, the day of the slides. This places the event in the extreme disaster area within the Lumb (1975) model.

The spine of highland (800–900 m a.s.l.) with similar slope conditions (commonly 32–38°) is marked by frequent slide scars, mostly taking the form of slump-slide features. These have been mapped from aerial photographs (1:12 500) and some key sites visited in the field. Many of these proved to be major landslides, involving large masses of material that had moved up to 1.5 km from the backwall, carrying large blocks across nearly flat terrain. All were forested and some toe areas settled and farmed. There seemed to be no local knowledge of the formative events. Similar features were seen inland in Sierra Leone, where some isolated groups of hills are ringed by landslide scars and debris. The plateau of the so-called Sula Mountains, which is underlain by at least 100 m of weathered metavolcanic rocks and supracrustal sediments, is also surrounded by large slump scars (Thomas, 1974, 1994), similar to those described by Modenesi (1988) from Brazil.

The conclusion from these observations is that landslides occur during heavy monsoon rains, potentially on an annual basis. But in most cases these are shallow translational slides involving weathered debris on steep, rocky inselberg slopes, slumps being confined to undercut slopes. Most of the larger slides may, therefore, be due to even wetter conditions in the past, or to the effects of rapidly changing climatic regimes in the Early Holocene. Some of these may be older than the late Quaternary, because the debris has become ferricreted; others may be much more recent.

Some large landslides and associated fans have also been identified during aerial reconnaissance in equatorial Kalimantan, in an area of supposedly inactive tectonics, and it appears that these features have a very widespread occurrence in tropical areas.

CONCLUSIONS

In the absence of dates it is difficult to deduce too much from these observations, but some conclusions seem valid.

- Large palaeolandslides are evident in tropical environments, ranging from the seasonal savannas to perhumid, rainforest areas.
- Many appear to be major, deep-seated events unknown (or at least very rare) within the contemporary process regime.
- Their spatial distribution and comparable age characteristics indicate responses to events taking place within 10^2–10^3 years.
- Their morphology and sedimentology indicate relatively rapid mass movements in the presence of excess moisture.
- There can be little doubt that those found in savanna areas are relict.
- In rainforest areas landslides form part of the present-day process system, but there are indications in some areas of past activity on a scale not apparent today.

It is hypothesised that unusual features of this kind may result from one or more of three situations:

(1) enhanced rainfall such as may have been experienced during the Holocene pluvial;
(2) the effects of rapidly increasing rainfall after the prolonged dry climates of the last glacial (21 000–12 500 yr BP);
(3) seismic shocks associated with eustatic and isostatic changes at the end of the last glacial.

The possibility of increased seismicity in coastal areas during the Holocene sea level rise remains largely untested in aseismic areas. Nevertheless, there is an increasing interest in palaeoseismology (reflected in the *Journal of Geophysical Research* symposium in March 1996), and in the links between global climate changes and active tectonics (Bull, 1991, 1996). However, the existence of large landslides inland from the Sierra Leone coastal areas and in central Africa, where in both cases there is no record of active Quaternary tectonics, makes this an unlikely global factor.

Enhanced annual rainfalls, associated with larger storm sizes, might bring a greater incidence of landsliding, but most geomorphic systems adjust regolith, slope and stream frequency characteristics to minimise instability. In many humid tropical mountain areas mass movement takes place almost continuously, deforming soils and displacing trees in the forest without total disruption or the generation of rapid landslides. This adjustment is less effective in climates containing frequent events of high magnitude, such as occur in hurricane and typhoon areas, and places such as SE Brazil where cold frontal troughs can produce copious rainfall.

However, in many areas, such as central Africa, high magnitude events are very infrequent and, in the absence of seismic shocks, the occurrence of landslides can best be explained by the effect of combining a rapid change in climate conditions with enhanced rainfalls.

One interesting observation of the Zambian slides is that they took place in deep, highly altered saprolite, and this has almost certainly been forming throughout the Cenozoic on the 'High Veld' of the central African Plateau. Once disrupted, these profiles do not reform rapidly, and their involvement in landsliding suggests that the events which produced these features are quite separate from those which cause debris flows in shallow regolith. Moreover, it is unlikely that a repetition of the formative events could take place frequently. We may therefore be dealing with a very infrequent combination of events, which might take place no more often than once in a glacial–interglacial cycle.

If this reasoning can be verified, it would imply quite unique hydrological conditions during the last Termination. Unfortunately the dating of landslides is in its infancy, at least in the tropics. However, the association of the slides with torrential stream deposits may allow the effective use of thermoluminescence/optically stimulated luminescence techniques.

In this chapter, it can only be argued that the highly unusual events which brought about the documented erosion and sedimentation rhythms in tropical rivers may well have been accompanied by equally dramatic changes to hillslope hydrology and to the erosion and sedimentation patterns across whole landscapes in the tropics. During the middle and later Holocene, continuing climate fluctuation is likely to have led to further adjustments of the relatively delicate landforms produced in the manner described, and this adjustment remains incomplete. It is therefore a matter of some concern for the future stability of the land resource that we should understand the nature and distribution of these late Quaternary deposits.

REFERENCES

Bull, W.B., 1991. *Geomorphic Responses to Climatic Change*, OUP, New York.

Bull, W.B., 1996. Global climate change and active tectonics: effective tools for teaching and research. *Geomorphology*, **16**, 217–232.

Bush, M.B., Colinvaux, P., Weimann, M.C., Piperno, D.R. and Lui, K-B., 1990. Late Pleistocene temperature depression and vegetation change in Ecuadorian Amazonia. *Quaternary Research*, **34**, 300–345.

Colinvaux, P.A., 1987. Environmental history of the Amazon Basin. In Rabassa (ed.) *Journal of The Quaternary of South America and Antarctic Peninsula*, **5**, 223–237.

Colinvaux, P., 1991. A commentary on: palaeoecological background: neotropics. *Climate Change*, **19**, 49–51 (special issue: Tropical Forests and Climate).

De Moura, J.R. da Silva, Mello, C.L., De Barros, M.A. and Barth-Fiocruz, O.M., 1993. O limite Pleistoceno-Holoceno no médio Rio Paraíba do Sul. *IV Congresso da ABEQUA*, São Paulo, July 1993, Associação Brasileira de Estudos do Quaternário, Resumos, 15–16.

Dong, B, Valdes, P.J. and Hall, N.M.J., 1996. The changes of monsoonal climates due to Earth's orbital perturbations and Ice Age boundary conditions. *Palaeoclimates*, **1**, 203–240.

Ferreira, R.C. and Monteiro, L.B., 1985. Identification and evaluation of collapsibility of colluvial soils that occur in the São Paulo State. *First International Conference on Geomechanics in Tropical Lateritic and Saprolitic Soils*, Brasilia, 1985, Volume 1, 269–280.

González Díez, A., Salas, L., Ramon Diaz de Terán, J. and Cendrero, A., 1996. Late Quaternary climate changes and mass movement frequency and magnitude in the Cantabrian region, Spain. *Geomorphology*, **15**, 291–309.

Hall, A.M., Thomas, M.F. and Thorp, M.B., 1985. Late Quaternary alluvial placer development in the humid tropics: the case of the Birim Diamond Placer, Ghana. *Journal of the Geological Society*, **142**, 777–787.

Iriondo, M. and Latrubesse, E.M., 1994. A probable scenario for a dry climate in central Amazonia during the late Quaternary. *Quaternary International*, **21**, 121–128.

Latrubesse, E.M. and Ramonell, C.G., 1996. A climatic model for southwestern Amazonia in last glacial times. *Quaternary International*, **21**, 163–169.

Ledru, M-P., 1993. Late Quaternary environmental and climatic changes in central Brazil. *Quaternary Research*, **39**, 90–98.

Linsley, B.K., 1996. Oxygen-isotope record of sea level and climate variations in the Sulu Sea over the past 150,000 years. *Nature*, **380**, 234–237.

Lumb, P., 1975. Slope failures in Hong Kong. *Quarterly Journal of Engineering Geology*, **8**, 31–65.

Meadows, M.E., 1983. Past and present environments of the Nyika Plateau, Malawi. *Palaeoecology of Africa*, **16**, 353–390.

Meadows, M.E., 1985. Dambos and environmental change in Malawi, central Africa. In M.F. Thomas and A.S. Goudie (eds), *Dambos: small channelless valleys in the tropics. Zeitschrift für Geomorphologie Supplementband*, **52**, 147–169.

Modenesi, M.C., 1988. Quaternary mass movements in a tropical plateau (Campos do Jordão, São Paulo, Brazil). *Zeitschrift für Geomorphologie N.F.*, **32**, 425–440.

Murray, L.M. and Olsen, M.T., 1990. Colluvial slopes – A geotechnical and climatic study. *Geomechanics in Tropical Soils, Proceedings of the Second International Conference on Geomechanics in Tropical Soils*, Singapore, 1988, Balkema, Rotterdam, Volume 2, 573–579.

Nott, J. and Price, D., 1994. Plunge pools and palaeoprecipitation. *Geology*, **22**, 1047–1050.

Nott, J., Price, D.M. and Bryant, E.A., 1996. A 30,000 year record of extreme floods in tropical Australia from relict plunge-pool deposits: Implications for future climate change. *Geophysical Research Letters*, **23**, 379–382.

Peters, M. and Tetzlaff, G., 1990. West African palaeosynoptic patterns at the Last Glacial Maximum. *Theoretical and Applied Climatology*, **42**, 67–79.

Pflug, R., 1969. Quaternary lakes of eastern Brazil. *Photogrammetria*, **24**, 29–35.

Reneau, S.L., Dietrich, W.E., Dorn, R.I., Berger, C.R. and Rubin, M., 1986. Geomorphic and Paleoclimatic implications of latest Pleistocene radiocarbon dates from colluvium-mantle hollows, California. *Geology*, **14**, 655–658.

Rhind, D., 1995. Drying out the tropics. *New Scientist*, **6**, 36–40.

Riccomini, C., Peloggia, A.U.G., Saloni, J.C.L., Kohnke, M.W. and Figueira, R.M., 1989. Neotectonic activity in the Serra do Mar rift system (southeastern Brazil). *Journal of South American Earth Sciences*, **2**, 191–197.

Servant, M., Soubiés, F., Suguio, K., Turcq, B. and Fournier, M., 1989. Alluviual fans in southeastern Brazil as an evidence for early Holocene dry climatic period. *International Symposium on Global Changes in South America During the Quaternary*, São Paulo, 1989, Special Publication 1, 75–77.

Shroder, Jr J.F., 1976. Mass movements on the Nyika Plateau, Malawi. *Zeitschrift für Geomorphologie, N.F.*, **20**, 56–77.

Tamura, T., 1986. Regolith-stratigraphic study of Late Quaternary environmental history in the West Cameroon Highlands and the Adamaoua Plateau. In H. Kadomura (ed.), *Geomorphology and Environmental Changes in Tropical Africa: Case Studies in Cameroon and Kenya*, Laboratory of Fundamental Research, Graduate School of Environmental Science, Hokkaido University, Special Publication 4, 69–93.

Thomas, M.F., 1974. *Tropical Geomorphology*, Macmillan, London.

Thomas, M.F., 1994. *Geomorphology in the Tropics*, John Wiley, Chichester.

Thomas, M.F. and Goudie, A.S. (eds), 1985. Dambos: small channelless valleys in the tropics. *Zeitschrift für Geomorphologie, Supplementband*, **52**.

Thomas, M.F. and Thorp, M.B., 1980. Some aspects of the geomorphological interpretation of Quaternary alluvial sediments in Sierra Leone. *Zeitschrift für Geomorphologie, N.F., Supplementband*, **36**, 140–161.

Thomas, M.F. and Thorp, M.B., 1995. Geomorphic response to rapid climatic and hydrologic change during the Late Pleistocene and Early Holocene in the humid and sub-humid tropics. *Quaternary Science Reviews*, **14**, 193–207.

Thomas, M.F. and Thorp, M.B., 1996. The response of geomorphic systems to climatic and hydrologic changes during the Late Glacial and early Holocene in the humid and sub-humid tropics. In J. Branson, A.G. Brown and K.J. Gregory (eds), *Global Continental Change: the Context of Palaeohydrology*, Geological Society, London, Special Publication 115, 139–153.

Thomas, M.F., Thorp, M.B. and Teeuw, R.M., 1985. Palaeogeomorphology and the occurrence of diamondiferous placer deposits in Koidu, Sierra Leone. *Journal of the Geological Society*, **142**, 789–802.

Thunell, R.C. and Miao, Q., 1996. Sea surface temperatures of the western Equatorial Pacific Ocean during the Younger Dryas. *Quaternary Research*, **46**, 72–77.

Van der Hammen, T. and Absy, M.C., 1994. Amazonia during the last glacial. *Palaeogeography, Palaeoclimatology, Palaeoecology*, **109**, 247–261.

Verstappen, H.Th., 1994. Climatic change and geomorphology in south and south-east Asia. *Geo-Eco-Trop*, **16**, 101–147.

Williams, M.A.J., 1980. Late Quaternary depositional history of the Blue and White Nile rivers in central Sudan. In M.A.J. Williams and H. Faure (eds), *The Sahara and the Nile*, Balkema, Rotterdam, 37–62.

Zeese, R., 1991. Paleosols of different age in central and northeast Nigeria. *Journal of African Earth Sciences*, **12**, 311–318.

CHAPTER 19

Late Quaternary Alluvial Sedimentation in the Upper Rio Negro Basin, Amazonia, Brazil: Palaeohydrological Implications

EDGARDO M. LATRUBESSE

Universidade Federal do Acre, Rio Branco, Acre, Brazil

AND

ELENA FRANZINELLI

Universidade Federal do Amazonas, Manaus, Brazil

INTRODUCTION

Multidisciplinary data obtained in the Brazilian Amazon have shown that the rainforest experienced significant biogeographic and climatic changes during the last glaciation. Recently, palynological and vertebrate palaeontological studies have presented evidence of alternating dry and wet climates with forest replacement by savanna vegetation during the dry periods and forest advancement during the wet periods (Absy et al., 1991; Rancy 1991, 1992). New geomorphological and sedimentological data indicate the presence of widespread areas covered by aeolian sediments (Santos et al., 1993; Iriondo and La-trubesse, 1994). Based on multidisciplinary data, Latrubesse and Ramonell (1994) have estimated wind patterns for the late Pleistocene.

In this chapter we focus on the Quaternary alluvial records of the Upper Rio Negro basin, and their correlation with the records of other Amazon fluvial sub-basins. The data presented here are the first that consider the relationship of alluvial sedimentation and Quaternary climatic changes with control of absolute dating for a cratonic fluvial basin of the Amazon fluvial system.

LOCATION

The region studied is located in the western part of the Guyana Shield. It is traversed by the upper course of the Rio Negro and is bordered by Colombia and Venezuela. The area is covered by tropical rainforest, has a very low population density, lacks roads, and the

---

Palaeohydrology and Environmental Change. Edited by G. Benito, V. R. Baker and K. J. Gregory
© 1998 John Wiley & Sons Ltd.

rivers are seasonally unnavigable. The most important town is São Gabriel da Cachoeira which is about 800 km northwest of Manaus.

The climate of this region is humid tropical, characterized by annual precipitation varying from 2900 to 3600 mm. The rainfall is evenly distributed during the year, without a dry season. The monthly mean temperature varies between 24 and 26°C (Radambrasil, 1976).

GEOMORPHOLOGICAL FRAMEWORK

According to Radambrasil (1976), rocks of the Guyanese Complex (early Precambrian in age) and subordinately metamorphosed sedimentary rocks of the Tunui Group (middle Precambrian) and of the Roraima Group (late Precambrian in age) outcrop in this area. Porphyroblastic granulites, mostly of granite composition but with some adamelite and quartzdiorite, form most of the Brazilian Shield rocks. In addition, Cretaceous carbonatite (in Serra dos Seis Lagos) and Quaternary sediments are found in the area (Figure 19.1).

Regionally, the Upper Rio Negro basin is a flat area of low relief with height varying between 60 and 160 m a.s.l. The most important features of the relief are numerous inselbergs that emerge locally on the plain, isolated or grouped, built in the Precambrian rocks of the Guyana Shield. Their tops occur at two levels, about 460 and 700 m a.s.l. These inselbergs have varied shapes: gentle slopes with exfoliation sheets as well as steep slopes with solution channels which are formed by acids from the plants growing at their tops. Extensive isolated areas on the plain are covered by residual white sands 0.5 to 1 m in thickness.

The Upper Rio Negro and its tributaries, the most notable being the Xié, Içana, Vaupés and Curicuriari on the right side, are entrenched in the flat surface formed on the Guyana Shield. The channels of these rivers present straight sections alternating with rapids and waterfalls less than 8 m high.

On the basis of the average discharge at its mouth (about 30 000 m^3 s^{-1}) the Rio Negro ranks as the largest tributary of the Amazon River and as the fifth largest river of the world (Meade et al., 1991). The drainage basin is about 600 000 km^2. The name 'Negro' is due to the large quantity of dissolved organic matter that gives a dark brown colour to the water. Paradoxically, the Negro transport only 6 million tons per year of suspended sediments to the Amazon (Forsberg et al., 1988). The tributaries of its upper basin also carry little suspended load and a small amount of bed load formed by quartz sand derived from the Guyana Shield. The drainage area at Serrinha gauging station, approximately 700 km above the mouth, is 250 000 km^2 (40% of the total drainage area) and includes the whole studied area. At Serrinha, discharges vary between approximately 4000 m^3 s^{-1} to more than 27 000 m^3 s^{-1}.

QUATERNARY SEDIMENTATION

The fluvial belts of the rivers are complex and formed by different geomorphological and sedimentological Quaternary units. In this study three geomorphological and sedimentological units were identified: upper terrace sediments, lower terrace sediments and present floodplain.

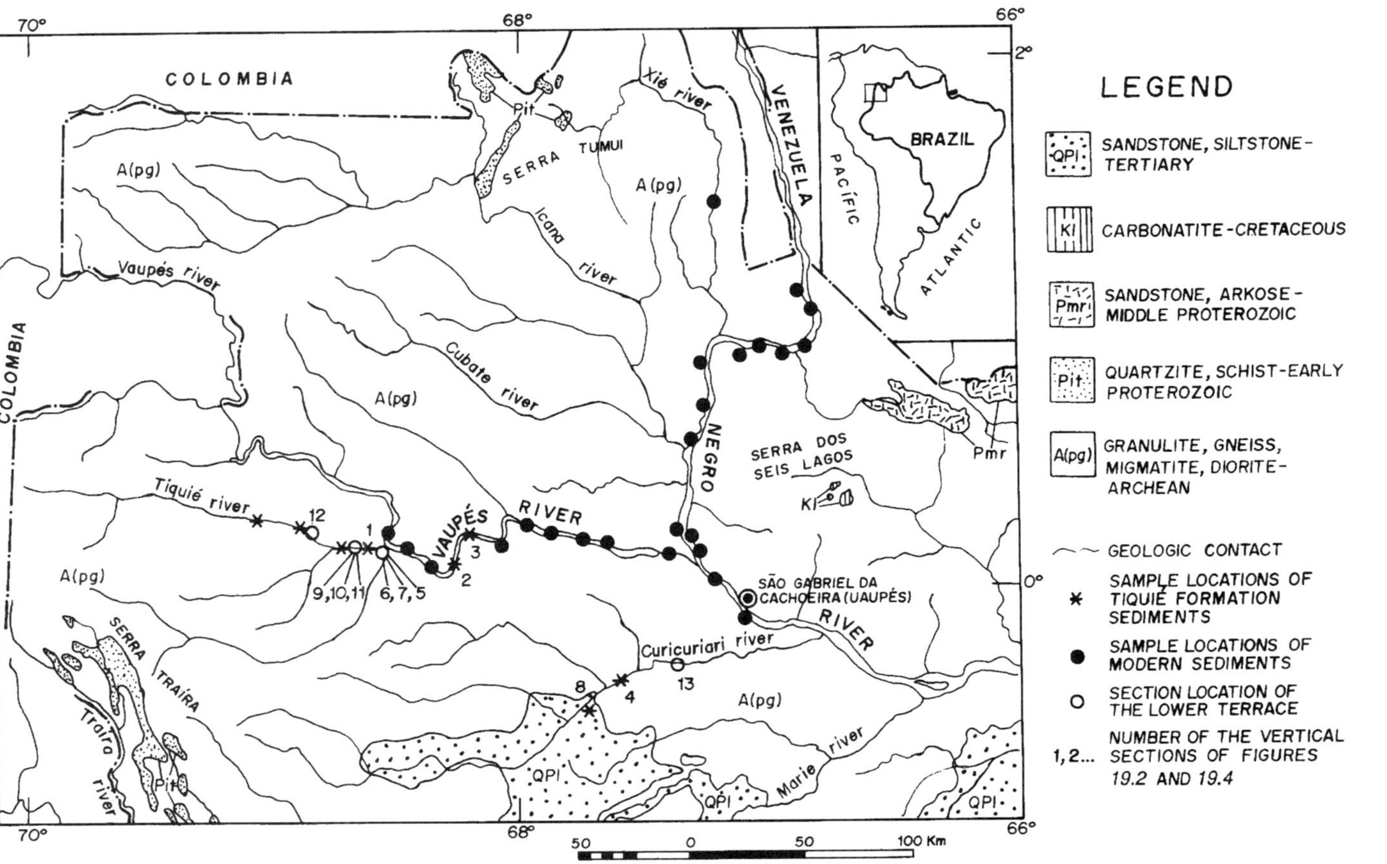

Figure 19.1 Geological map of the Upper Rio Negro region, sample locations of ancient (upper terrace, Tiquié Formation and lower terrace sediments) and modern sediments, and profile location (see Figures 19.2 and 19.4)

Upper terrace sediments

In the rivers of the Upper Rio Negro basin an ancient terrace level at a maximum height of 14 m above the water level of the rivers in the dry season was found. The Quaternary fluvial sediments of this upper terrace were defined here as the Tiquié Formation. Outcrops of this upper terrace sediment along the Vaupés, Tiquié and Curicuriari rivers are scarce, relatively small, occur in isolated patches and are generally several kilometres apart. However, this unit shows a remarkable textural similarity and equates to sedimentary structures of the sediments along the studied rivers.

The sediments of this unit were first described by Paiva (1929) as ferruginous conglomeratic sandstones impregnated by organic material that gives a characteristic dark brown to black colour. These alluvial sediments also occur in Venezuela where they are called Negrito in the Brazo Casiquiare River region. The geologists of Radambrasil's project (1976) have also briefly described the characteristics of these deposits.

The sediments are typically sandy with predominant planar cross-bedding and trough cross-bedding structures. Fine-grained sediments are subordinated, present as thin beds which are sometimes lens-shaped and not more than 40–50 cm thick. Gravels are of secondary importance compared to the sandy facies. Pebbles are found associated with the planar and trough cross-bedding structures. The sediments are impregnated with organic matter and iron oxide.

In these deposits it is common to find vegetal fragments, especially of stems and leaves and even some fragments of trunks. At the uppermost part of some outcrops, organic matter and oxides were leached leaving *in situ* residual, white friable sand (profiles 1 and 2, Figure 19.2). Grain size data of these sediments show that the deposits along the Curicuriari River are finer and more homogeneous than in the other rivers. Mean size (*Mz* of Folk and Ward, 1957) in this case ranges between 2.6 and 1.35ϕ. They are also well sorted. In comparison, the sediments of the Tiquié and Vaupés outcrops have mean sizes varying between 0ϕ and 1.24ϕ, and are poorly to moderately sorted. In all cases the particles, mostly of quartz (> 95%), are subrounded to angular in shape. X-ray diffraction of the total grain size fraction indicates that kaolinite and feldspar are present with traces of chlorite and illite/mica. Radiocarbon dates of logs and organic matter found in these sediments range between 27 ka and older than 40 ka.

The type section of the Tiquié Formation was defined in the Tiquié River. The outcrop, nearly 14 m high, occurs on the left bank of the river in the reach between Igarapé Ira and Maloca Matapí (point 1 of Figure 19.1 and profile 1 of Figure 19.2). It is formed principally of gravels up to 3 cm in diameter with sandy matrix and sandy sediments, dark brown to black in colour. Two radiocarbon samples gave ages of $27\,220 \pm 200$ [14]C yr BP (Beta 52467) and $37\,240 \pm 520$ [14]C yr BP (Beta 52465).

In the Vaupés River at Monte Alegre (point 2 of Figure 19.1 and profile 2 of Figure 19.2), a radiocarbon dating of wood indicates an age older than 38 450 ^{14}C yr BP (Beta 52470). In the Curicuriari River another radiocarbon dating of wood indicates an age older than 40 000 ^{14}C yr BP (Beta 52462) (point 4 of Figure 19.1 and profile 4 of Figure 19.2). Complementary, selected profiles of Vaupés and Curicuriari rivers are presented in Figure 19.2.

Considering these chronological data we can correlate this episode of sedimentation with the Middle Pleniglacial. However, most of the ages are beyond the range of ^{14}C dating techniques or are almost at the reliability limit of the method. Apparently,

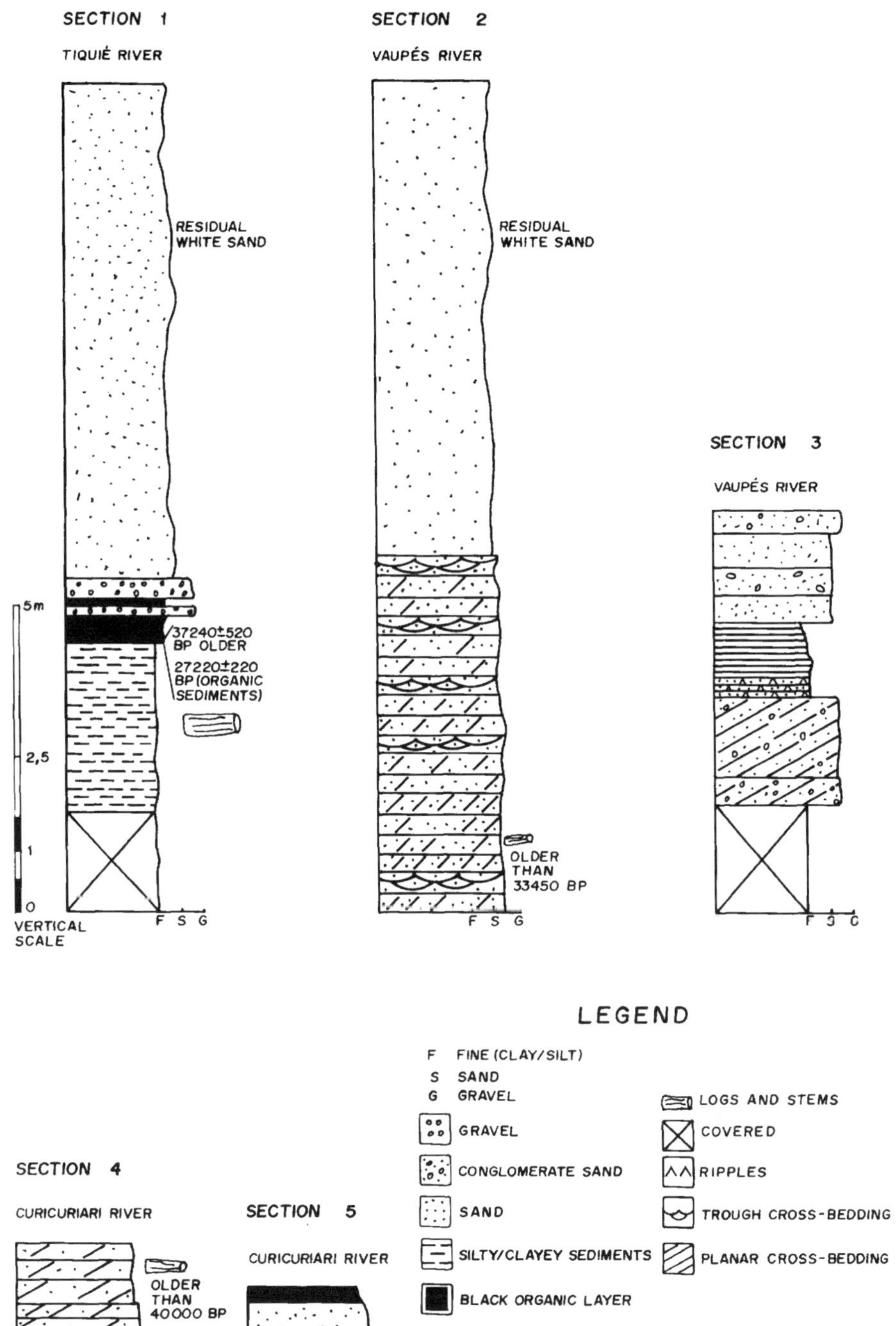

Figure 19.2   Selected vertical sections of the Tiquié Formation. The locations are indicated in Figure 19.1

contamination can be frequent in the tropics (Heine, 1994; Rasanen et al., 1992). For these reasons, the suggested correlation of the Tiquié Formation with the Middle Pleniglacial should be interpreted with caution.

Lower terrace sediments

In the Upper Rio Negro basin a lower terrace 2 to 4 m in height above the lowest water level occurs extensively along the rivers. It is composed of fine deposits that are defined here as lower terrace sediments. In general the base of this unit is not visible, but at some points it was possible to identify that it is lying in unconformity, either on the Brazilian Shield basement or on the Tiquié Formation (Figure 19.3).

The sediments of the lower terrace can be clearly differentiated from the Tiquié Formation. In the lower terrace, fine sediments (silt and clay) are predominant. Sands (principally fine and very fine sand) are scarce and the typical black to dark brown colour

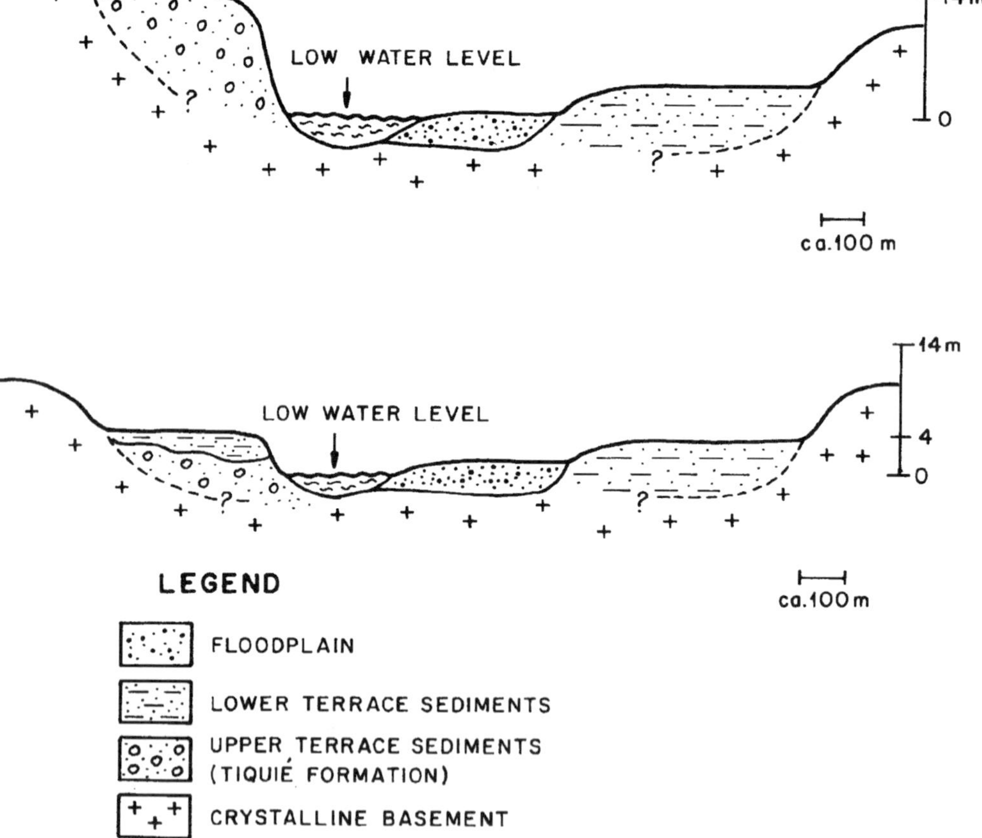

Figure 19.3   Schematic transverse sections showing the most frequent type of occurrences of the upper and lower terraces

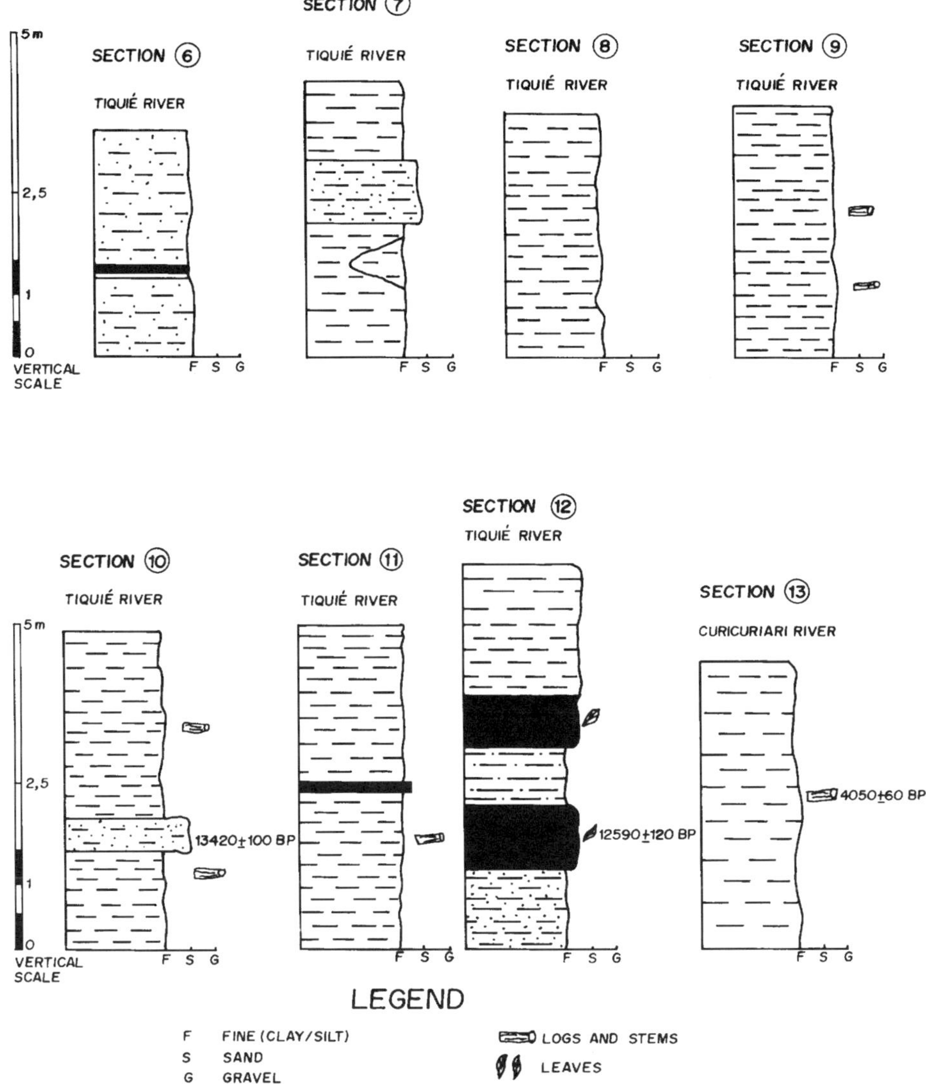

Figure 19.4   Selected vertical sections of the lower terrace. The locations of the sections are indicated in Figure 19.1

of the Tiquié Formation is absent. Well-developed beds of leaves, in the form of organic strata about 30–50 cm thick, and a considerable quantity of logs are characteristic in this unit.

The dominant lithofacies are fine laminated or massive sediments, characteristic of overbank and channel-related facies indicating lateral accretion deposition. Different vertical sections of the Tiquié, Curicuriari and Vaupés rivers are depicted in Figure 19.4. The older ^{14}C dates are Late Pleistocene ranging between *c.* 13.5 and 12.5 ka (Beta 52464 and Beta 52469; profiles 10 and 12 of Figure 19.4). The younger chronologies are Holocene (4050 ± 60 ^{14}C yr BP; Beta 52466, profile 13 of Figure 19.4). Additional datings are necessary for a more precise differentiation among the different Holocene episodes of sedimentation. X-ray diffraction analysis shows that quartz is dominant followed by considerable quantities of kaolinite. Chlorite and feldspar are scarce, while illite and mica are insignificant.

The modern floodplain

In the Amazon, black water rivers such as the Negro, which carries a small quantity of suspended load, are typical of basins that drain cratonic areas. Today, lateral migration and bank erosion of the rivers are almost insignificant in the tributaries of the Upper Rio Negro basin, as well as in the Rio Negro itself.

The bedload of the rivers produces channel bars, stable point bars and sand waves. These structures are covered by water during the high water season and are more frequent and larger in the Vaupés River. Frequently, the bars developed along or close to a Guyana Shield bedrock outcrop.

The petrographic and textural compositions of these modern sands were discussed by Franzinelli and Potter (1983), Potter and Franzinelli (1985) and Venzo et al. (1992). Mineralogically, they are pure quartz sand of more than 98% quartz. From the textural point of view, these sediments are made up of medium-sized sand with a very low percentage of silts, at the most 1%. According to Venzo et al. (1992), the mean size (Mz; Folk and Ward, 1957) varies between 0.38 and 2.81ϕ, with dominant values ranging between 1.0 and 1.5ϕ. The sorting ranges between 0.47 (well-sorted) and 1.07 (poorly sorted). The grains are subangular to subrounded.

DISCUSSION

If we consider that the Upper Rio Negro basin is located in a cratonic low relief area, the neotectonic factor can be ignored as a variable affecting the amount and type of sediment load of the fluvial system. Therefore, climate was the main variable capable of modifying fluvial behaviour during the Late Quaternary.

Data obtained from other rivers of the western Amazon basin suggest that aggradation probably dominated during the Middle Pleniglacial (*c.* 65–25 ka) in the fluvial belts. This episode of sedimentation was recognized in the Caquetá (Van der Hammen et al., 1992a), Madre de Dios (Rasanen et al., 1990) and the Ucayali rivers (Dumont et al., 1988, 1992). According to many authors (Dumont et al., 1992; Van der Hammen et al., 1992a; Rasanen and Linna, 1993; Van der Hammen and Absy, 1994) the alluvial sedimentation in rivers of western Amazonia increased simultaneously with the glacial advances in the central and northern Andes. The maximum advance of the Late Pleniglacial appears to have occurred between *c.* 34 and *c.* 27 ka (Van der Hammen et al., 1981; Clapperton, 1993). This glacial advance probably occurred in response to atmospheric cooling while humidity remained high and produced early glacier advances in tropical latitudes with respect to the last Global Glacial Maximum (Clapperton, 1993). Terraces composed of sand and even gravel suggest that the high production of coarse sediments was associated with heavy rains produced in the Andes (Van der Hammen et al., 1992a; Dumont et al., 1992; Rasanen and Linna, 1993). This interpretation, however, could be applied to rivers with headwaters in the Andes, but not for rivers whose headwaters are in the lowlands (Latrubesse and Ramonell, 1994) or in cratonic areas. The Upper Rio Negro has its headwaters in the cratonic area. Some tributaries flow in the shield area (Tiquié, Cubaté, Xié, etc.) and other tributaries run either on Tertiary sediments in their upper basins or in the shield area in their lower courses (Curucuriari, Marié).

The modern sedimentation in the Upper Rio Negro basin can be used as a key for the interpretation of the Tiquié Formation sediments. Petrographic studies on the modern sand of the Rio Negro fluvial system confirmed that granitic rocks under tropical weathering and low relief produce a small amount of fluvial sediment load made up of pure quartz sand (Franzinelli and Potter, 1983; Potter and Franzinelli, 1985). Two conditions would be necessary to produce pure quartz sands: an environment of intense chemical weathering and a mechanism to provide sufficient time for weathering action (Johnson et al., 1988).

The modern fluvial sediments of the Upper Rio Negro basin show some differences with respect to the sediments of the Tiquié Formation. Furthermore, the ancient sediments of the Tiquié Formation have a larger amount of gravel facies and organic matter (logs and leaves) that do not occur in the modern fluvial system. For these reasons, the two sedimentary units under study could indicate some minor variations in the environmental conditions and in the dynamics of the fluvial system during the episodes of sedimentation. During the fluvial activity that deposited the Tiquié Formation, the climate was tropical in the Upper Rio Negro basin, permitting the production of pure quartz sands and the input of a low quantity of suspended load to the system (Table 19.1). This climate would have had stronger seasonal contrasts of heavy rains than the present one, resulting in the occurrence of rivers with more active channels than the present-day ones. For this reason, a greater abundance of organic material (organic strata, leaves, trunks) and pebbles were produced in the ancient sediments than in the modern ones. The presence of organic detritus in the sandy deposits indicates lateral erosion and active deposition in the channel. For example, the current vegetation plays an important role for some Amazonian rivers (white water rivers, or rivers with abundant suspended load and more active channels) in accounting for channel variation and sedimentary deposits (Baker, 1978; Latrubesse, 1992). In addition, palynological data obtained by Van der Hammen et al. (1992a) in fluvial sediments of the middle Caquetá, a neighbouring basin of the Upper Rio Negro area, indicate that during the Middle Pleniglacial, drier and more open vegetation covered a larger area than nowadays, but that Amazonian forest was the dominant type of vegetation.

Considering that the radiocarbon datings of the Tiquié Formation should be considered with caution, the Middle Pleniglacial age attributed to this formation is not definitive.

A different environmental scenario can be drafted for the lower terrace that occurs extensively along the banks of the rivers in the Upper Rio Negro basin.

Lower and complex terraces with terminal Pleistocene (Late Glacial) and Middle

Table 19.1 Morpho-sedimentary evolution of the Upper Rio Negro basin during the Late Quaternary

Morpho-sedimentological unit	Dominant sediments	Processes	River type	Climate	Radiocarbon dating yr (BP: max. and min. ages)	Age
Present floodplain	Pure quartz sands	Aggradation (channel-related facies dominant)	Black water river	Tropical humid		Upper Holocene
		Dissection (formation of the lower terrace level)				Terminal Middle Holocene–Upper Holocene
Lower terrace sediments	Silty-clayey	Aggradation (floodplain-related facies dominant)	White water river	Drier than present	4050 ± 60 13 420 ± 100	Late Pleistocene (Late Glacial) to Middle Holocene.
		Dissection (formation of the upper terrace level)				Late Pleistocene (Last Glacial Maximum)
Upper terrace sediments (Tiquié Formation)	Pure quartz sands	Aggradation (channel-related facies dominant)	Black water river-like	Humid tropical, more seasonal than present	27 220 ± 200 Older than 40 000	Late Pleistocene (Middle Pleniglacial)

Holocene sediments were described in some Amazonian rivers (Iriondo, 1988; Rasanen et al., 1990, 1992; Dumont et al., 1992; Latrubesse, 1992; Van der Hammen et al., 1992b; Latrubesse and Ramonell, 1994). During the Late Glacial–Middle Holocene the discharges of some rivers could have been smaller than those found at present (Dumont et al., 1992; Van der Hammen et al., 1992b).

Radiocarbon datings in the lower terrace sediments of the Upper Rio Negro basin suggest that they were deposited between the terminal Pleistocene (Late Glacial) and the Lower–Middle Holocene. During this period, the Upper Rio Negro basin provided abundant fine sediments to the river system. The older sedimentation phase of the lower terrace occurred during the last deglaciation after approximately 14–13 ka (Table 19.1). In order to differentiate one or more phases of Holocene sedimentation, more data on absolute chronology are necessary. During this interval (14–4 ka), the Upper Rio Negro basin had a white water river behaviour, transporting abundant suspended load.

The present behaviour of black water rivers carrying pure quartz sands and very little suspended load started recently, after the Hypsithermal (c. 8–4 ka).

CONCLUSIONS

Considering the geological and chronological data presented in this study and their correlation with those of other Amazon basins, we can conclude that drastic hydrological changes occurred in the Upper Rio Negro basin during the Late Pleistocene–Holocene. The Rio Negro is considered to be a typical black water river of Amazonia. Today this giant river carries an insignificant quantity of suspended load when compared with its great water discharge. The sediments of the Tiquié Formation, comprising principally pure quartz sands, indicate that during the Middle Pleniglacial the palaeohydrological regime of the basin was similar to that of the present, but it was morphogenetically more active. We have not yet found a sedimentological record of the Last Glacial Maximum (c. 24–14 ka). Another complex phase of erosion/deposition occurred between 14 ka and 4 ka, when the lower complex terrace was deposited. The older sediments of the lower terrace, which are younger than 14 ka, are related to climatic changes associated with the last global deglaciation. During the Late Glacial and most of the Holocene, the rivers of the Upper Rio Negro basin transported abundant suspended load. At this time, most of the rivers were mainly of white water type. The present-day fluvial dynamics of black water rivers, which transport very little suspended load and have a bed load consisting of white quartz sands, are very recent. The maximum age of this black water river behaviour could be approximately 4 ka. It is likely that these palaeohydrological data of the Upper Rio Negro basin can help biologists and ecologists in explaining and understanding the biodiversity and complexity of the aquatic ecosystems of the Amazon basin.

ACKNOWLEDGEMENTS

We are grateful to Dr Antonio Rossi, of the Università degli Studi di Modena, Italy, for performing mineralogical analysis. We thank Dr Guillermina Garzón and Dr Tomasz Kalicki for critical review of the manuscript and for their suggestions that enriched this chapter.

REFERENCES

Absy, M.L., Cleef, A.L.M., Fournier, M., Martin, L., Servant, M., Sifeddine, A., da Silva, X.F., Soubies, F., Suguio, K., Turq, B. and Van der Hammen, T., 1991. Mise en évidence de quatre phases d'ouverture de la forêt dense dans le sud-est de l'Amazonie au cours des 60,000 dernières années. Première comparásion avec d'autre regions tropicales. *C. R. Academie Sciences, Paris, II*, **312**, 673–678.

Baker, V.R., 1978. Adjustment of fluvial systems to climate and source terrain in tropical and subtropical environments. In A.D. Miall (ed.), *Fluvial Sedimentology*, Canadian Society of Petroleum Geologist, Memoir 5, 211–230.

Clapperton, Ch., 1993. Nature of environmental changes in South America at Last Glacial maximum. *Palaeogeogeography, Palaeoecology, Palaeoclimatology*, **101**, 189–102.

Dumont, J.F., Lamotte, S. and Fournier, M., 1988. Neotectónica del Arco de Iquitos (Jenaro Herrera, Perú). *Boletín de la Sociedad Geológica del Perú*, **77**, 717.

Dumont, J.F., García, F. and Fournier, M., 1992. Registros de cambios climáticos para los depósitos y morfologías fluviales en la Amazonia Occidental. In L. Ortlieb and J. Macharé (eds), *Paleo ENSO records* Extended Abstracts, ORSTOM—CONCYTEC, Lima, Peru, *International Symposium*, 87–92.

Folk, R.R. and Ward, W.C., 1957. Brazos River bar: a study in the significance of grain size parameters. *Journal of Sedimentary Petrology*, **27**, 3–26.

Forsberg, B., Martinelli, E. and Richey, J., 1988. Sediment delivery rates for the Amazon river and its principal Brazilian tributaries. *Chapman Conference American Geophysical Union*, 77–81.

Franzinelli, E. and Potter, P., 1983. Petrology, chemistry and texture of modern river sands, Amazon river system. *Journal of Geology*, **91**, 23–39.

Heine, K., 1994. The Mera site revisited: Ice-Age Amazon in the light of the new evidence. *Quaternary International*, **21**, 113–119.

Iriondo, M., 1988. A comparison between the Amazon and Paraná river systems. *Mittelungen des Geologish-Paleontologisches Institut*, **66**, 77–92.

Iriondo, M. and Latrubesse, E., 1994. A probable scenario for a dry climate in Central Amazonia during the late Quaternary. *Quaternary International*, **21**, 121–128.

Johnson, M., Stallard, R. and Meade, R., 1988. First cycle quartz arenites in the Orinoco basin, Venezuela and Colombia. *Journal of Geology*, **96**, 263–276.

Latrubesse, E., 1992. *El Cuaternario fluvial de la cuenca del Purus en el Estado de Acre, Brasil*, PhD Thesis, Universidad Nacional de San Luis, Argentina.

Latrubesse, E. and Ramonell, C., 1994. A climatic model for Southwestern Amazonía at Last Glacial times. *Quaternary International*, **21**, 163–169.

Meade, R., Rayol, J., da Conceição, S. and Natividade, J., 1991. Backwater effects in the Amazon River basin of Brazil. *Environmental Geology and Water Sciences*, **18**, 105–114.

Paiva, 1929. *Vale do Río Negro: Phísiografia e Geologia*, Servico Geologico e Mineral, Rio de Janeiro, 40–62.

Potter, P. and Franzinelli, E. 1985. Fraction analyses of modern river sand of rios Negro and Solimões, Brazil, implications for the origin of quartz-rich sandstones. *Revista Brasileira de Geociências*, **15**, 31–35.

Radambrasil, 1976. Departamento Nacional de Produção mineral-DNPM. *Geologia*, **11**, 1–137.

Rancy, A., 1991. *Pleistocene mammals and paleoecology of the western Amazon*, PhD thesis, University of Florida.

Rancy, A., 1992. Western Amazon – Paleomammals and the forest refugia model. In E. Franzinelli and E. Latrubesse (eds), *International Symposium on the Quaternary of Amazonia*, UFAM, Manaus, Abstracts and Scientific Contributions, 45–48.

Rasanen, M. and Linna, A., 1993. Late Pleistocene alluvial terrace deposits and their dating in Peruvian Southwestern Amazonia. In E. Franzinelli and E. Latrubesse (eds), *International Symposium on the Quaternary of Amazonia*, UFAM, Manaus, Abstracts and Scientific Contributions.

Rasanen, M., Salo, J.S., Jungens, H. and Romero Pitman, L., 1990. Evolution of western Amazon lowland relief: impact of Andean Foreland dynamics. *Terra Nova*, **2**, 320–332.

Rasanen, M., Neller, R., Salo, J. and Jungens, H., 1992. Recent and ancient fluvial deposition system in the Amazonian foreland basin Peru. *Geological Magazine*, **129**, 293–306.

Santos, O., Nelson, B. and Giovannini, C.A., 1993. Campos de dunas: Corpos de areia sob leitos abandonados de grandes ríos. *Ciência Hoje*, **16**(93), 22–25.

Van der Hammen, T. and Absy, M.L., 1994. Amazonia during the Last Glacial. *Palaeogeography, Palaeoclimatology, Palaeoecology*, **109**, 247–261.

Van der Hammen, T., de Jong, E. and de Veer, A., 1981. Glacial sequence and environmental history in the Sierra Nevada del Cocuy (Colombia). *Paleogeography, Palaeoclimatology, Paleocology*, **32**, 247–340.

Van der Hammen, T., Duivenvoorden, J., Lips, J.X., Espejo, N. and Rego, L. 1992a. The late Quaternary of the middle Caquetá River area (Colombian Amazonia). *Journal of Quaternary Science*, **7**, 45–55.

Van der Hammen, T., Urrego, L., Espejo, N., Duivenvoorden, J. and Lips, J., 1992b. Late Glacial and Holocene sedimentation and fluctuation of river water level in the Caquetá river area (Colombia, Amazonia). *Journal of Quaternary Science*, **7**, 57–67.

Venzo, G., Franzinelli, E., Lenardon, G. and Marocco, R., 1992. Mineralogy, grain size, roundness and surface microstructures of modern sands in the Alto rio Negro partial basin (Brasil–Colombia–Venezuela). *Studi Trentini di Scienze Natural. Acta Geologica*, **67**, 31–49.

CHAPTER 20

Palaeohydrological Changes in the Upper Paraná River, Brazil, During the Late Quaternary: A Facies Approach

JOSÉ C. STEVAUX AND MANOEL L. DOS SANTOS

Department of Geography, University of Maringá, Brazil

INTRODUCTION

The Upper Paraná River basin area is about 820 000 km^2 and includes the most industrial and populated cities of Brazil. The principal rivers in the basin – the Paraná, Paranaiba, Paranapanema, Grande and Peixe, for example – are almost totally impounded (Figure 20.1). Two of the largest dams in the world – the Itaipu and Porto Primavera Hydroelectricity Plants, with 12 600 MW and 1814 MW, respectively (Amarante, 1982) – are located in the upstream and downstream limits of the study area: the last reach of the Paraná River in its natural condition.

Despite its high density of occupation, the Paraná River has only a short period of historical data for discharge, water level, flow velocity and flood magnitude. For example, in the study area, the fluvial station of Porto São José, located between the Itaipu, Porto Primavera and Rosana dams, has operated only since 1964.

Presently, the area is undergoing rapid economic development and growth involving expansion in agrobusiness, urbanization, transportation networks and hydroelectric dams, that require a more extensive knowledge of river hydrology over a longer time than afforded by the limited gauge records. These factors emphasize the value of the palaeohydrologic information for guiding occupation and management of the region.

In this study, a change in hydrological characteristics of the Paraná River during the Holocene is interpreted from facies analysis of alluvial deposits.

CHARACTERISTICS OF THE UPPER PARANÁ FLUVIAL SYSTEM

The study area is located near the town of Porto Rico (lat. 22°43′S, long. 53°10′W) just downstream from the mouth of the Paranapanema River (Figure 20.1). Here, the Paraná River has a multichannelled system 3.4–4.0 km wide with an extensive flood plain (5.0 to 6.0 km wide) on its right (northwest) margin, where there are many lakes, back swamps and secondary streams that form a mixed braided–anastomosing alluvial plain.

The annual range of temperature varies from 10.3 to 33.6 °C, averaging 22 °C, and there

Palaeohydrology and Environmental Change. Edited by G. Benito, V. R. Baker and K. J. Gregory
© 1998 John Wiley & Sons Ltd.

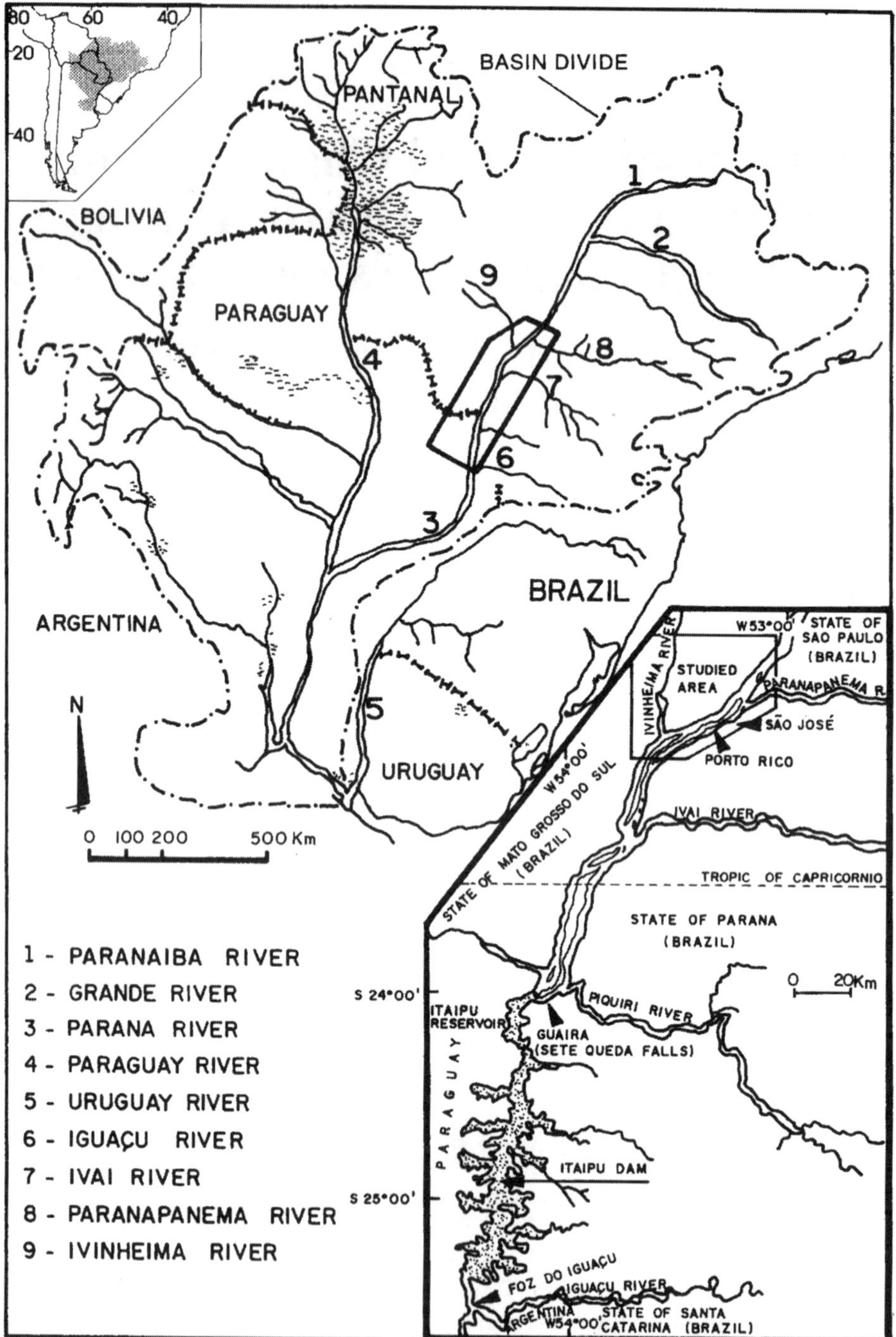

Figure 20.1 Del Plata Basin showing relative areas drained by the Paraná, Paraguay and Uruguay rivers (modified from Bonetto et al., 1989).The Paraná–Paraguay river basin covers about $2.8 \times 10^6 \, \text{km}^2$

is an annual rainfall of 1200 mm, so that the region has a tropical–subtropical climate. In the fluvial station of Porto São José, the Paraná River has a discharge that varies from 8400 m^3 s^{-1} (June to August) to 13 000 m^3 s^{-1} (November to March), with the highest discharge recorded as 33 470 m^3 s^{-1} in 1983 and the lowest 2550 m^3 s^{-1} in 1969. The river slope in this area is 0.096 m km^{-1} (Stevaux, 1994, pp. 143–144).

The Paraná River valley in the study area has four geomorphological units (Figure 20.2, Table 20.1), where the two youngest can be directly correlated with Paraná River activity (Stevaux, 1994).

Table 20.1   Geomorphological units of the Upper Paraná River area

Geomorphological unit	Age	Deposits	Height above river water level (m)
Porto Rico	Pleistocene	Soil and colluvium from Caiuá Formation	20 to 80
Taquaruçu	Middle Pleistocene	Sandy colluvium and gravel in the base	30 to 50
Fazenda Boa Vista	40–30 ka (base)	Alluvial sand and gravel (Paraná River's terrace)	8 to 20
Rio Paraná	6 ka (base)	Sandy mud (flood plain deposits), sand (channel deposits)	0 to 4

Modified from Stevaux (1994)

PREVIOUS WORKS

Except for some reports about gravel availability in dam construction areas, studies of sedimentation of the Paraná River are very few and limited in scope. Studies regarding palaeohydrology are practically absent. Suguio et al. (1984) studied the sedimentology of alluvial deposits of the Paraná River in the area of the Porto Primavera Dam, and presented the first hypothesis of its hydrological changes during the Quaternary. Nogueira Jr (1988) studied gravel deposits at the same locality from the point of view of their geotechnical features. Fernandez (1990) focused on geotechnical characteristics of bank deposits related to marginal erosion and proposed three types of channel margin: stable, accretional and erosional. Santos (1991) presented the first study on sand bar morphology and facies and distinguished between lateral and central bars. Stevaux (1994) made an extensive description of alluvial deposits. Using Miall's (1977, 1978, 1985a,b) facies classification, he recognized 13 facies in Paraná sediments. He also presented some hypotheses for climatic evolution of the area at the end of the Quaternary. The first study on Paraná River palaeohydrology was made by Stevaux et al. (1995). The authors identified a strong change in river hydrology at the beginning of the Holocene. At that time, the river pattern changed from braided to anastomosing.

FACIES ANALYSIS AND PALAEOHYDROLOGICAL INFERENCES

The Paraná River alluvial deposits have a strong asymmetrical shape and their thickness varies from 4 to 20 m. The terrace shows a thickness of 10 to 15 m, the flood plain 4 to 6 m

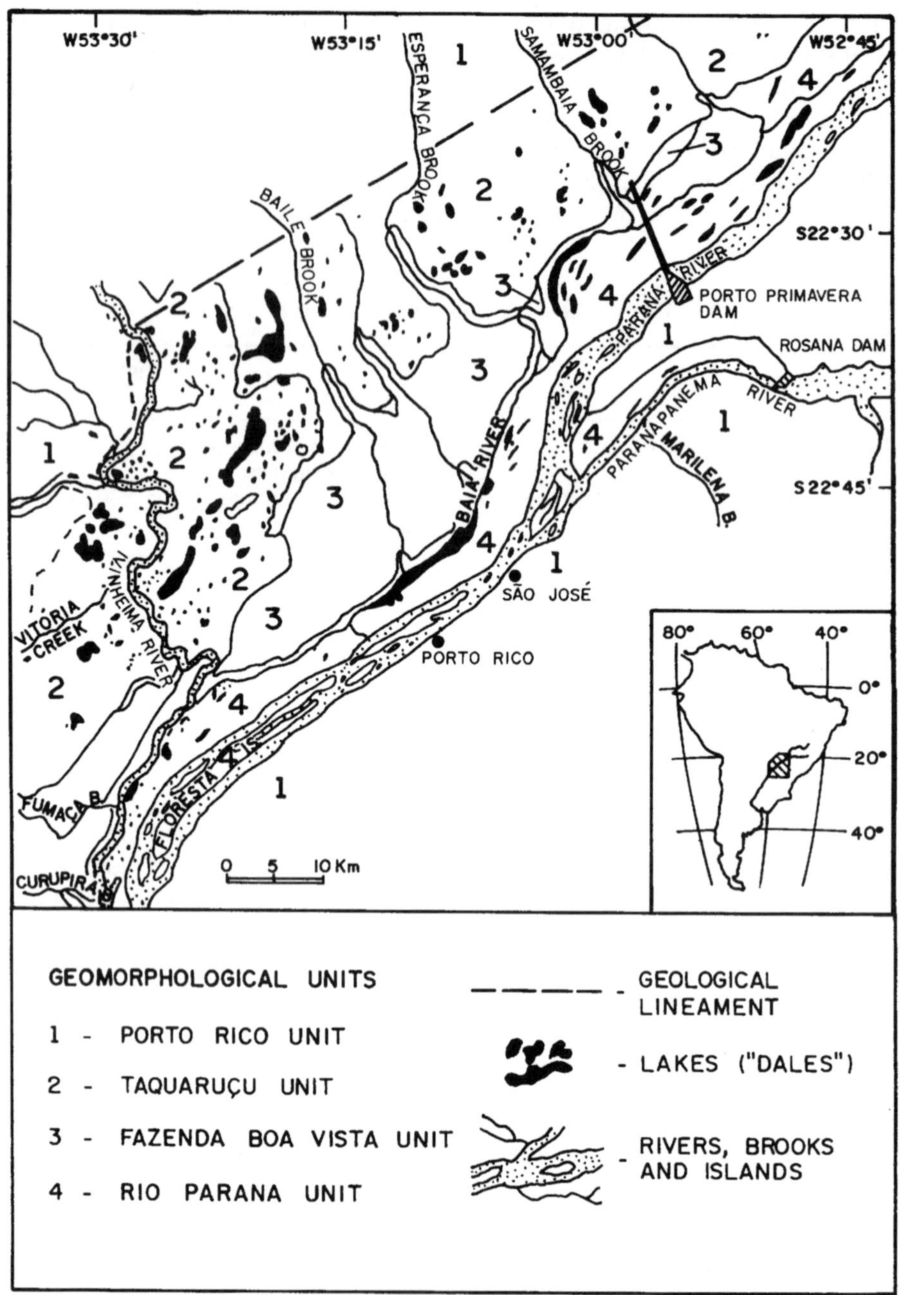

Figure 20.2   Location and geomorphological map of the Upper Paraná River near Porto Rico

and channel deposits from 10 to 20 m. The alluvial deposits belong to Facies Associations 1 and 2 whose units are described and classified using Miall's (1978, 1985a,b) lithofacies classification (Figure 20.3, Tables 20.2 and 20.3).

Facies Association 1

Polymictic gravel facies unit

The lowermost unit consists of 0.5 to 1.0 m of crudely stratified to massive pebble to cobble gravel (Gm, Gms facies) with lenses of coarse sand and trough cross-bedded sandy gravel (Gt facies). Gravels are clast-supported and sandy matrix-supported. Most of the clasts are well-rounded, locally imbricated and exhibit normal to inverse grading. Clast lithologies include agate, chalcedony, chert, sandstone, quartz and quartzite. This unit is practically continuous at the base of alluvial deposits. In this unit, grain flow and debris flow textures are common. Sieve and lag deposits are also present. Pebble imbrication and trough cross-stratified gravelly sand indicates deposition in longitudinal bars in a braided channel (Miall, 1977, 1978, 1985a,b; Stevaux, 1994). These processes are common in high-energy fluvial or fan-braided systems in semi-arid to arid environments (Nilsen, 1982). An example of this unit can be observed in the base of the core shown in Figure 20.4.

Gravelly sand facies unit

This unit is 7 to 8 m thick (about 80% of the total thickness of Association 1) in terrace deposits and about 2 to 3 m thick in the channel. It is characterized by interbedding of fine to medium gravelly massive or cross-stratified sand (Sm, Sp, Sg facies) and lenticular bodies of massive or cross-stratified sandy gravel with very fine to coarse pebbles (Gms, Gt facies). This facies unit was deposited in a sandy gravel braided system.

Fine sand to muddy sand facies unit

This unit consists of massive to stratified, fine sand, moderately sorted and 2 to 3 m thick. In its base, this unit develops a thin layer of organic sandy mud interpreted as a palaeosoil. Some lenses of massive sandy mud occur in the sandy sequence. Given the stratigraphic position of this unit, deposited over the gravelly sand unit, it is best interpreted as a deposit of an anastomosed fluvial system developed at the end of Association 1 deposition. This hydrological change, however, is not well studied.

Facies Association 2

Stratified sand facies unit

This unit is present in the upper part of the channel and in the flood plain deposits. In the upper part of the channel, the thickness varies from 10 to 12 m. Lenses and blankets of coarse to fine sand and gravelly sand are 100 m wide and 0.8 to 1.0 m thick, with local levels of limonite cementation. Planar and trough cross-stratified sand (Sp and St) are very common and were generated by mega-ripple and dune bedforms in low to medium

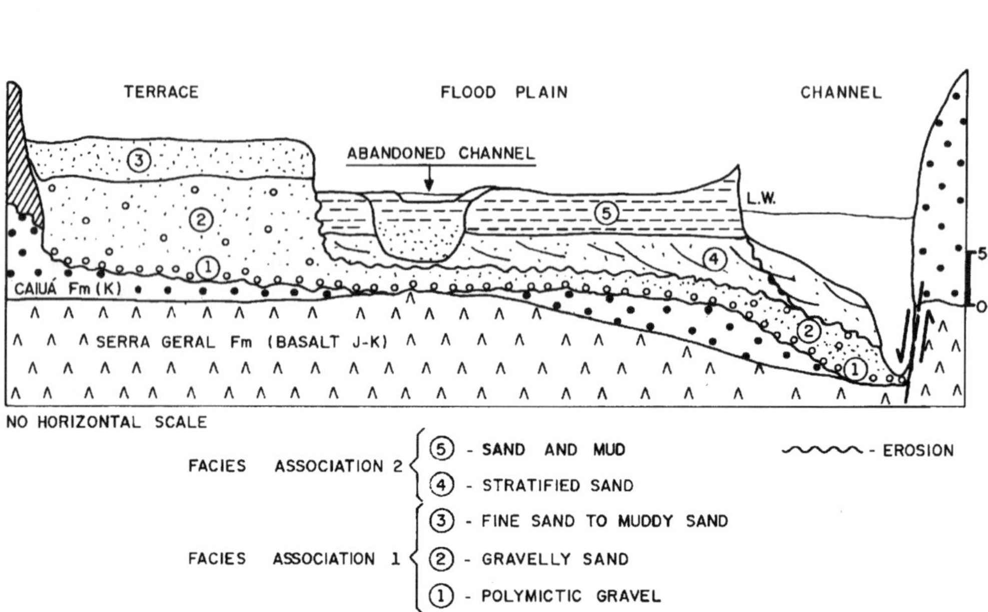

Figure 20.3   Facies relation and geological schematic section of the Paraná River deposits

flow regimes. Sigmoidal stratified sand (Ss) is present in layers 0.8 m thick erosionally truncated by reactivation surfaces. In the flood plain this unit is no more than 1.0 m in thickness and presents the same features as in the channel. This facies corresponds to the present anastomosing channel processes (Smith, 1986, pp. 184–186; Stevaux, 1994, p. 152).

Sand and mud facies unit

This unit consists of reddish brown (10YR5/6 to 2YR5/8 to 2.5YR4/8, Munsell Soil Colour Chart, 1995) and very dark brown to medium grey (7.5YR2/0 to 7.5YR5/0), massive or crudely laminated sandy mud or silt (Fm facies). Sandy silt layers are patchy; lamination disrupted by burrowing organisms can be identified in X-ray radiographs. Some intervals have pelletal fabric and pedogenetic structures. Organic mud or vegetation remnants are uncommon. Sand in this unit is white (10YR8/2), fine-grained and occurs in beds from millimetres to 5 cm thick, with climbing ripple lamination and lenses (Sr facies). This unit is interpreted as forming through vertical accretion in a flood plain environment, where sand layers were deposited as crevasse splays or natural levees. About nine ^{14}C dates were obtained from cores of the three facies units (Figure 20.4, Table 20.4).

PALAEOHYDROLOGY IN THE CONTEXT OF ENVIRONMENTAL CHANGE

Stevaux (1994) inferred four climatic events for the Upper Paraná River area during the Late Pleistocene and Holocene (Table 20.5) based on facies analysis in lake and fluvial

Table 20.2   Lithofacies classification

Facies code	Lithofacies	Sedimentary structures	Interpretation
Gm	Massive matrix-supported gravels	Grading	Debris flow deposits
Gms	Massive or crudely bedded gravel	Horizontal bedding, imbrication	Longitudinal bars, lag and sieve deposits
Gt	Gravel	Trough cross-bedding	Minor channel fills
Br	Resedimented blocks	Massive	–
Sp	Medium to coarse sand, pebbly	Planar cross-bedding	Dune and megaripples
Sr	Very fine to fine sand, mud	Ripple drift and climbing, drape and lenses	Low velocity flow
So	Fine to very fine sand, organic-rich mud	Laminations, bioturbation and mottled	–
Sh	Fine to medium sand	Planar stratification	Main channel bed form
Sl	Fine sand to mud	Planar lamination	Low velocity flow
Sm	Fine to medium sand	None	Sandy gravel braided system
St	Fine to coarse sand	Trough cross-stratification	Channel fills
Ss	Fine to coarse sand	Broad, shallow scours	Scour fills
Fm	Silt and clay	Massive	Overbank deposits
Fl	Mud, fine to very fine sand	Planar and rhythmic lamination	Levee deposits

Modified from Miall (1978, 1985a,b) and Stevaux (1994)

Table 20.3   Facies associations of Paraná River alluvial deposits

Faces association	Faces unit	Lithofacies assemblage	Age	Environment
Association 1	Polymictic gravel	Gm, Gms (Gt)	no date	Alluvial fan, gravel bar
	Gravelly sand	Sg, Sm, Sp (Gms, Gt)	> 40 ka	Longitudinal bar, braided channel
	Fine sand to muddy sand	Sr, Sm, Sh	?	Ripple bed form
Association 2	Stratified sand	Sp, St (Ss)	c. 8.0 ka	Megaripple and dune bedforms, anastomosing channel
	Sand and mud	Fm, Fl, Sr, So	4.4 ka	Flood plain/levee

deposits plus palynological analysis. These climatic events show a close correlation with the palaeohydrology of the Paraná River described previously. The braided system under dry climatic conditions (debris and grain flow in sandy deposits) of Association 1 was developed during the First Drier Event. The First Wetter Event, initiated c. 8.0 ka, induced an intensive change in river hydrology in which the Paraná changed from a braided pattern to an anastomosing pattern with a wide flood plain. Large islands, swamps, natural levees and many secondary channels were developed during this period. The short time interval of climatic transition back to drier conditions (Second Drier

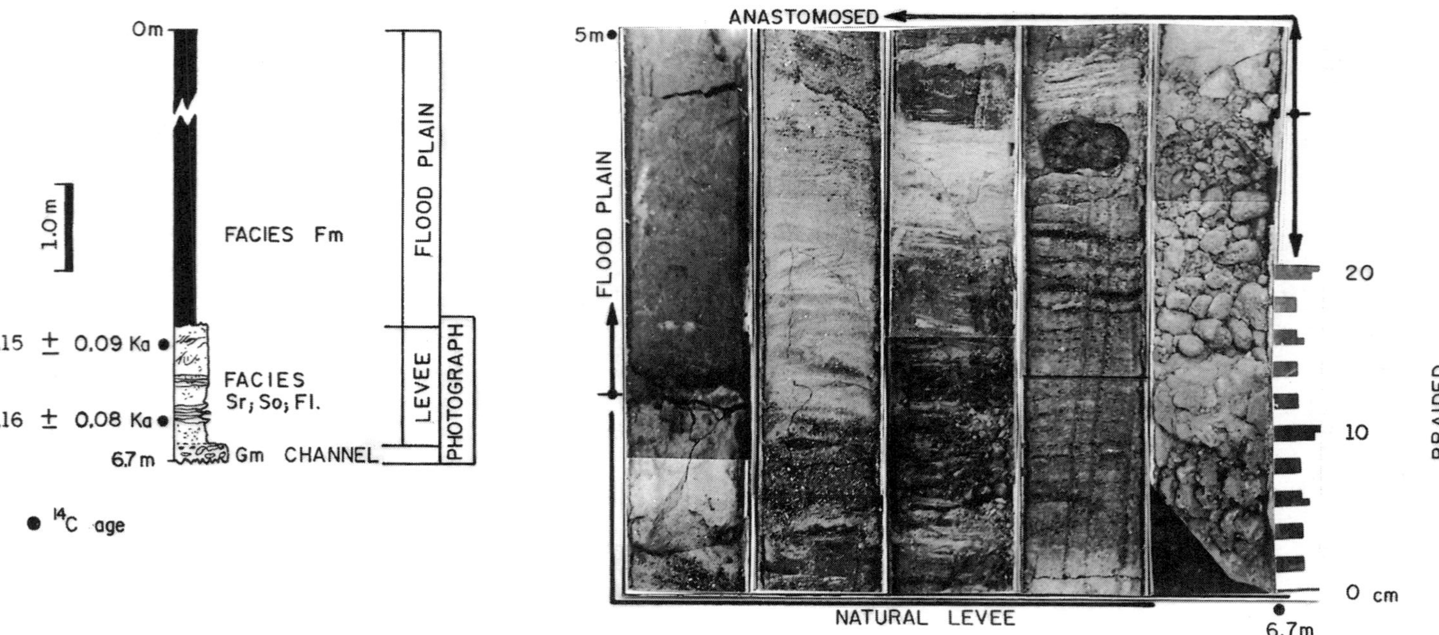

Figure 20.4   Geological profile of the Paraná River deposits at Porto Primavera. Profile presents the hydrological change from a braided system (polymictic gravel facies unit in the base) to the anastomosed system (sand and mud facies unit)

Table 20.4 Facies units and ^{14}C dating survey

Facies unit	Dating process	Number of dates	Age (ka)	Sample
Polymictic gravel	–	0	?	–
Gravelly sand	^{14}C	4	> 40[1,2]; 42.52 ± 1.70[3]; 31.14 ± 1.7[4]	Leaves, wood
Stratified sand	correlation	0	c. 8	–
Sand and mud	^{14}C	5	4.8 ± 0.1[5]; 3.16 ± 0.08[6]; 3.15 ± 0.09[7]; 1.99 ± 0.08[8]; 0.27 ± 0.06[9]	Leaves, peat and wood

[1] In Nogueira (1988); [2] CENA #110, standard; [3] BETA #92911, standard; [4] BETA #92912, AMS; [5] CENA #112, standard; [6] BETA #92910, standard; [7] BETA #92909, standard; [8] BETA #92909, standard; [9] BETA #92908, standard. BETA: Beta Analytic, Florida, USA; CENA: Universidade de São Paulo, Brazil.

Table 20.5   Climatic changes in the Upper Paraná River area during the Holocene

Climatic event	Time interval (ka)	Comparison with present climate
First Drier Event	> 8.0	Drier
First Wetter Event	8.0–3.5	Wetter
Second Drier Event	3.5 to 2.5	Drier
Second Wetter Event	2.5 to present	Present climate

Modified from Stevaux (1994)

Event: 3.5 to 2.5 ka) is not identified from facies analysis but from a series of abandoned channels along the west margin of the Paraná River. Stevaux et al. (1997) identified other evidence for this climatic change such as alluvial fan development and aeolian features in the Fazenda Boa Vista and Taquaruçu geomorphological units (Figure 20.2). Neotectonics was also postulated by Stevaux et al. (1997) in order to explain channel abandonment. The present climatic condition promoted an incision in the river thalweg that has reached the Cretaceous sandstone of the Caiuá Formation in some sections.

Large amounts of ancient fluvial sand and gravel were entrained in the channel by the last thalweg incision. The increase in bed load rate induced a disequilibrium condition in the Paraná River resulting in sand and gravel bar forms that are uncommon in an anastomosing river. This fact suggests that a new classification is needed for the Paraná River: a mixed braided–anastomosing channel pattern.

Figure 20.5 shows the stratigraphic relations between alluvial deposits of the ancient braided system and the deposits of the anastomosed system developed since the beginning of the Holocene.

CONCLUDING DISCUSSION

The evidence of a strong climatic change at the Pleistocene–Holocene transition and lesser changes during the Holocene is recognized throughout southern, central southern and northern Brazil and northeastern Argentina (Fairbridge, 1976; Tricart, 1982; Markgraf, 1989; Servant et al., 1989; Clapperton 1993; Iriondo and Garcia, 1993; Ledru, 1993; Thomas and Thorp, 1995, 1996; Stevaux et al., 1997). Based on generalizations by

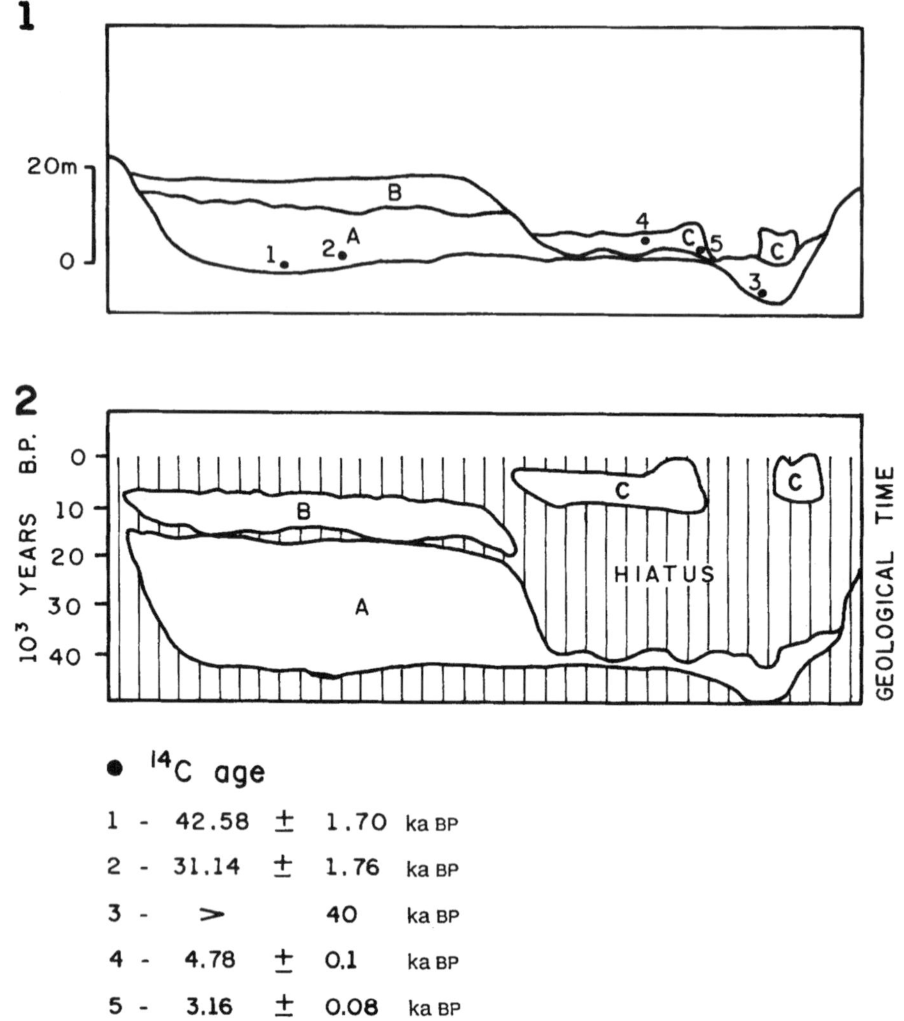

Figure 20.5   Geological (1) and chronostratigraphic (2) cross-sections of the Paraná deposits. Legend: A, ancient braided system; B, ancient anastomosed system; C, modern anastomosed system

Thomas and Thorp (1995, 1996), it was possible to recognize the changing processes in the study area during the Holocene which are summarized below (Figure 20.6).

(1) During the First Drier Event, the Paraná River developed a typical braided channel with gravelly and sandy deposits (Facies Association 1); the entire river valley was covered by braided channels during this interval.
(2) During the climatic change around 8.5 ka (First Wetter Event), the alluvial plain was vertically and laterally excavated producing a terrace on its west margin (Fazenda Boa Vista Geomorphological Unit); at this time ancient braided deposits were almost

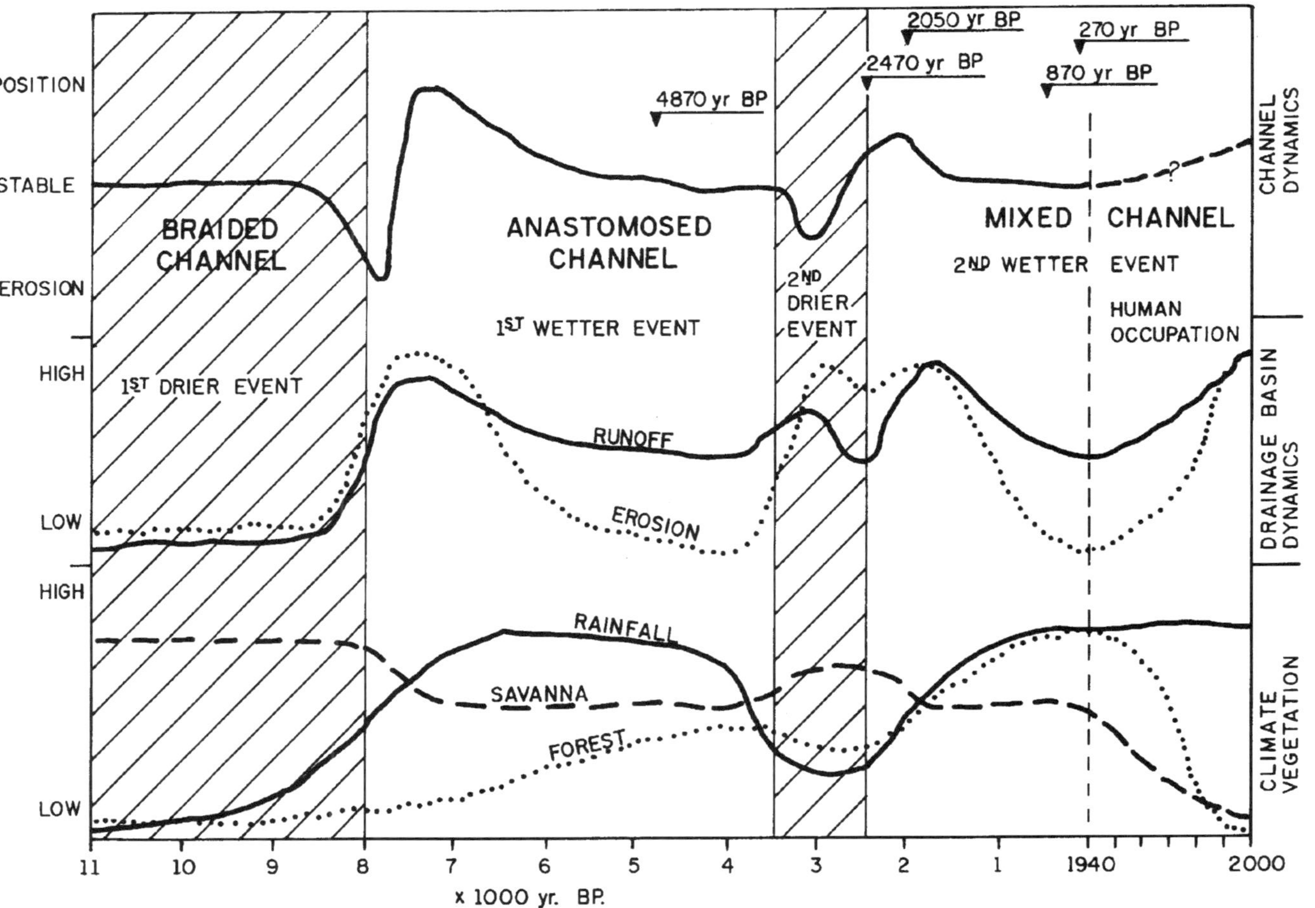

Figure 20.6   Vegetation, climate and geomorphic processes in the Upper Paraná River during the Holocene (adapted from Thomas and Thorp, 1995)

completely eroded and preserved only on the terrace and in the deepest parts of the channel. An anastomosing pattern was developed and the wide flood plain began to be constructed (Facies Association 2). The pluvial subtropical forest began to develop in the area (Klein, 1975).

(3) Between 3.5 and 2.5 ka (Second Drier Event) there was a short period of drier climate, not as intensive as the first one, that slightly altered the state of the river equilibrium; many secondary channels were abandoned and produced many ox-bow lakes and wet areas in the flood plain, and the main channel migrated to its margin.

(4) From 2.5 ka to the present, the channel thalweg incised the ancient deposits and continued to migrate towards the left margin. The river, although still having an anastomosing pattern with a large number of islands, developed an uncommon mixed anastomosing–braided channel with sand and gravel bars. The present pluvial subtropical forest reached its maximum during this interval (Klein, 1975).

(5) Since 1940 the area has been extremely impacted by human activity. The forest cover has been reduced to less than 1% and many farms and towns have developed in the area. Surface and gully erosion, soil degradation and desertification are the main problems in the area. The impact of these activities on the river dynamics is not yet understood.

ACKNOWLEDGEMENTS

We gratefully acknowledge the financial support granted by the CNPq (Conselho Nacional de Desenvolvimento Científico e Tecnológico). We also acknowledge the encouragement and suggestions of Dr Vera Markgraf (University of Colorado, USA), Dr Martín Iriondo (Universidad del Litoral, Argentina) and Dr Paul Potter (Universidade Federal do Rio Grande do Sul, Brazil). We also thank the anonymous reviewer for invaluable comments and suggestions for improving the manuscript.

REFERENCES

Amarante, C.A.P., 1982. *Dams in Brazil*, Brazilian Committee on Large Dams, São Paulo, Brazil.

Bonetto, A.A., Wais, J.R. and Castello, H.P., 1989. The increasing damming of the Paraná Basin and its effects on the lower reaches. *Regulated Rivers*, **4**, 333–346.

Clapperton, C.M., 1993. *Quaternary Geology and Geomorphology of South America*, Balkema, Rotterdam.

Fairbridge, R.W., 1976. Effects of Holocene climatic change on some tropical geomorphic processes. *Quaternary Research*, **6**, 529–566.

Fernandez, O.V.Q., 1990. *Mudanças no Canal Fluvial do Rio Paraná e Processos de Erosão nas Margens: Região de Porto Rico, PR*, Master's dissertation, UNESP, Instituto de Geociências e Ciências Exatas, Rio Claro, Brazil.

Iriondo, M.H. and Garcia, N.O., 1993. Climatic variation in the Argentine plains during the last 18,999 years. *Palaeogeography, Palaeoclimatology, Palaeoecology*, **66**, 77–92.

Klein, R.M., 1975. Southern Brazilian phytogeographic features and probable influence of Upper Quaternary climatic changes in the floristic distribution. In J.J. Bigarella and R.D. Baker (eds), *International Symposium on the Quaternary, Special Contribution*, Curitiba, 1975, Boletim Paranaense de Geociências, **33**, 67–88.

Ledru, M.-P., 1993. Late Quaternary environmental and climatic changes in central Brazil. *Quaternary Research*, **39**, 90–98.

Markgraf, V., 1989. Paleoclimates in central and south America since 18 000 BP based on pollen and lake-level record. *Quaternary Science Reviews*, **18**, 1–24.

Miall, A.D., 1977. A review of the braided-river depositional environment. *Earth-Science Review*, **13**, 1–62.

Miall, A.D., 1978. Fluvial sedimentology: an historical review. In A.D. Miall (ed.), *Fluvial Sedimentology*, Canadian Society of Petroleum Geologists, Memoir 5, 1–47.

Miall, A.D., 1985a. Architectural-element analysis applied to fluvial deposits. *Earth-Science Review*, **22**, 261–308.

Miall, A.D., 1985b. Sedimentation on an early Proterozoic continental margin under glacial influence: the Gowganda Formation (Huronian), Elliot Lake area, Ontario, Canada. *Sedimentology*, **32**, 763–788.

Nilsen, H., 1982. Alluvial fan deposits. In P.A. Scholle and D. Spearing (eds), *Sandstone Depositional Environments*, The American Association of Petroleum Geologists, Memoir 31, 49–86.

Nogueira Jr, J., 1988. *Possibilidades de Colmatação Química dos Filtros da Barragem de Porto Primavera (SP) por Compostos de Ferro*, Master's dissertation, Universidade de São Paulo, Instituto de Geociências, São Paulo, Brazil.

Santos, M.L., 1991. *Faciologia e Evolução de Barras de Canal do Rio Paraná na Região de Porto Rico (PR)*, Master's dissertation, UNESP, Instituto de Geociências e Ciências Exatas, Rio Claro, Brazil.

Servant, M., Soubiés, F., Suguio, K. Turcq, B. and Fournier, M., 1989. Alluvial fans in south-eastern Brazil as an evidence for early Holocene dry climatic period. *International Symposium on Global Changes in South America During the Quaternary*, São Paulo, 1989, Special Publication, 1, 75–77.

Smith, G.E., 1986. Anastomosing river deposits, sedimentation rates and basin subsidence, Magdalena River, NW Colombia, South America. *Sedimentary Geology*, **46**, 177–196.

Stevaux, J.C., 1994. The Upper Paraná River (Brazil): geomorphology, sedimentology and paleoclimatology. *Quaternary International*, **21**, 143–161.

Stevaux, J.C., Takeda, A. and Moraes, M., 1995. Dinâmica sedimentar no canal do rio Paraná. *Anais VI Simpósio Sul-Brasileiro de Geologia*, Porto Alegre, Brazil, 284–287.

Stevaux, J.C., Souza Filho, E.E. and Jabur, I.C., 1997. A história quaternária do rio Paraná em seu alto curso. In A.E.A.M Vazzoler, A.A. Agostinho and N.S. Hahn (eds), *A Planíce de Inundação do Alto Rio Paraná*, EDUEM, UEM – NUPELIA. Maringá, Brazil, 47–72.

Suguio, K., Nogueira Jr, J. Tanaguchi, H. and Vasconcelos, M.L., 1984. Quaternário do rio Paraná em Pontal do Paranapanema: proposta de um modelo de sedimentação. *Anais do XXXIII Congresso Brasileiro de Geologia*, Rio de Janeiro, **1**, 10–18.

Thomas, M.F. and Thorp, M.B., 1995. Geomorphic response to rapid climatic and hydrologic change during the Late Pleistocene and Early Holocene in the humid and sub-humid tropics. *Quaternary Science Review*, **14**, 193–207.

Thomas, M.F. and Thorp, M.B., 1996. The response of geomorphic systems to climatic and hydrological change during the Late Glacial and early Holocene in humid and sub-humid tropics. In J. Branson, A.G. Brown and K.J. Gregory (eds), *Global Continental Changes: the Context of Paleohydrology*, Geological Society, London, Special Publication 115, 139–153.

Tricart, J., 1982. El Pantanal: Un ejemplo del impacto geomorfológico sobre el ambiente. *Informaciones Geográficas*, Universidad de Chile, Santiago, Chile, **29**, 81–97.

PART VI

Response of Extreme Events to Climate Change

Extreme Events in the Last 200 Years in the Upper Vistula Basin: Their Palaeohydrological Implications

LESZEK STARKEL

Polish Academy of Sciences, Cracow, Poland

INTRODUCTION

Continuous meteorological and hydrological records, old maps as well as monitoring of floods, soil erosion, landslides and other phenomena give the possibility of studying the causes, course and effects of various extreme events. By comparison of the hydrological phenomena and their sedimentological and geomorphic effects, as well as their frequency, it is possible to calibrate them in order to reconstruct water circulation during the Holocene based on inherited forms and sediments. This methodology is well elaborated (Baker et al., 1988), but still the problem of changes in frequency on a scale of a century and the role of these changes in the transformation of fluvial systems are not sufficiently recognized.

During the Holocene there were both wetter and drier phases (alternating) as well as phases with higher and lower frequency of extreme events (Starkel, 1983). Both types coincide and correlate in time in various parts of the globe, but not everywhere (cf. Knox, 1983; Starkel et al., 1991; Magny, 1993; Gregory et al., 1995).

In the temperate zone the alternation of wetter and drier phases is well reflected in the evolution and succession of plant communities, the rate of peat growth and lake level changes (Ralska-Jasiewiczowa and Starkel, 1988; Berglund et al., 1996) as well as in fluctuations of mountain glaciers (Grove, 1979; Karlen, 1991).

This concept, adapted to the evolution of alluvial plains with sequences of cuts and fills (Starkel, 1966), was later modified by the recognition of phases with various frequencies of extreme events, especially heavy rainfalls of different totals, intensities and durations, which are responsible for the formation of extensive floods, landslides, debris flows etc. (Starkel, 1983, 1985). Particularly good evidence of these phases with frequent floods is registered by the clustering of fossil oak trunks in alluvia (Becker, 1982).

To provide characteristics of present-day extreme events which might serve as a key to better recognition of past hydrological changes, the upper Vistula basin was taken as an example (Figure 21.1). This area is a 'motherland' of my concept and has especially well-dated various Holocene phases with a higher frequency of extreme events (Starkel, 1983, 1994, 1996b; Kalicki, 1991).

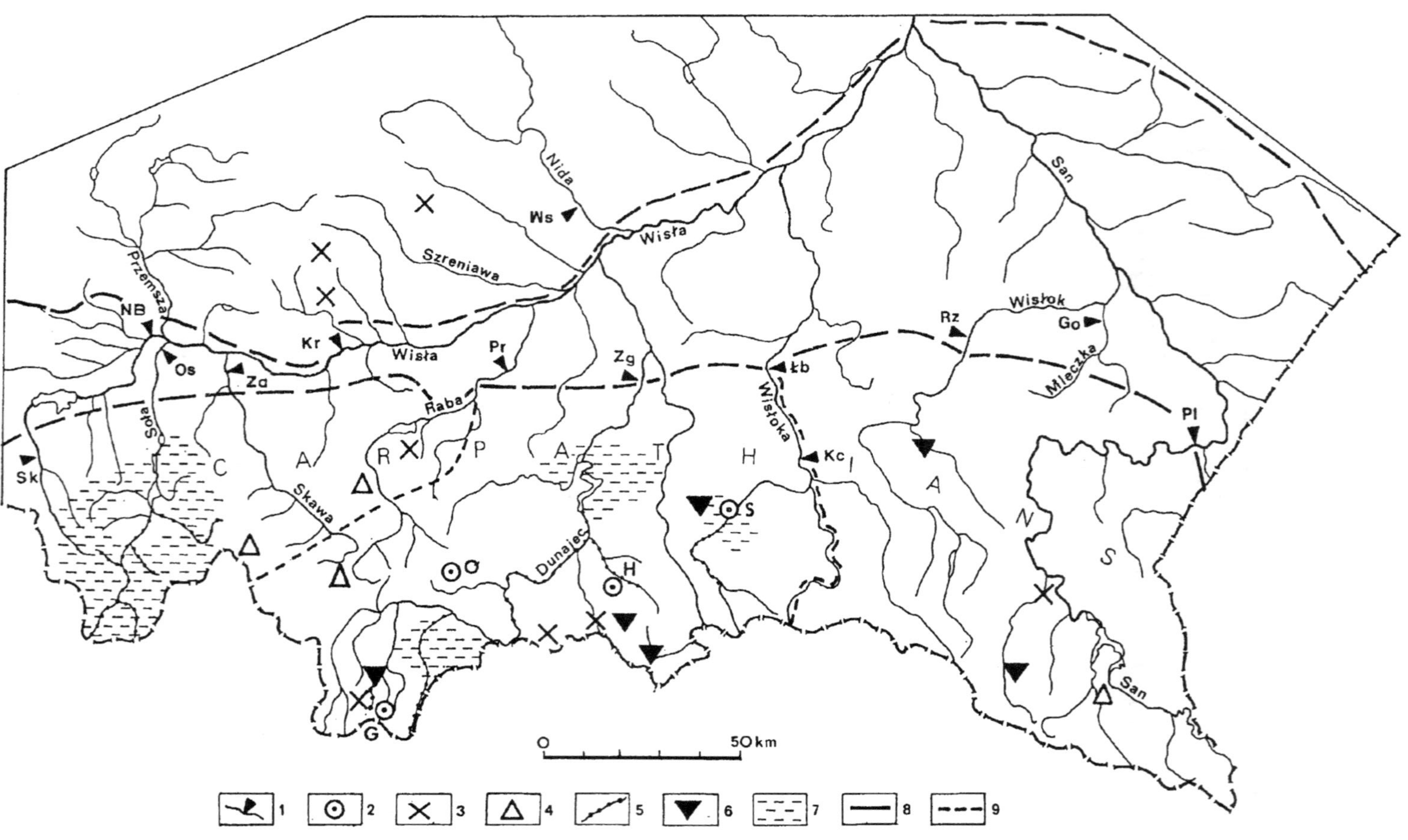
Nida
Ws
Wisła
Szreniawa
Przemsza
NB
San
Rz
Wisłok
Go
Kr
Wisła
Pr
Zg
łb
Mleczka
Pl
Os
Za
Raba
Wisłoka
Kc
Sk
Soła
C
A
R
P
A
T
H
Kc
A
Skawa
Dunajec
S
H
N
S
G
X
San
50km
1
2
3
4
5
6
7
8
9

Figure 21.1   Areas and sites of studies on extreme events in the upper Vistula basin (based on figure 1 in Starkel, 1996a): Legend: 1, monitoring stations of water level and river discharge; 2, monitoring stations of extreme geomorphic processes: G, Hala Gasienicowa (Kotarba, 1994); H, Homerka Basin (Froehlich, 1982); S, Szymbark (Gil, 1976, 1994); O, Ochotnica (Niemirowski, 1972); 3, localities with registered effects of heavy downpours; 4, localities with registered effects of continuous rains; 5, river channels with registered flood effects; 6, landslides created during rainy seasons; 7, regions with survey of active landslides; 8, boundaries of three main geomorphic units: Carpathians; subcarpathian basins; S, Polish uplands; 9, boundaries of four main Carpathian regions presented on Figure 21.6

The upper Vistula drains the Carpathian mountains, built mainly of flysch sandstones and shales in the south, limestones in the low uplands, marls and loess covers in the north, while the Carpathian foredeep is filled with unconsolidated Quaternary deposits. The annual precipitation fluctuates between 600–700 mm in the north and 800–1500 mm in the mountains rising in the Tatras up to the alpine vertical belt. Flood events are mainly connected with summer rains, but snowmelt floods cannot be excluded, and they are especially frequent in the eastern part of the basin.

WHAT DO THE RECORDS OF THE 19th AND 20th CENTURIES REPRESENT?

A general cooling in the first decades of the 19th century followed by a distinct warming trend are well recognised in Europe (Lamb, 1977; Grove, 1988). The onset of the Little Ice Age is also characterised by frequent extreme events such as heavy frosts, floods, debris flows, avalanches etc. (Figure 21.2).

In the case of the upper Vistula basin, the climatic characteristics are based on temperature and rainfall records going back at Cracow station to the late 18th century (Trepińska, 1982; Obrębska-Starkel et al., 1995), as well as on dendro-climatological reconstructions (Bednarz, 1984). The first half of the 19th century was cooler, accompanied by glacial advances in the Alps. Later, there was a warming trend interrupted by cooler waves between AD 1905 and 1930 as well as between 1965 and 1975. The precipitation regime shows various trends in the 10-year moving averages, with a general tendency to decrease. Annual precipitation totals in the upper Vistula basin may fluctuate in an individual location between 600 and 1500 mm and may have a very random distribution of extremes. Frequently two or three wetter years form clusters that are separated by single dry years. Summer precipitation dominates and shows a distinct spatial distribution. If the highest daily precipitation in the western part is brought about by continuous rains, so in the eastern part it is caused by local frontal storms (Cebulak, 1992a,b). The 100–150 mm daily precipitation at Łabowa Station has a recurrence interval of 10 years on average (Froehlich and Starkel, 1995).

Regarding the runoff regime there are distinct differences in space. The calculated maximum discharge of 1% probability varies with height from 0.5 m^3 s^{-1} to 2–4 m^3 s^{-1}, although in the western part of the mountains it is higher (2–4 m^3 s^{-1}) than in the eastern part (1.5–2.5 m^3 s^{-1}) (cf. Punzet, 1991). There are also clear, seasonal differences in the flow pattern. In the western part of the Carpathians summer high discharges predominate (60–75%), while winter and spring high discharges (65 to > 85%) take place in the eastern part of the mountains and in their foreland (cf. Ziemońska, 1973; Punzet, 1991; Figures 21.3 and 21.4).

Another very important change in time is the observed lowering of high water levels connected with regulation and straightening of channels and incision of river beds (Punzet, 1981). In most of the Carpathian river valleys, the former braided channels of the 18th and early 19th centuries were probably controlled by the frequent floods and ice-jams during the Little Ice Age (Klimek, 1974; Szumański, 1977). Starting from the second half of the 19th century, these wide channels, after their dissection, were transformed into low floodplains. The next incision, after the Second World War, in the range of 1–2 m was connected with exploitation of gravel and sand directly from the channels, construction of water reservoirs in several valleys (Klimek, 1983), and in the upland areas with abandonment or destruction of hundreds of water mills. Therefore, the extreme events in the river valleys have different causes during the second half of this century.

Regarding the characteristics of the extreme events during the last 200 years in the upper Vistula basin, the length of observation series and number of recorded places should also be considered. Most of the rainfall and water level measurements were initiated in the 1860s and 70s. The first descriptions of geomorphic or sedimentological effects go back about 90 years, and detailed monitoring of slope and channel processes at experimental field stations started no more than 30–45 years ago (Figure 21.2).

VARIOUS TYPES OF EXTREME EVENTS IN THE LAST TWO CENTURIES

Among the events recorded in sediments and forms, and observed by various authors, there were heavy downpours, continuous rains, rainy seasons and rapid snowmelts (Table 21.1, Figure 21.3; Starkel, 1996a).

Heavy downpours are local and short-lasting events with total rainfall of 30–200 mm, duration of 30–40 min up to 4 h and rainfall intensity of 1–3 mm min^{-1}. Their characteristic feature is a very random distribution both in space and time. During such rainfalls the threshold of infiltration rate is passed, and overland flow combined with splash erosion starts. Several downpours were described at the Szymbark station (Gil and Słupik 1972a; Słupik 1973) as well as at the Homerka station (Froehlich, 1982). During such events, rills develop over arable land and earthflows of the arable soil layer occur, carrying several tons of soil from 1 ha (Figuła, 1960; Gil, 1976). Deepening of gullies and cart-roads takes place (Froehlich, 1982), because the specific runoff may exceed $10 \, \text{m}^3 \, \text{s}^{-1} \, \text{km}^{-2}$. Erosion is accompanied by the formation of debris flows (Soja, 1981). On the steeper slopes and on scarps of cultivated terraces, hundreds of slumps and earth flows may develop, such as those formed near Postołów after 110.7 mm rainfall in 1953 (Starkel, 1960) or in Jaworki after a downpour exceeding 100 mm in 1958 (Gerlach, 1966).

In the alpine belt of the granite Tatras, 95 mm of rain with maximum intensity of over 1 mm min^{-1} gave rise to the formation of debris flow troughs up to 600 m long (Kotarba, 1994).

Heavy downpours may cause rapid flooding of the valley floors in first to fourth order basins and deposition of a sediment layer up to 20–50 cm thick. Such floods were recorded in the Tatra mountains, where they transported heavy bedload (Kaszowski, 1973), as well as on the loess plateaus north of Cracow in 1937 (Kondracki, 1937) and in September 1995 (Czyżowska, 1996), and in limestone uplands (Kaszowski and Kotarba, 1970).

Debris flows in the Tatra mountains during the last two centuries were investigated using the lichenometric method (Kotarba, 1989, 1995). The most frequent and largest ones coincide with wet years of the final phase of the Little Ice Age (AD 1826–35, AD 1843–52, 1860–70) (cf. Bednarz, 1984).

Continuous rainfalls are regional phenomena and are related to travelling cyclonic systems (Niedźwiedź, 1972; Cebulak 1992b). The precipitation totals during two to five days reach 150–400 mm and the rainfall intensity does not exceed 3–12 mm h^{-1}. Continuous rainfalls occur every five to 20 years in the western Carpathians, but they are rarer in the eastern part and in the foreland.

Sometimes such rainfalls may occur in two or three consecutive years. During the period 1970–1980 such rainfalls caused extreme floods in various parts of the upper

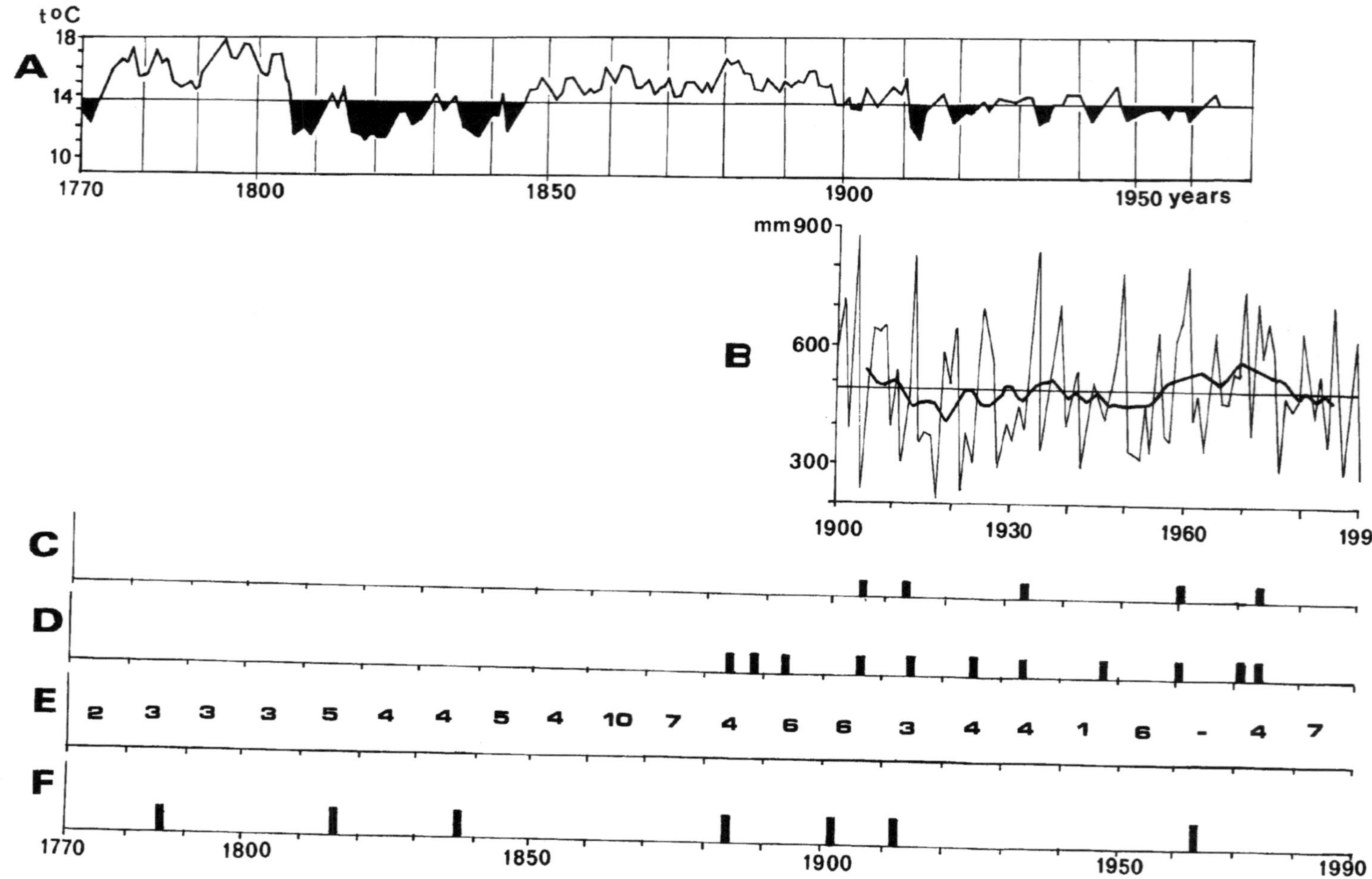

A
t°C
18
14
10
1770
1800
1850
1900
1950 years
B
mm 900
600
300
1900
1930
1960
1990
C
D
E
2 3 3 3 5 4 4 5 4 10 7 4 6 6 3 4 4 1 6 - 4 7
F
1770
1800
1850
1900
1950
1990

Figure 21.2   Temperatures, precipitations, debris flow tracks in the Tatra mountains and main volcanic eruptions (based on Kotarba (1995) slightly changed). (A) Mean VI–VII temperatures in Zakopane reconstructed by Bednarz (1984); (B) mean summer precipitation (VI–VIII) in Zakopane and 10 year moving averages (Obrębska-Starkel et al., 1995); (C) years with frequent landslides (Starkel, 1996a); (D) years with extreme floods in Dunajec valley (Starkel, 1996a); (E) number of debris flow tracks in Gasienicowa valley (Kotarba 1995); (F) years of great volcanic eruptions

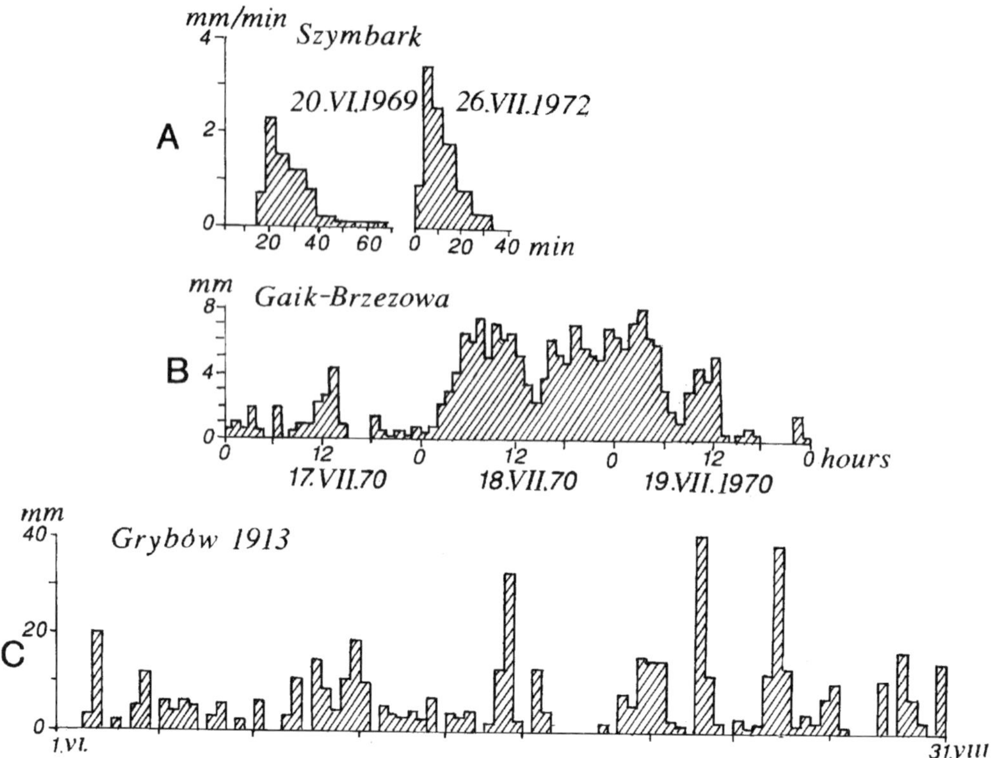

Figure 21.3 Rainfall intensity during various types of extreme rainfalls (after Starkel, 1996a). (A) Heavy downpours in Szymbark (Gil and Słupik, 1972a); (B) continuous rain at Gaik Brzezowa (Niedźwiedź, 1972); (C) rainy season in 1913, when great landslide in Szymbark was formed

Vistula basin but the whole basin was never occupied by a single flood (Figures 21.3, 21.4 and 21.5).

The morphological effects over the slopes depend on the lithology of the substratum and slope gradient. Slope wash is registered on steeper cultivated slopes (Gil, 1976), and downcutting of cart-roads is intensive (Froehlich, 1982). Many earth slides were formed or reactivated, especially during the series of continuous rainfalls in 1958–60 in the western part of the flysch Carpathians (Ziętara, 1968). In the case of debris covers on the sandstone slopes, the 400 mm heavy rainfall in 1970 caused the formation of episodic springs; their density reached up to 600 per square kilometre (Brykowicz et al., 1973).

Floods caused by continuous rains are the main transforming factors of the river channels and floodplains in the Carpathian valleys. The specific runoff exceeds $1–2\,\mathrm{m}^3$ $\mathrm{s}^{-1}\,\mathrm{km}^{-2}$ (Punzet, 1991) and in smaller basins is above $3\,\mathrm{m}^3\,\mathrm{s}^{-1}\,\mathrm{km}^{-2}$. During the flood of 1970 the suspended load and bedload accounted for 90% of the annual sediment load and were 20 times greater than in a normal year without floods (in the Homerka experimental basin; Froehlich, 1975). In the Białka Tatrzańska drainage basin with braided channel, a five-year flood with discharge of $200–250\,\mathrm{m}^3\,\mathrm{s}^{-1}$ (specific runoff $c.$ $1\,\mathrm{m}^3\,\mathrm{s}^{-1}\,\mathrm{km}^2$) may change the system of bars and pools and a 50 year flood with

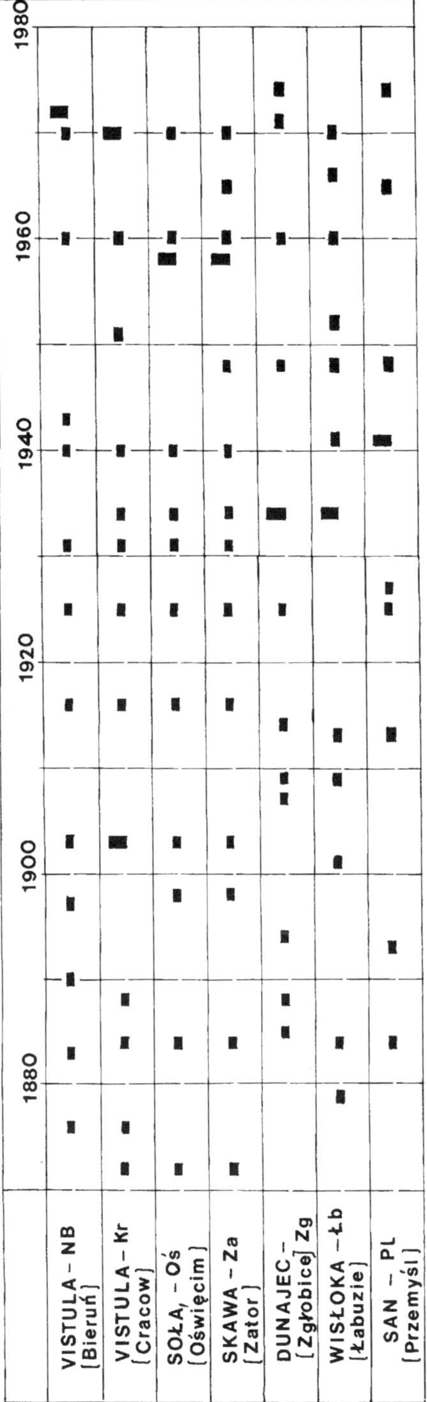

Figure 21.4 Differences in extreme floods in W–E transect of Carpathians; the highest floods during the century are indicated by large squares (based on Osuch, 1991, and Punzet, 1991)

Table 21.1   Active processes during various extreme rainfalls and floods on slope and channel systems

Type of extreme	Alpine rockwall–talus slopes	Steep sandstone, debris slopes forested	Steeper shale–sandstone slopes with thick soil/colluvia	Gentle slopes, thick loess or colluvia	Steep gradient mountain channels	Low gradient channels (uplands and basins)
Heavy downpours	Debris flows	Small slips and debris flows	Small slips and slumps, soil flows	Slope wash, soil flows, gully erosion	High load, local erosion, gravel bars, debris flows	Lateral erosion, overbank deposition
Continuous rains	Subsurface wash, piping	Linear erosion, piping	Earthslides, linear erosion	Earthslides, earthflows, piping, gully erosion	Deep and lateral erosion, shifting of bars	Lateral erosion, avulsions, shift of bars, overbank deposition
Rainy seasons		Deep rockslides	Deep earth and rockslides	Earthslides	Lateral erosion	Lateral erosion
Rapid snowmelts	Avalanches, rockfalls		Slope wash, piping	Slope wash, piping, solifluction	Ice jams, high load, lateral erosion	Overbank deposition, shifting of channel

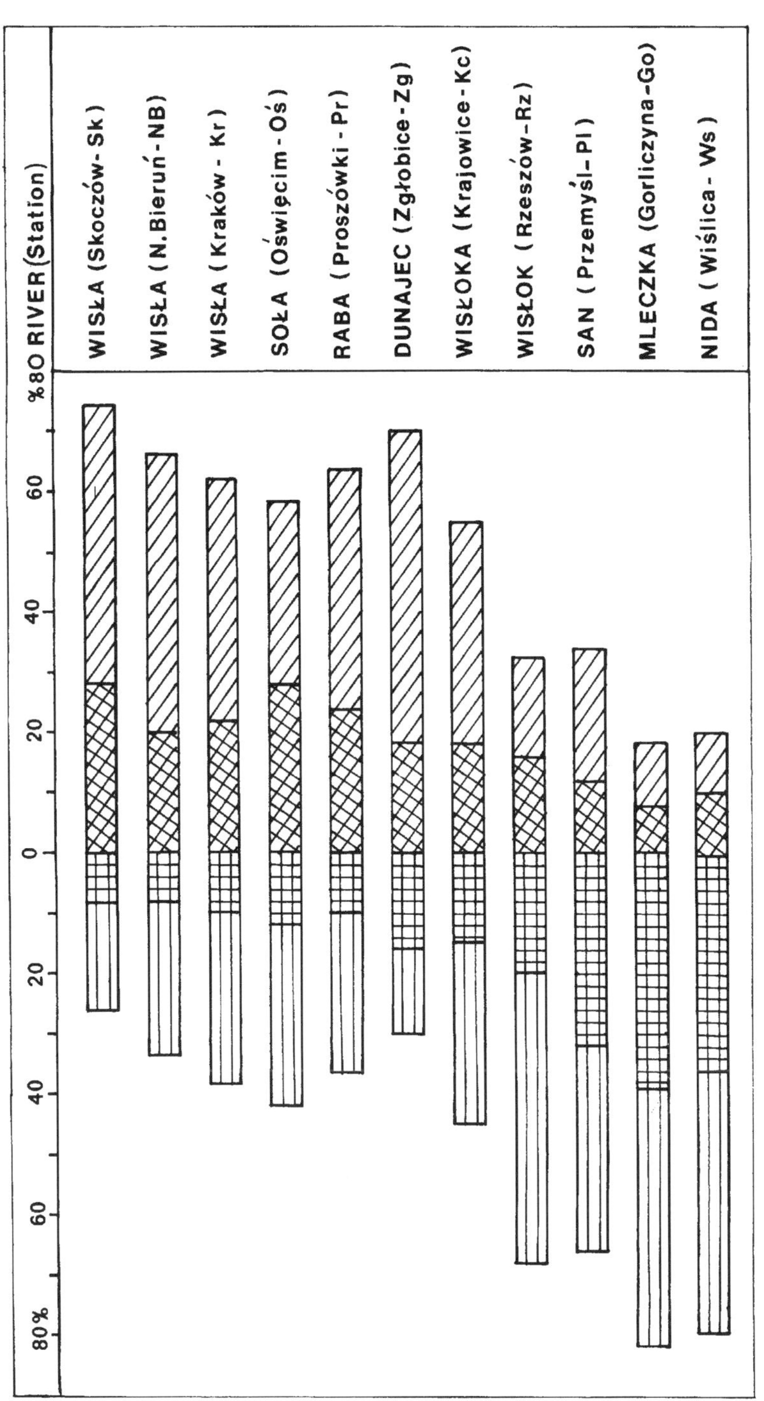

Figure 21.5 Frequency of highest annual discharges in half-years and selected months in various rivers (based on Punzet, 1991). Legend: 1, summer half-year; 2, July or June; 3, winter half-year; 4, March or April (snowmelt flood)

discharge of over $400 \, \text{m}^3 \, \text{s}^{-1}$ is able to transform the whole river bed (Baumgart-Kotarba, 1983).

The rain floods may occur several times during the following years. Floods in the Ropa valley in 1970, 1972, 1973 and 1974 resulted in the removal of debris from the channel and finally caused the incision of the bedrock (Soja, 1977). In the case of a larger river, the Dunajec, it was documented that large floods led to step-like changes in the channel depth, either to scouring in the gravels or to aggradation associated with large migrating bars (Froehlich, 1975). Continuous rainfalls seldom lead to the simultaneous passing of thresholds and transformation of slope and channel systems. This was observed by Ziętara (1968) in the upper Soła river basin in the years 1958–60.

Rainy seasons are typical of moist years when the annual precipitation in the Carpathians reaches 1200–1500 mm. In an individual summer month the rainfall may reach 200–500 mm and 100–200 mm in autumn and winter months (after a vegetation season). Excluding the continuous rainfalls triggering floods, the rainfall intensity does not exceed $1–1.5 \, \text{mm} \, \text{h}^{-1}$ (Gil and Starkel, 1979). Here, a characteristic feature lies in the great number of rainy days. Not only are the loose deposits saturated with water, but rain water infiltrates into the bedrock along cracks. Seepage pressure increases, and some beds become plasticised. Experimental studies in Szymbark have shown that in 1974, as well as to a lesser extent in 1980 and 1985, precipitation exceeding 250 mm during several weeks was able to trigger mass movements over the slopes (Thiel, 1989; Gil, 1994).

The largest flysch landslides in the current century are connected with such wet years. In April 1907 the rock slide in Duszatyn transferred $10 \times 10^6 \, \text{m}^3$ of land mass after the rainy year of 1906 (1474 mm recorded) and heavy precipitation (371 mm) in the first months of 1907 (Schramm, 1925).

In 1913 near Szymbark, after precipitation that amounted to 500–600 mm in two summer months, a valley rock slide formed that transferred $3.5 \times 10^6 \, \text{m}^3$ of land mass (Sawicki, 1917). In such rainy seasons, numerous landslides are reactivated and play an important role in slope evolution, even though the distance of transfer is very short.

Snowmelts are common phenomena in the winter season in the upper Vistula basin. But it is only in very cold winters with deeply frozen ground (to 1–1.5 m depth) and heavy snowfall, occurring in the late season, that rapid snowmelt accompanied by rain may cause intensive runoff and slope wash as well as the flooding of valley floors. Under such circumstances, the water cannot infiltrate the frozen layer and the saturated upper part is exposed to rill erosion, soil flows and solifluction. Such events were described from the uplands covered with loessic soils by Ziemnicki and Orlik (1971), from the Cracow region in 1956 (Starkel, 1976) and from the research station in Szymbark in 1969 (Gil and Słupik, 1972b). The spring of 1997 was similar.

Snowmelt floods are typical of the upland areas and of the eastern parts of the Carpathian foothills which are exposed to advection of continental air masses (Figure 21.6). The ice-jam floods in the last decades were not recorded on larger rivers, but they were probably much more frequent before the regulation, at the onset of the Little Ice Age.

LESSONS WHICH MAY BE LEARNT FOR THE PAST

During different types of extreme rainfalls and snowmelts, the thresholds for various slope and fluvial processes may be passed: overland flow and wash, piping, various types

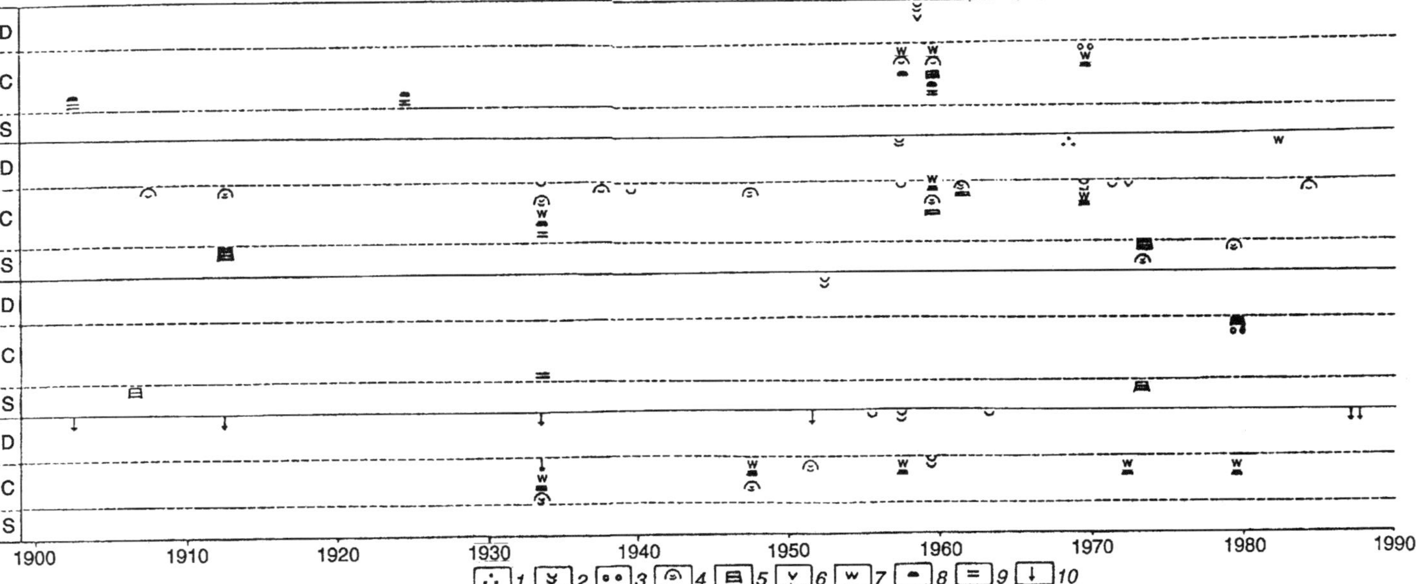

Figure 21.6   Geomorphic effects of extreme rainfalls in various parts of the Polish Carpathians registered in the 20th century (after Starkel, 1996a). Legend: D, downpours; C, continuous rains; S, rainy seasons. 1, Heavy slope wash; 2, soil flows and earth slumps; 3, piping; 4, earth slides; 5, great rockslides; 6, channel erosion (local); 7, channel erosion (intensive, regional); 8, shifting of bars and channels; 9, overbank deposition; 10, debris flows in high mountains. From the top to the bottom are presented: western part, central part, eastern part and the Tatra region (cf. Figure 21.1)

of mass movements, overbank flooding, channel bedload, etc. (Figure 21.6). The effects of all these processes are reflected in forms and sediments. Their size, thickness and areal extent correspond to various rainfall and runoff parameters and are the basic information for the reconstruction procedures (cf. Baker et al., 1988; Starkel 1995). By studying the sequences of overbank or channel deposits from various phases of the Holocene (Starkel 1983, 1994) it is possible to reconstruct high frequency floods connected with continuous rainfalls, or by studying the debris flow tracks, proluvial and deluvial deposits, to document periods with frequent heavy downpours (Kotarba, 1995; Śnieszko, 1995; Czyżowska 1997).

The careful examination of extreme events registered in the last two centuries shows a random distribution in the 20th century with only few clusterings of events, e.g. in 1958–60 (described by Ziętara, 1968) and in 1970–74 (described by Soja, 1977), leading to direct changes in the trends in slope or channel evolution. In most cases, after an individual event a recovery time of 10–50 years is sufficient for the system to return to the previous slope or channel stability (e.g. by revegetation or development of the mature meandering channel). But the number of debris flows in the Tatra mountains (Kotarba, 1989, 1995) which developed in the first half of the 19th century indicates a distinct rise in heavy downpours (and probably avalanches). This was also a time with extensive floods probably connected with heavy continuous rainfalls (Mikulski, 1963; Klimek, 1974).

Observations of the geomorphic effects of clusterings of floods, mentioned above, are of great importance for understanding the phases with high flood frequency during the Holocene, when the rate of deposition and lateral channel shift increased, as well as channel avulsions and new landslides being registered (Starkel 1983, 1985, 1996b). Observations of present-day channels have also shown that a flood phase may work in opposite directions in very closely located reaches of river channels, e.g. a parallel tendency to downcutting or to aggradation may be observed. Therefore, there is a need for very careful evaluation of the position of the investigated section and its consideration on a wide spatial and temporal background.

The present-day regional differences in type and frequency of events between the Carpathians and their foreland, in vertical belts in the western and eastern parts, probably existed during the whole Holocene, although a shift of these phenomena in space cannot be excluded.

Summing up, the examination of the geomorphic and sedimentological effects of present-day extreme events should help to evaluate the main causes (rainfall totals, intensity, specific runoff) and mechanism of past events and to distinguish the role of individual events from clusterings of several events and from longer phases with higher frequency. Any reconstruction should be made very carefully, because human intervention in the fluvial regime is very great, especially in relation to accelerated runoff and soil erosion as well as the regulation of river channels and sediment load.

REFERENCES

Baker, V.R., Kochel, R.C. and Patton, P.C. (eds), 1988. *Flood Geomorphology*, Wiley, New York.
Baumgart-Kotarba, M., 1983. Kształtowanie koryt i teras rzecznych w warunkach zróżnicowanych ruchów tektonicznych (na przykładzie wschodniego Podhala). *Prace Geograficzne IGiPZ PAN*, **145**, 1–145.
Becker, B., 1982. Dendrochronologie und Paläoekologie subfossilen Baumstämme aus Flussab-

lagerungen; ein Beitrag zur nacheiszeitlichen Auenentwicklung im südlichen Mitteleuropa. *Mitteilungen der Kommission für Quartärforschung der Osterreichischen Akademie der Wissenschaften*, **5**, 1–120.

Bednarz, Z., 1984. The comparison of dendroclimatological reconstructions of summer temperatures from the Alps and Tatra Mountains from 1741–1965. *Dendrologia*, Verona, **2**, 63–72.

Berglund, B.E., Birks, H.J.B., Ralska-Jasiewiczowa, M. and Wright, H.E., 1996. *Palaeoecological Events During the Last 15 000 Years*, Wiley, Chichester.

Brykowicz, K., Rotter, A. and Waksmundzki, K., 1973. Hydrograficzne i morfologiczne skutki katastrofalnego opadu i wezbrania w lipcu 1970 roku w zródłowej części zlewni Wisly. *Folia Geographica, Series Geographica Physica*, **7**, 115–129.

Cebulak, E., 1992a. Maksymalne opady dobowe w dorzeczu górnej Wisly, *Zeszyty Naukowe UJ, Prace Geograficzne*, **90**, 79–96.

Cebulak, E., 1992b. Wpływ sytuacji synoptycznej na maksymalne opady dobowe w dorzeczu górnej Wisly. *Folia Geographica, Series Geographica Physica*, **23**, 81–95.

Czyżowska, E., 1996. Skutki geomorfologiczne i sedymentologiczne gwałtownej ulewy w dolinie Kalinki 15 wrzeœnia 1995 r (Wyżyna Miechowska). *Przeglad Geologiczny*, **44**(8), 818–816.

Czyżowska, E., 1997. Zapis zdarzeń powodziowych na pograniczu boreału i atlantyku w osadach stożka napływowego w Podgrodziu. *Dokumentacja Geograficzna IGiPZ PAN*, **5**.

Figuła, K., 1960. Erozja w terenach górskich. *Wiadomości IMUZ I. Warszawa*, **4**, 109–147.

Froehlich, W., 1975. Dynamika transportu fluwialnego Kamienicy Nawojowskiej. *Prace Geograficzne IGiPZ PAN*, **114**, 1–122.

Froehlich, W., 1982. Mechanizm transportu fluwialnego i dostawy zwietrzelin do koryta w górskiej zlewni fliszowej, *Prace Geograficzne IGiPZ PAN*, Warsaw, **143**.

Froehlich, W. and Starkel, L., 1995. The response of slope and channel systems to various types of extreme rainfalls (temperate zone-humid tropics comparison). *Geomorphology*, **11**(4), 337–346.

Gerlach, T., 1966. Współczesny rozwój stoków w dorzeczu górnego Grajcarka (Beskid Wysoki – Karpaty Zachodnie). *Prace Geograficzne IG PAN*, Warszaw, **52**.

Gil, E., 1976. Spłukiwanie gleby na stokach fliszowych w rejonie Szymbarku. *Dokumentacja Geograficzna IG PAN*, Warszawa, **2**.

Gil, E., 1994. Meteorologiczne i hydrologiczne warunki ruchów osuwiskowych. *Conference Papers IGiPZ PAN, Warszawa*, **20**, 89–102.

Gil, E. and Słupik J., 1972a. The influence of the plant cover and land use on the surface run-off and wash down during heavy rain. *Studia Geomorphologica Carpatho-Balcanica*, **6**, 181–190.

Gil, E. and Slupik, J., 1972b. Hydroclimatic conditions of slope wash during snowmelt in Flysch Carpathians. *Symposium International de Geomorphologie*, Universite de Liege, Vol. 67, 75–90.

Gil, E. and Starkel, L., 1979. Long-term extreme rainfalls and their role in the modelling of flysch slopes, *Studia Geomorphologica Carpatho-Balcanica*, **13**, 207–220.

Gregory, K.J., Starkel, L. and Baker, V.R. (eds), 1995. *Global Continental Palaeohydrology*, Wiley, Chichester.

Grove, J., 1979. The glacial history of the Holocene. *Progress in Physical Geography*, **3**(1), 1–54.

Grove, J.M., 1988. *The Little Ice Age*, Methuen, London.

Kalicki, T., 1991. The evolution of the Vistula river valley between Cracow and Niepołomice in Late Vistulian and Holocene times. Evolution of the Vistula river valley during the last 15 000 years, part IV. *Geographical Studies*, Special Issue, **6**, 11–37.

Karlen, W., 1991. Glacier fluctuations in Scandinavia during the last 9 000 years. In L. Starkel, K.J. Gregory and J.B. Thornes (eds), *Temperate Palaeohydrology*, Wiley, Chichester, 395–412.

Kaszowski, L., 1973. Morphological activity of the mountain stream (with Biały Potok in the Tatra Mts. as example). *Zeszyty Naukowe UJ, Prace Geograficzne*, **31**.

Kaszowski, L. and Kotarba, A., 1970. Wpływ katastrofalnych wezbran na przebieg procesów fluwialnych (na przykładzie Potoku Kobylanka na Wyżynie Krakowskiej). *Prace Geograficzne IG PAN*, Warszaw, **80**, 5–87.

Klimek, K., 1974. The retreat of alluvial rivers banks in the Wisłoka Valley (South Poland). *Geographia Polonica*, **28**, 59–75.

Klimek, K., 1983. Erozja wgłębna dopływów Wisly na przedpolu Karpat. In Z. Kajak (ed.), *Ekologiczne podstawy zagospodarowania Wisly i jej dorzecza*, PWN, Warszaw, Łódź.

Knox, J.C., 1983. Responses of river systems to Holocene climates. In H.E. Wright, Jr (ed.), *Late*

Quaternary Environment of the United States, Vol. 2, The Holocene, University of Minnesota Press, 26–41.

Kondracki, J., 1937. Skutki ulewy w dniu 22 maja 1937 roku w dolinie Prądnika. *Prązeglad Geograficzny*, **16**, 161–165.

Kotarba, A., 1989. On the age of debris flows in the Tatra Mountains. *Studia Geomorphologica Carpatho-Balcanica*, **23**, 139–152.

Kotarba, A., 1994. Geomorphologiczne skutki katastrofalnych, letnich ulew w Tatrach Wysokich. *Acta Univ. N. Copernici, Geografia 27*, Nauki Mat.-Przyr. 92, Toruń, 21–34.

Kotarba, A., 1995. Rapid mass wasting over the last 500 years in the High Tatra Mountains, *Questiones Geographicae*, Special Issue, **4**, 177–183.

Lamb, H.H., 1977. *Climate. Present, Past and Future, Vol. 2. Climatic History and the Future*, Methuen, London.

Magny, M., 1993. Holocene fluctuations of lake levels in the French Jura and sub-Alpine ranges and their implications for past general circulation pattern. *The Holocene*, **3**(4), 306–313.

Mikulski, Z., 1963. O najstarszych systematycznych obserwacjach wodowskazowych na ziemiach polskich. *Przeglad Geofizy*, **10**(2), 130–140.

Niedźwiedź, T., 1972. Heavy rainfall in the Polish Carpathians during the flood of July 1970. *Studia Geomorphologica Carpatho-Balcanica*, **6**, 194–198.

Niemirowski, M., 1972. Comparison of the effects of floods in two catchment basins of the Gorce Mts. (Beskid Sadecki). *Studia Geomorphologica Carpatho-Balcanica*, **6**, 201–203.

Obrębska-Starkel, B., Bednarz, Z., Niedźwiedź, T. and Trepińska, J., 1995. On the trends of the climate changes in the higher parts of the Carpathian mountains. *Zeszyty Naukowe UJ, Prace Geograficzne*, **98**, 123–151.

Osuch, B., 1991. Stany wód. In J. Dynowska and M. Maciejewski (eds) *Dorzecze górnej Wisły, part. I*, PWN, Warszaw.

Punzet, J., 1981. Zmiany w przebiegu stanów wody w dorzeczu górnej Wisły na przestrzeni 100 lat (1871–1970). *Folia Geographica, Series Geographica Physica*, **14**, 5–28.

Punzet, J., 1991. Przepływy charakterystyczne. In J. Dynowska and M. Maciejewski (eds), *Dorzecze górnej Wisły, part. I*, PWN, Warszaw, 167–215.

Ralska-Jasiewiczowa, M. and Starkel, L., 1988. Record of the hydrological changes during the Holocene in the lake, mire and fluvial deposits, Folia Quaternara 57, Cracow, 91–127.

Sawicki, L., 1917. Osuwiska ziemne w Szymbarku i inne zsuwy powstale w 1913 roku w Galicji zachodniej. Rozprawy. *Wydziatu Matematyczno–Przyrodniczego. PAU*, **3**, 13, dz.A., 227–313.

Schramm, W., 1925. Zsuwiska stoków górskich w Beskidzie. Wielkie zsuwisko w lesie wsi Duszatyn ziemi sanockiej. Kosmos, **50**, Lwów.

Słupik, J., 1973. Zróżnicowanie spływu powierzchniowego na fliszowych stokach górskich. *Dokumentacja Geograficzna IG PAN*, Warszaw, **2**.

Śnieszko, Z., 1995. *Ewolucja obszarów lessowych Wyżyn Polskich w czasie ostatnich 15 000 lat*. Prace Naukowe Uniw. Śląskiego 1496, Sosnowiec.

Soja, R., 1977. Deepening of channel in the light of the cross profile analysis Carpathian river as example. *Studia Geomorphologica Carpatho-Balcanica*, **11**, 127–138.

Soja, R., 1981. Analiza odpływu z fliszowych zlewni Bystrzanki i Ropy (Beskid Niski). *Dokumentacja Geograficzna, IGiPZ PAN*, **1**, 1–91.

Starkel, L., 1960. Rozwóf rzeźby polskich Karpat fliszowych w holocenie, Prace Geograficzne IG PAN, 220.

Starkel, L., 1966. Post-glacial climate and the moulding of Europe relief. *Proceedings of Symposium on World Climate from 8000–0 B.C.*, London, 15–33.

Starkel, L., 1976. The role of extreme (catastrophic) meteorological events in the contemporary evolution of slopes. In *Geomorphology and Climate*, Wiley, Chichester, 203–246.

Starkel, L., 1983. The reflection of hydrologic changes in the fluvial environment of the temperate zone during the last 15 000 years, In K.J. Gregory (ed.), *Background to Palaeohydrology: A Perspective*, Wiley, Chichester, 213–235.

Starkel, L., 1985. The reflection of the Holocene climatic variations in the slope and fluvial deposits and forms in the European mountains. *Ecologia Mediterranea*, **11**(1), 91–98.

Starkel, L., 1994. Frequency of floods during the Holocene in the Upper Vistula Basin. *Studia Geomorphologica Carpatho-Balcanica*, **27–28**, 3–13.

Starkel, L., 1995. Palaeohydrology of the temperate zone. In K.J. Gregory, L. Starkel and V.R. Baker (eds), *Global Continental Palaeohydrology*, Wiley, Chichester, 233–257.

Starkel, L., 1996a. Geomorphic role of extreme rainfalls in the Polish Carpathians. *Studia Geomorphologica Carpatho-Balcanica*, **30**, 21–38.

Starkel, L., (ed.), 1996b. Evolution of the Vistula river valley during the last 15 000 years, *Geographical Studies*, Special Issue, **9**.

Starkel, L., Gregory, K.J. and Thornes, J.B. (eds.), 1991. *Temperate Palaeohydrology*, Wiley, Chichester.

Szumański, A., 1977. Zmiany układu koryta dolnego Sanu w XIX i XX wieku oraz ich wpływ na morfogenezę terasu łęgowego. *Studia Geomorphologica Carpatho-Balcanica*, **11**,139–154.

Thiel, K. (ed.), 1989. *Kształtowanie fliszowych stoków karpackich przez ruchy masowe*, Instytut Budownictwa Wodnego Gdańsk.

Trepińska, J., 1982. Characterization of the measurements of climatological series at the Jagiellonian University. *Zeszyty Naukowe UJ, Prace Geograficzne*, Cracow, **55**.

Ziemnicki, S. and Orlik T., 1971. Charakterystyka okresowych spływów z falistej zlewni lessowej. *Zeszyt Problemy Postę pów Nauk Rolniczyc*, **119**, 7–22.

Ziemońska, Z., 1973. Stosunki wodne w Polskich Karpatach Zachodnich. *Prace Geograficzne IG PAN*, **103**, 1–126.

Ziętara, T., 1968. Rola gwałtownych ulew i powodzi w modelowaniu rzeżby Beskidów [Role of extreme rainfalls and floods in the modeling of relief of the Beskidy Mts]. *Prace Geograficzne IG PAN*.

CHAPTER 22

Mid-Holocene Floods of the Syrian Euphrates Inferred from 'Tell' Sediments

TAKASHI OGUCHI

Center for Spatial Information Science, University of Tokyo, Japan

AND

CHIAKI T. OGUCHI

Institute of Geoscience, University of Tsukuba, Ibaraki, Japan

INTRODUCTION

Palaeoflood reconstruction is a major component of continental palaeohydrology (Baker et al., 1988). Various methods are used for palaeoflood reconstruction based on historical records, channel geometry, palaeocompetence relations, geobotanical records, and palaeostage indicators (Wohl and Enzel, 1995). Among the palaeostage indicators, slackwater deposits (Kochel and Baker, 1988) have received considerable attention, particularly in arid regions (Baker et al., 1995), where flood deposits are well preserved because of slow weathering processes and weak bioturbation (Baker, 1987; Wohl et al., 1994a). Although slackwater deposits have been surveyed in the arid regions of the American Southwest (e.g. Kochel and Baker, 1982; Ely et al., 1993), Australia (e.g. Patton et al., 1993; Wohl et al., 1994a), and Spain (Benito et al., 1996), research in the Near and Middle East has been confined to small river basins in Israel (Wohl et al., 1994b; Greenbaum and Frumkin, 1996). Accordingly, we have limited knowledge about the palaeofloods of large rivers in the Near and Middle East, although palaeofloods of the Tigris and the Euphrates have attracted special attention in relation to the Biblical Deluge (e.g. Wooley, 1929; Hallo and Simpson, 1971; Issar, 1990).

In the Near and Middle East, mounds resulting from the accumulation of debris on long-lived settlements can be found in many places. These mounds are called 'tells', an Arabic term. Tells often occur along large perennial rivers such as the Tigris and the Euphrates, because high water availability was favourable for human settlement. Numerous archaeologists have excavated tells because their sediments give valuable information about past human activities, especially those during the Holocene. In addition, if tells at river sides were subjected to occasional flood inundation, fluvial sediments accumulated between human-induced deposits (e.g. Blackburn and Fortin, 1994; Peltenburg, 1997). Such flood deposits are generally well preserved because they were rigidly covered with human-induced sediments. Thus, analyses of tell sediments provide useful information about the palaeofloods of large rivers in the Near and Middle East.

The purpose of this chapter is to describe flood deposits at some tells along the Syrian Euphrates and to introduce geoarchaeological techniques for understanding hydrological and environmental changes during the mid-Holocene.

SURVEYED TELLS

This study concerns three tells located along a 20 km stretch of the Syrian Euphrates near the Turkish border (Figure 22.1). They are called Kosak (Tell No.1), Abr (Tell No.2) and Ahmar (Tell No.3). All the tells were rapidly excavated in recent years, because they will be under the backwater of the Tishrin Dam by the end of the 20th century.

Figure 22.2A shows the settings of Tell No.1 and the Euphrates river. The tell is about

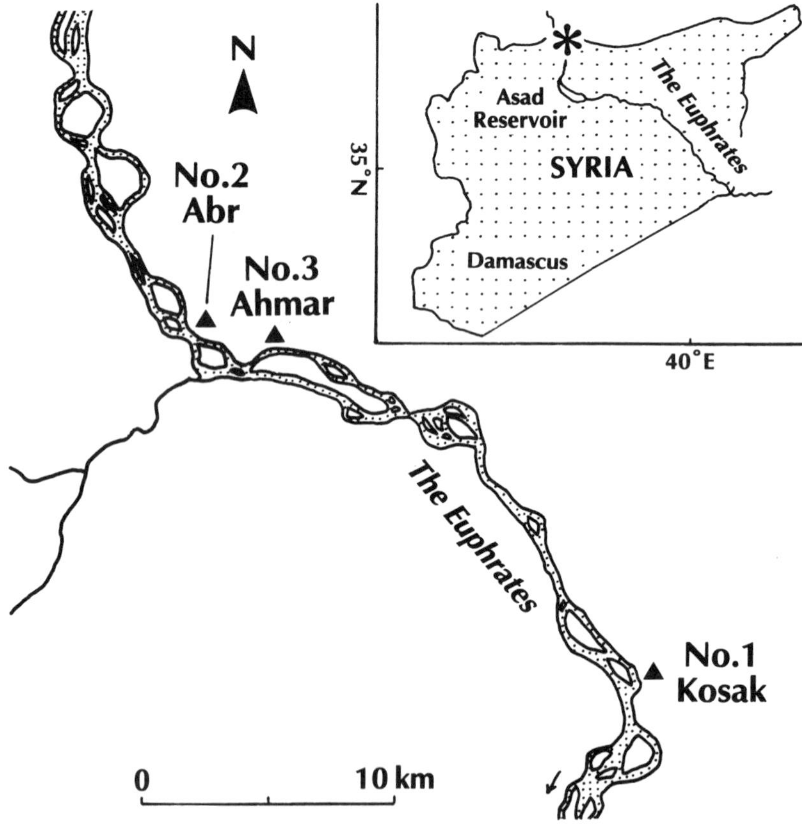

Figure 22.1   Location of studied tells

500 m away from the channel of the Euphrates. The tell is located on a flood plain, which is slightly dissected by a tributary (the Wadi Sarrin) running by the tell (Figure 22.2B). The maximum height of the tell above the flood plain is about 10 m, and the diameter of the tell is about 60 m. Tell No.2 and Tell No.3 are also located within a short distance from the channel of the Euphrates (Figure 22.1) and have dimensions similar to Tell No.1.

Archaeological missions from Japan, Syria and Australia excavated these tells (e.g. Hammade and Yamazaki, 1992; Matsutani and Nishiaki, 1995; Bunnens et al., 1997). Latest findings from the excavation of some other Syrian tells have been reported via the Internet, using World Wide Web pages. We have quoted two such pages for new and important information (Bunnens et al., 1997; Peltenburg, 1997), although the pages may be revised in the future. In these archaeological missions, they trenched portions of the tells to a depth of a few to several metres. The periods of tell construction were estimated from the types of excavated artifacts. Tell No.1 was mainly formed during the Ubaidian period between c. 6 and 8 ka. Tell No.2 was also formed during the Ubaidian period, although the uppermost parts were constructed in the Uruk period between c. 5.5 and 6 ka. In contrast, most of Tell No.3 was constructed during the Iron Age between c. 2.6 and 3.2 ka, when the Assyrians governed the site and the surrounding areas.

BEDDED DEPOSITS ON TELLS

The excavation of the three tells revealed that most deposits on the tells were introduced by human activities. Excavated walls made of stones and bricks indicate that ancient people built houses to live on the tells. Abundant pottery, stone tools and kilns were also discovered during excavation (e.g. Hammade and Yamazaki, 1992; Matsutani and Nishiaki, 1995).

At Tell No.1 and Tell No.2, however, horizontally bedded deposits with no housing construction and few artifacts are intercalated between the human-related deposits. Figure 22.3A shows the bedded deposits exposed on an excavated section of Tell No.1. The bedded deposits are about 70 cm in thickness, more than 12 m in width, and stand about 15 m above the usual water level of the Euphrates near the tell. They are mostly composed of sand and silt with brown to grey colours. Dry sieving of the deposits has shown that the dominant particle size ranges from 0.5ϕ to 3.0ϕ. The bedded deposits occur between human-related sediments of the Ubaidian period, indicating that they were accumulated in the mid-Holocene between c. 6 and 8 ka.

Bedded layers of sand and silt with rounded cobbles are also intercalated at some sedimentological levels of Tell No.2 (Figure 22.3B). The bedded deposits stand about 8 to 12 m from the usual water level of the Euphrates. Like Tell No.1, the bedded deposits occur between the human-related deposits of the Ubaidian period. The mineral composition of the bedded deposits was analysed using X-ray diffraction (XRD). Figure 22.4 shows the analytical results for sediment samples taken from Tell No.1. The samples were collected from six levels of the bedded deposits in Figure 22.3A with a 10 cm vertical interval. Before sampling, the surface sediment on the excavated wall was removed to a depth of a few centimetres to ensure that sediment samples were not contaminated by other materials during excavation.

Figure 22.4 shows that all the samples contain abundant calcite and quartz as well as moderate amounts of feldspar. The calcite is present because the area surrounding the tell is underlain by limestone. Microscopic observation revealed that the roundness of quartz

Figure 22.2   (A) Tell No.1 (Kosak, below the arrow) seen from the western side of the Euphrates; (B) Tell No. 1 (below the arrow) and a neighbouring wadi which dissects the flood plain of the Euphrates

Figure 22.3   Intercalated bedded deposits at excavated sections: (A) Tell No.1; (B) Tell No.2

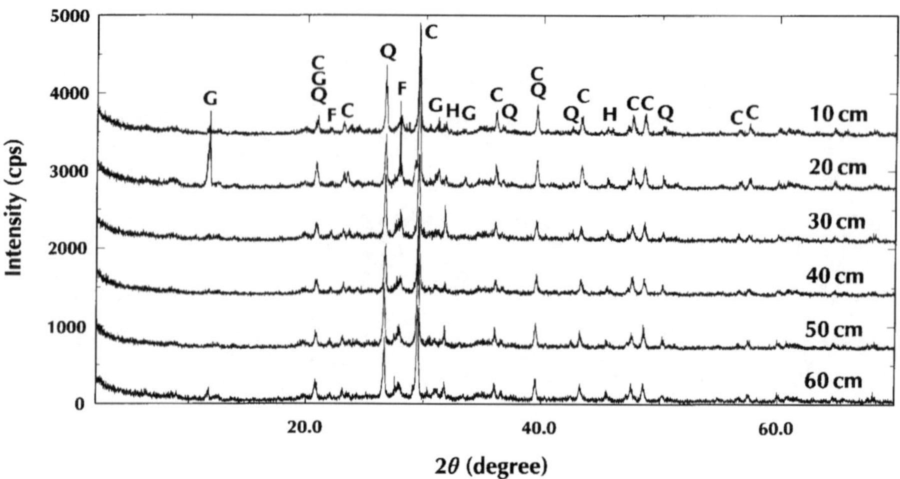

Figure 22.4   The result of XRD analyses for bedded deposits at Tell No.1. Six soil samples were taken from 10 to 60 cm below the top of the deposits, at 10 cm intervals. Legend: C, calcite; F, feldspar; G, gypsum; H, halite; Q, quartz

grains is medium to low, suggesting that they were derived not by aeolian processes but by fluvial transportation. The XRD charts show that soil samples also contain halite and gypsum. Halite occurs at all six levels: most abundantly at 30 cm from the top of the bedded deposits, moderately at 50 and 60 cm, and slightly at 10, 20 and 40 cm. Gypsum occurs abundantly at 20 cm from the top and slightly at 30 and 60 cm, whereas it is negligible at the other levels. Abundant evaporite zones in the middle parts of the deposits indicate the breaks of sedimentation, because evaporite tends to concentrate on the ground surface by repeated wetting and drying.

As noted, intercalated bedded deposits occur more frequently at Tell No.2 than at Tell No.1. The results of XRD analyses showed that gypsum and halite also occur in the bedded deposits of Tell No.2 (Oguchi and Oguchi, in preparation). This indicates that bedded deposits at Tell No.2 also accumulated intermittently.

By contrast, Tell No.3 has no well-bedded intercalated deposits comparable to those at Tells No.1 and No.2. Instead, there is evidence for enhanced human activity during the Assyrian occupation. They constructed large buildings with many rooms and courtyards with mosaic pavements (Bunnens et al., 1997).

DISCUSSION

The well-bedded deposits found at Tell No.1 and Tell No.2 can be attributed to palaeo-floods from the Euphrates river. The deposits were accumulated on small mounds with a very limited catchment area, where wide water flow could not be generated by local storms. For the same reason, sediment supply from the other parts of the tell was limited, indicating that most sediments comprising the bedded layers came from beyond the tell itself. As shown in Figure 22.2B, horizontally bedded deposits of fine materials accumulated on the flood plain of the Euphrates beside Tell No.1. These flood plain deposits consist mostly of sand and silt, which look very similar to the bedded deposits on the tell.

These facts indicate that the bedded deposits on the tell were induced by large-magnitude floods of the Euphrates. The shape of quartz grains and cobbles in the deposits also indicates that they were derived from fluvial sedimentation.

Abundant evaporites in the middle of the bedded deposits at Tell No.1 suggest that more than one flood was responsible for their deposition. Nevertheless, no marked disconformity can be recognized within the bedded deposits, implying that time intervals between flood events were not long. In summary, flood water from the Euphrates occasionally reached c. 15 m above the present river level some time in the Ubaidian period, but such a condition persisted for a relatively short time.

Tell No.2 was subjected to flood inundation more often than Tell No.1, because bedded flood deposits with evaporites occur more frequently. Such differing flood frequencies probably reflect the height of the tells from the river. As noted, flood deposits at Tell No.2 stand c. 8 to 12 m above the present river. Floods with this magnitude must have taken place more often than floods that reached 15 m above the present river. Therefore, Tell No.2 experienced more intercalation of flood deposits than Tell No.1.

Although Tell No.3 is also located on a flood plain of the Euphrates within a short distance from and at a low level above the river, there were no flood-related deposits on the tell. This contrast between Tell No.3 and the other two tells probably reflects changes in flood magnitude during the Holocene. Most deposits on Tell No.3 were accumulated during the Iron Age, whereas flood deposits on Tell No.1 and Tell No.2 were accumulated during the mid-Holocene. Therefore, it can be hypothesized that the magnitude of floods of the Euphrates was relatively large in the mid-Holocene but decreased in the late Holocene.

The above hypothesis can be tested and validated by previous research on the palaeoclimate in the Near and Middle East. The flood magnitude of the Syrian Euphrates is controlled by two factors: (1) river discharge from the upper source areas of the Anatolian mountains; and (2) flash floods from tributaries in the deserts of southern Turkey and Syria. Concerning Anatolian mountains, Erol (1981) inferred that climate was most humid during the Climatic Optimum around 6 to 7 ka. In the deserts of the Near and Middle East, climate also tended to be very humid in the mid-Holocene, whereas drier conditions have prevailed since 4 to 5 ka (e.g. Bull, 1991, p. 127; Flohn, 1991). In particular, conspicuous climatic drying took place at the beginning of the Middle Bronze Age at c. 4.2 ka (e.g. Neev and Emery, 1967; Issar, 1995).

Evidence for mid-Holocene floods from the Syrian Euphrates was also found at a tell of Jerablus-Tahtani, which is located to the north of the three surveyed tells. The tell is situated immediately beside the Euphrates, standing several metres above the river. According to Peltenburg (1997), the tell experienced flood inundation at least twice between 4 and 6 ka. Furthermore, the lower reaches of the Tigris and the Euphrates also underwent frequent floods around 5 ka (Issar, 1995). Thus, we can conclude that the mid-Holocene wet climate was responsible for the large-magnitude floods of the Syrian Euphrates.

Although a climatic drying since c. 4.2 ka was prominent in the Near and Middle East, there were also fluctuations of moisture conditions during the mid-Holocene between 4.2 and 8 ka (e.g. Stiller et al., 1984; Issar, 1995). Floods of the Syrian Euphrates may have been affected by such second-order climatic changes. Precise dating of flood sediments on the tells needs to be performed to discuss the relation between mid-Holocene climatic fluctuations and palaeoflood occurrence.

ACKNOWLEDGEMENTS

We would like to thank Professor V.R. Baker, University of Arizona, and an anonymous reviewer for constructive comments on the manuscript. We would also like to thank Professor T. Matsutani and Professor Y. Nishiaki, University of Tokyo, and Dr Y. Yamazaki, Aleppo Museum, for fruitful discussion, although we are solely responsible for any errors.

REFERENCES

Baker, V.R., 1987. Paleoflood hydrology and extraordinary flood events. *Journal of Hydrology*, **96**, 79–99.

Baker, V.R., Kochel, R.C. and Patton, P.C. (eds), 1988. *Flood Geomorphology*, Wiley, New York.

Baker, V.R., Bowler, J.M., Enzel, Y. and Lancaster, N., 1995. Late Quaternary palaeohydrology of arid and semi-arid regions. In K.J. Gregory, L. Starkel and V.R. Baker (eds), *Global Continental Palaeohydrology*, Wiley, Chichester, 203–231.

Benito, G., Machado, M.J., Pérez-González, A. and Sopeña, A., 1996. Palaeoflood analysis of the Tagus river in the El Puente del Arzobispo gorge (Central Spain). *Palaeohydrology in Spain: Field Excursion Guide*, Second International Meeting on Global Continental Palaeohydrology, GLOCOPH'96, Toledo, Spain, 5–16.

Blackburn, M. and Fortin, M., 1994. Geomorphology of Tell 'Atij, Northern Syria. *Geoarchaeology*, **9**, 57–74.

Bull, W.B., 1991. *Geomorphic Responses to Climatic Change*, Oxford University Press, New York.

Bunnens, G., Myers, S., Glynn, M. and Smith, J., 1997. *Tell Ahmar Excavations*, Internet WWW pages, University of Melbourne (http://adhocalypse.arts.unimelb.edu.au/ Dept/Arch/Tell—Ahmar/), with seven pictures and three figures (as of August 1997).

Ely, L.L., Enzel, Y., Baker, V.R. and Cayan, D.R., 1993. A 5000-yr record of extreme floods and climate change in the southwestern United States. *Science*, **262**, 410–412.

Erol, O., 1981. Quaternary pluvial and interpluvial conditions in Anatolia and environmental changes especially in south-central Anatolia since the Last Glaciation. In W. Frey and H.-P. Eerpman (eds), *Beiträge zur Umweltgeschichte des Vorderen Orients*. In Kommission bei Dr Ludwig Reichert Verlag, Wiesbaden, 101–109.

Flohn, H., 1991. Towards a physical interpretation of the end of the Holocene moist period in the Near East. *Erdkunde*, **45**, 163–167.

Greenbaum, N. and Frumkin, A., 1996. Surface and cave paleofloods: evidence for climatic fluctuation in the northern Negev (Nahal Zin and Mount Sedon), Israel. *Palaeohydrology and Modelling of Environmental Change*, Second International Meeting on Global Continental Palaeohydrology, GLOCOPH'96, Toledo, Spain, 34–35.

Hallo, W.W. and Simpson, W.K., 1971. *The Ancient Near East: A History*, Harcourt Brace Jovanovich, New York.

Hammade, H. and Yamazaki, Y., 1992. Syrian archaeological expedition in the Tishreen Dam Basin: excavations at Tell al-'Abr 1990 and 1991. *Damaszener Mitteilungen*, **6**, 109–137.

Issar, A.S., 1990. *Water Shall Flow from the Rock: Hydrogeology and Climate in the Lands of the Bible*, Springer Verlag, Berlin Heidelberg.

Issar, A.S., 1995. *Impacts of Climate Variation on Water Management and Related Socio-Economic Systems*. Technical Documents in Hydrology, UNESCO, Paris.

Kochel, R.C. and Baker, V.R., 1982. Paleoflood hydrology. *Science*, **215**, 353–361.

Kochel, R.C. and Baker, V.R., 1988. Paleoflood analysis using slack water deposits. In V.R. Baker, R.C. Kochel and P.C. Patton (eds), *Flood Geomorphology*, Wiley, New York, 357–376.

Matsutani, T. and Nishiaki, Y., 1995. Preliminary report on the archaeological investigations at Tell Kosak Shamali, the Upper Euphrates, Syria: the 1994 season. *Akkadica*, **93**, 11–20.

Neev, D. and Emery, K.O., 1967. The Dead Sea, depositional processes and environments of deposition. *Bulletin of the Geological Survey of Israel*, **41**, 1–147.

Oguchi, T. and Oguchi, C.T. (in preparation): Holocene sediments and Palaeoenvironment, Tell

al-'Abr, Syria.
Patton, P.C., Pickup, G. and Price, D.M., 1993. Holocene paleofloods of the Ross River, central Australia. *Quaternary Research*, **40**, 201–212.
Peltenburg, E., 1997. *The Jerablus-Tahtani Project, Syria*, Internet WWW pages, University of Edinburgh (http://www.geo.ed.ac.uk/arch/jerablus/jerahome.html), with two pictures and one figure (as of August 1997).
Stiller, M., Ehrlich, A., Pollinger, U., Baruch, U. and Kaufman, A., 1984. *The late Holocene sediments of Lake Kinneret (Israel) – multidisciplinary study of a 5 m core*, Geological Survey of Israel, Ministry of Energy and Infrastructure, Jerusalem.
Wohl, E.E. and Enzel, Y., 1995. Data for palaeohydrology. In K.J. Gregory, L. Starkel and V.R. Baker (eds), *Global Continental Palaeohydrology*, Wiley, Chichester, 23–59.
Wohl, E.E., Fuertsch, S.J. and Baker, V.R., 1994a. Sedimentary records of late Holocene floods along the Fitzloy and Margaret rivers, Western Australia. *Australian Journal of Earth Sciences*, **41**, 273–280.
Wohl, E.E., Greenbaum, N., Schick, A.P. and Baker, V.R., 1994b. Controls on bedrock channel incision along Nahal Paran, Israel. *Earth Surface Processes and Landforms*, **19**, 1–13.
Wooley, C.L., 1929. *Ur of the Chaldees: A Record of Seven Years Excavation*, E. Benn, London.

CHAPTER 23

Palaeoflood Hydrology of the Tagus River, Central Spain

G. BENITO AND M.J. MACHADO

CSIC-Centro de Ciencias Medioambientales, Madrid, Spain

A. PÉREZ-GONZÁLEZ

Departamento de Geodinámica, Universidad Complutense de Madrid, Spain

AND

A. SOPEÑA

Instituto Geología Económica, CSIC-Universidad Complutense de Madrid, Spain

INTRODUCTION

The risks associated with floods are estimated either by the probability of their causing material damage or by their possible effects on society, e.g. loss of human life and direct damage to populated areas (Yevjevich, 1994). Material damage produced by floods throughout the 1980s in Spain extended to the mean sum of $350 million/year (data from National Insurance Consortium). Such economic loss may be considered small in comparison with the social impact of the loss of human life in flood events as catastrophic as those of the Valencia region (1957 and 1982, 86 and 38 deaths, respectively), NW Spain (1959, 150 deaths), Catalonia, NE Spain (1962, 1971 and 1994, 1000, 400 and nine deaths, respectively) and SE Spain (1963, 1973 and 1989, > 300, 300 and 42 deaths, respectively).

In the last five years, Spain has been witness to a series of floods of immense social impact, associated more with the loss of human life than with material damage. Most recent examples include the floods of Yebra-Almoguera in central Spain (1995, 10 deaths), Biescas in the Pyrenees (1996, 87 deaths), Alicante (1997, five deaths and $75 million losses) and Badajoz (1997, 24 deaths and $300 million losses) (flood damage data from National Insurance Consortium). Most of the technical reports issued by the central administration have described these events as unique, natural phenomena of unforseeable nature. Such declarations were made since there is a lack of hydrological records of floods of similar magnitude in gauge stations. The gauge station record is used for flood zone planning purposes and assessment of the potential hazards of flooding.

The following conclusions may be drawn from the recent series of extreme floods: (1)

Palaeohydrology and Environmental Change. Edited by G. Benito, V. R. Baker and K. J. Gregory
© 1998 John Wiley & Sons Ltd.

extraordinary events follow periodicities which fall out of the scope of relatively short-term gauge station records; (2) human experience has shown that the damage produced by floods may be mitigated or alleviated but never fully avoided; (3) structural changes to rivers and early warning systems produce a false sense of security; (4) the channelling of rivers, building of dams for sediment retention, etc. may in fact increase flood damage when their capacity is surpassed, e.g. breakage of structures or operational and team errors; (5) the prediction of flood risks and the choice of measures to be taken require the adoption of new criteria to aid in the evaluation and the prevention of major flood hazards.

Generally speaking, there is a lack of knowledge of such extreme events. Tackling the problem of flood risk as a statistical problem of hydrological data may lead to considerable material and human loss (Baker, 1994). Extraordinary events may often be related to cycles which escape the conventional hydrological records. New methods are needed which involve the analysis of data of floods occurring over periods of thousands of years. Geomorphological and palaeohydrological studies provide the necessary tools and methodology for the evaluation of exceptional, large magnitude floods and for the estimation of discharge (Baker et al., 1983). Without doubt, such data would permit a more realistic interpretation of the natural environment, essential in the prediction of flood risks and long-term hydrological planning, and also take into account possible future climatic changes.

Techniques which estimate peak discharge based on palaeostage indicators have been successfully employed in several regions of the USA such as Texas (Kochel, 1981; Kochel et al., 1982), Arizona (Ely and Baker, 1985; Partridge and Baker, 1987; Ely et al., 1993; Martínez-Goytre et al., 1994), Utah (Webb et al., 1988; O'Connor and Webb, 1986), Colorado (Jarret, 1990, 1991), Wisconsin (Knox, 1985, 1995), Washington (Chatters and Hoover, 1986) and southern California (Enzel, 1992), and also in other countries including India (Kale et al., 1994, 1996), South Africa (Boshoff et al., 1993; Zawada, 1997), Israel (Greenbaum and Frumkin, 1996), Peru (Wells, 1990) and Australia (Pickup et al., 1988). In Europe, especially in the Mediterranean countries, palaeoflood hydrology shows great potential as a significant alternative for the investigation of exceptional events of great magnitude. Similarly, the study of spatial and temporal variability of floods of the past millennia involves several climatic settings and corresponding hydrological responses. These permit the understanding of the impact of global warming in extreme hydrological events. Such climate/flood relationships are particularly relevant in areas of high hydrological sensitivity to climatic variation such as the Iberian Peninsula.

HISTORICAL FLOODS AND CLIMATIC VARIABILITY IN SPAIN

The magnitude and frequency of floods differ in each basin depending upon the variability in size and morphometry of the drainage network and particularly upon the weather systems which produce flood events (Benito et al., 1996). The climatic characteristics of the Iberian Peninsula, and particularly of Spain, are largely influenced by the fact that it is surrounded by two large water masses: the Atlantic Ocean to the west and the Mediterranean Sea to the east (Figure 23.1). The influence of these large water masses on weather, in addition to the orographic characteristics of the Peninsula, give rise to a distinct division between the basins which drain into the Atlantic and those draining into the Mediterranean. Atlantic basins drain 69% of the Peninsula. The major rivers, of up to 1200 km, are

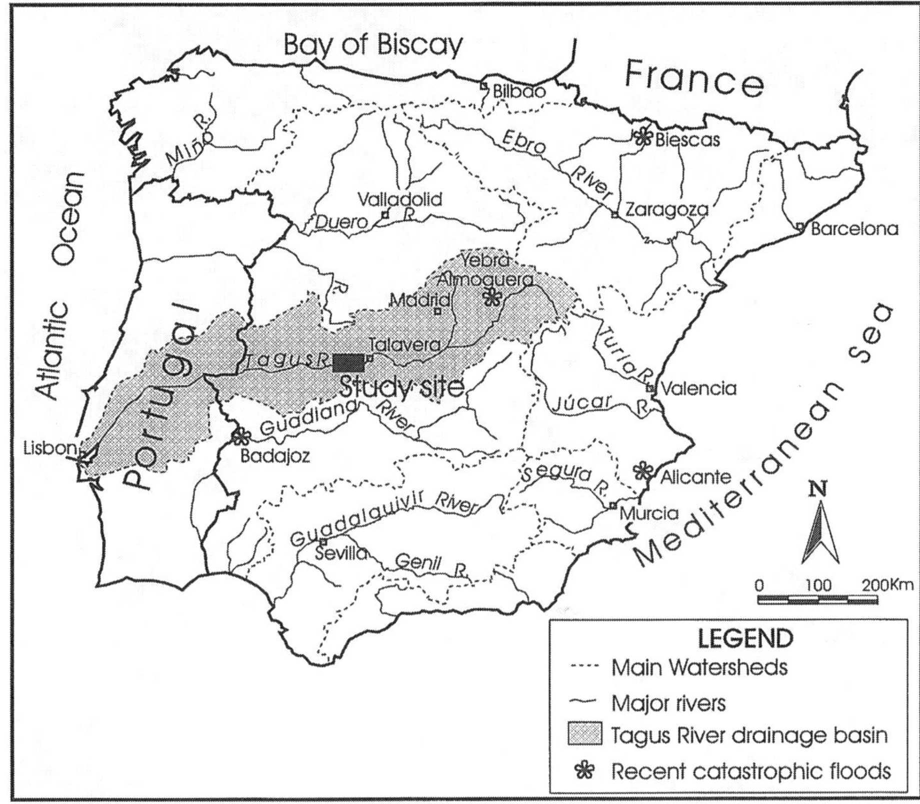

Figure 23.1 Main rivers and watersheds in Spain, and location of the study area and of the most recent catastrophic floods

the Duero, Tagus, Guadiana and Guadalquivir. Only one main river, the Ebro, drains into the Mediterranean. This division is related to the structure and the geological evolution of the Central Plateaux and the Alpine mountain systems. In addition to geographical and morphostructural controls, these drainage basins are also affected by different air masses which due to their characteristics are responsible for different rainfall events in terms of distribution, seasonality and duration.

The climate of the Iberian Peninsula is characterised by clear seasonal and monthly variability. Summers are hot and dry and winters are generally mild and relatively wet. This regime is controlled by two main systems: the subtropical anticyclone of the Azores during summer, and the westerlies, associated with cold fronts, in winter. However, certain major seasonal trends associated with the frequency of a given circulation pattern may be identified.

From the hydrological point of view and in relation to flood analysis in particular, attention must be drawn to the types of cyclonic circulation of the Iberian Peninsula which are related to most of the rainfall events occurring in this region. Capel (1981), in his study of the weather types of Spain, used climatological data for the period 1964 to 1979 to identify the circulation patterns associated with anticyclonic (53.5%) and cyclonic (40.8%) regimes. For the latter, he recognised the existence of five main types of circulation: northern, northeastern, western/northwestern, southern/southwestern flows and cold pools.

The present analysis of temporal flood variability during the last millennium in Spain was based on information extracted either directly from a wide range of written documents such as official ecclesiastical documents, local chronicles, etc., or from past compilation studies. This database was then cross-referenced. Floods affecting only one location were excluded.

Historical flood analysis permitted the identification of the different time intervals showing similar flood frequency and distribution in Spain (Benito et al., 1996). This flood variability seems to correspond to changes in the prevailing atmospheric circulation patterns affecting the Iberian Peninsula. Three periods of extensive, large-magnitude floods took place in AD 1150–1290, AD 1400–1500 and AD 1850–1910 corresponding to transitional climatic conditions. During these periods zonal flow, located probably at the lower latitudes, prevailed over meridional flow.

The flood distribution pattern for the period between AD 1290 and AD 1400 was characterised by a general decrease in the number of floods, particularly in the Atlantic basins. Two other periods, AD 1620–1690 and AD 1740–1760, presented a similar flood pattern. Historical records point to a higher frequency of severe winters with snow and frost even in SE Spain during these latter periods. Summers are generally described as very hot and dry. These phases seem to interrupt a rather large interval from AD 1500 until AD 1850. During these phases, the south and southeast basins showed the highest flood frequency due to cold pool conditions.

THE TAGUS RIVER

The Tagus river drains the central part of the 'Meseta' or Spanish Plateau (Figure 23.1). It is the longest river of the Iberian Peninsula (1200 km) and the third largest in catchment area (81 947 km^2). The Tagus headwaters drain the Iberian Range. Tributaries feed the river and drain the Central Range to the north and the Montes de Toledo to the south. Mean discharge close to the river mouth in Lisbon is 500 m^3 s^{-1}. The major contribution is made by the Central Range draining tributaries which open into the middle–low stretch of the Tagus. Hydrologically, the Tagus river is characterised by extreme seasonal and annual flow variability including severe floods with peak discharges up to 45 times the average discharge. Discharge characteristics are: (1) maximum discharge from February to March; (2) minimum discharge in August; (3) a peak in December; and (4) a discharge reduction in January.

Large floods recorded by the gauge station occurred mainly in December and January although some floods were also recorded in February and March. Historical records show a similar pattern in flood distribution with maximum values recorded in December and January.

Currently, the Tagus river is fully regulated by dams that have inundated most of the channel and floodplain areas downstream of Talavera de la Reina (central part of the basin) to the Portuguese border. The Azutan dam, approximately 10 km upstream of El Puente del Arzobispo (present study site; central part of the basin) was built in 1969 for hydroelectric purposes. The spillway was designed for flood control with a discharge capacity of 6000 m^3 s^{-1}. The longest gauge record upstream of El Puente del Arzobispo is that of Talavera de la Reina where the mean discharge is around 75 m^3 s^{-1}. The peak discharge record is incomplete and includes the periods 1942 to 1949 and 1971 to the present day (Figure 23.2). During this period only two floods exceeded 1500 m³ s⁻¹ and four floods surpassed 1000 m³ s⁻¹. The largest flood during this instrumental period occurred in March 1947. Estimated peak discharge was 7320 m³ s⁻¹ or approximately four times that of the second largest flood recorded.

Whether such flood events are taken into account or not, the analysis of flood frequency gives rise to large differences in peak discharge values for high recurrence intervals. The field data obtained suggests that the 1947 discharge values may not be all that accurate. Estimations of the discharge associated with the deposition of slackwater sediments may provide a better estimation of the 1947 flood peak discharge and a better understanding of the flood events occurring over longer periods of time.

STUDY AREA

The analysis of palaeofloods focused on the middle reach of the Tagus river (Figure 23.3), downstream of El Puente del Arzobispo, where the river is confined to a bedrock canyon (Figure 23.4). Geologically, the study area falls within the limits of two major morpho-structural units: a Hercynic basement and a Neogene cover. The Hercynic unit comprises both sedimentary and igneometamorphic rocks of Palaeozoic age. The Neogene cover mainly consists of arcosic rocks and shales with horizontal bedding lying discordantly on the Palaeozoic basement. This cover represents the most western part of the Tertiary Continental Tagus Basin.

Pliocene geological evolution was dominated by rapid uplifts (Iberomanchega phases) and climatic fluctuations giving rise to several levels of rañas, extensive alluvial piedmonts that exhibit a great complexity in terms of Quaternary geology and soil

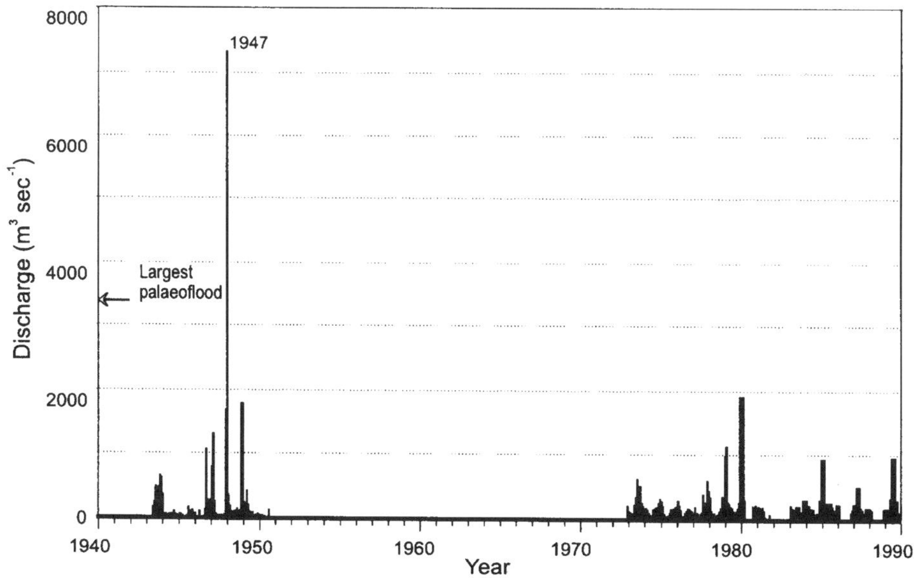

Figure 23.2   Monthly peak discharges recorded in the Talavera gauge station

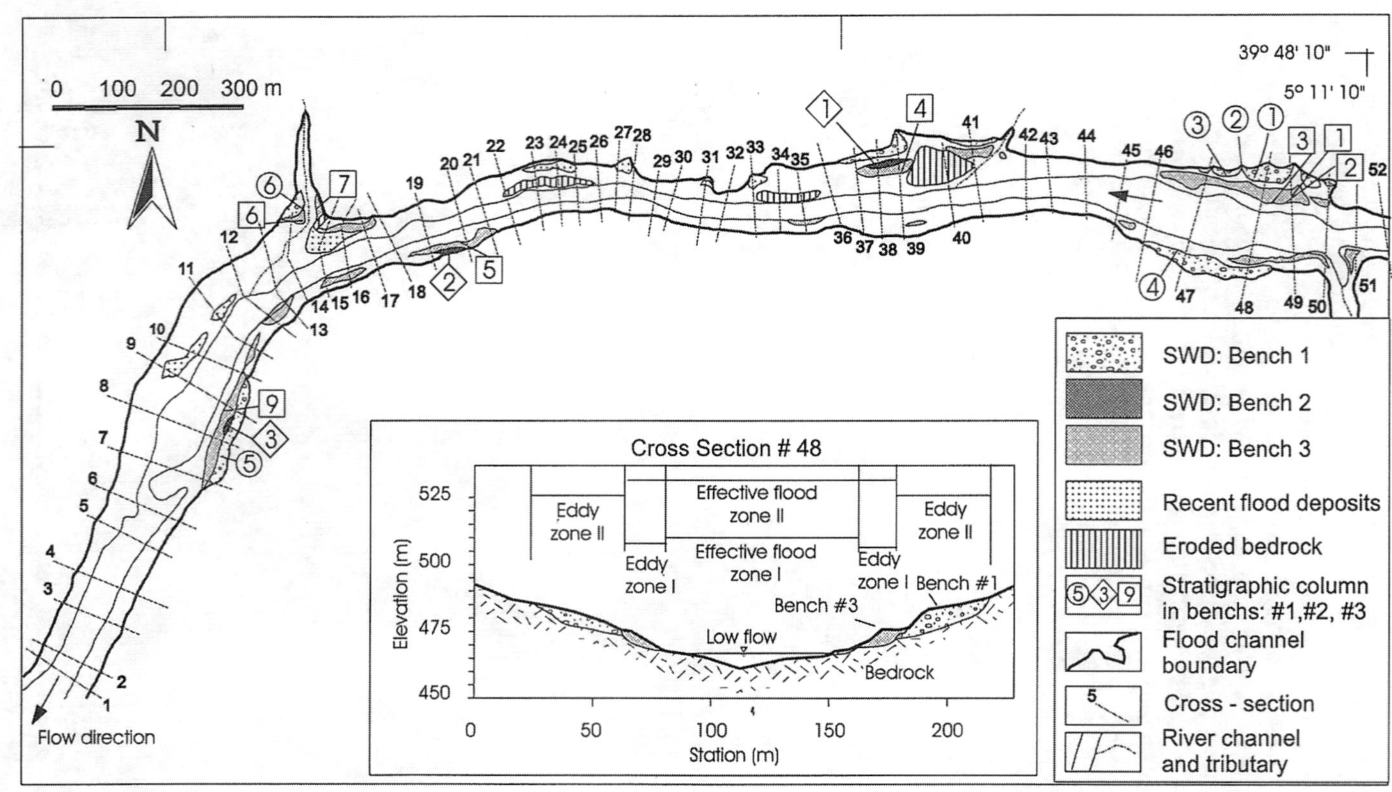

Figure 23.3   Location of the slackwater flood deposits along El Puente del Arzobispo gorge and the stratigraphic profiles within Bench 1 (circle), Bench 2 (diamond) and Bench 3 (square). Inset: cross-section 48 shows the stepped bench morphology of the flood deposits and their hydraulic interpretation with effective/ineffective zones for high magnitude (zone II) and for medium–low magnitude (zone I) flood events

Figure 23.4   Downstream view of El Puente del Arzobispo gorge and flood deposits forming
Bench 1 (B1) and Bench 3 (B3) on an expansion reach

development. During the Pleistocene, the Tagus river dissected these alluvial surfaces forming successive alluvial terraces. Late Pleistocene–Holocene terraces may be found lying on Neogene deposits east of El Puente del Arzobispo where the river valley opens. Westwards, the river cuts the granitic bedrock following tardihercinic structures and develops into a 125 m deep gorge. The remnants of eroded Pleistocene soils, only 6 m above the present thalweg, indicate that incision of the gorge during the Holocene was minor. Holocene sediments in the gorge are mostly related to the flood dynamics of the Tagus river and to local slope processes. Large floods through the El Puente del Arzobispo gorge have left a set of erosional and depositional features which allow reconstruction of the flood hydraulics and provide a long-term record of flood frequency and magnitude.

PALAEOFLOOD GEOMORPHOLOGY

The scattered location of flood deposits in the El Puente del Arzobispo gorge (Figure 23.3) reflects the potential conveyance capacity of the flood flow. Depositional areas were controlled by local hydraulic conditions such as flow separation at canyon expansions, tributary mouths or on the leeward side of bedrock obstacles. In these areas, the sharp change in energy flow conditions from high velocity (1.7–4 m s^{-1}) effective flow areas (main channel) to low velocity (0.1–0.6 m s^{-1}) ineffective flow areas, gave rise to the accumulation of thick sequences of slackwater deposits mainly of sand and silt. These types of sediments have been described by several researchers (e.g. Bretz, 1929; Baker, 1973; Kochel and Baker, 1982; Baker et al., 1983) as the most useful for the estimation of palaeoflood peak discharge.

A first step in understanding palaeoflood dynamics and the physics of the large floods of the El Puente del Arzobispo gorge, involves the generalised classification of erosional and depositional features. Two primary bar forms, eddy and pendant, can be recognised in the El Puente del Arzobispo gorge. The eddy bar displays bench or 'ridge' morphology. Bench morphology was found in canyon expansions (Figure 23.4) and in tributary mouths. Pendant bars are preserved downstream of bedrock obstructions.

The morphology of the eddy bars depends upon the geometry of the flow separation zone which controls the flow recirculation pattern. Ridge morphology is associated with more open geometries with secondary flow of relatively high velocity which continuously rework the bar form and prevent the accumulation of thick deposits. Bench morphology may be found in backflooded tributary valleys and in ineffective flow areas of channel expansion of standing or slow-moving water.

Three benches, each with multiple flood units, were found in the study area. These benches are formed by vertical accretion of slackwater sedimentation units deposited by successive floods of increasing magnitude. Inset benches are formed when smaller floods are unable to reach the upper bench and the eddy shear line shifts into the channel. The inset deposits may be flushed away periodically by extreme rare events corresponding to a shift of the eddy shear line into bedrock canyon margins. These deposits represent the selective preservation of evidence for the largest floods over the last 10 000 years. Deposits associated with high-frequency smaller floods were eroded periodically.

The El Puente del Arzobispo gorge bedrock channel can be described as a significant hydrological constraint. Despite the lack of measured erosion rates, there is clear field evidence of the resistance offered by granite bedrock to fluvial erosion. Only during extreme events and in the short reaches of high energy loss rates can the bedrock channel resistance be overcome. The vertically walled, granite bedrock channel is related to the largest amount of energy expenditure by the flood flow. This type of innerchannel erosion has been attributed to plucking and cavitation under macroturbulent conditions as well as to other processes requiring immense flow intensity (Baker, 1978; O'Connor, 1993; Benito, 1997).

In the channel margins, the erosional forms correspond to erosion bench or bare granite surfaces in which some erosional 'chute' or high-flow innerchannels were developed. In these reaches, the flow energy is dissipated by roughness adjustments with the appearance of macroturbulence that favours plucking erosion.

STRATIGRAPHY AND SEDIMENTOLOGY OF SLACKWATER DEPOSITS

The complete stratigraphical record of slackwater flood deposits along the Tagus gorge in the study area includes three benches (Figure 23.3). The upper deposit (Bench 1) is located 17 m above the present channel bottom. Inset within this bench the intermediate deposit (Bench 2) appears 12 m above the channel bottom. The lower slackwater bench (Bench 3) is either inset within Bench 1 and Bench 2 or attached directly to the channel margins at bedrock-protected sites. The Bench 3 elevation is about 10 m with different base elevations. The number of flood units preserved in each bench of the study area – particularly those of the oldest floods preserved only in a few sites – varies.

The flood sediment sequence of the uppermost part of the highest slackwater bench (Bench 1) is about 9 m in thickness and contains 30 slackwater sedimentation units. Three sets of flood units can be distinguished within this sequence. The lower set comprises six flood sediment units. In these units, upward fining of grain size is common. The facies mainly consist of sandy textures with a small amount of clay–silt deposits. Current ripples, large-scale trough cross-bedding due to the migration of bedfoms such as three-dimensional dunes, and parallel lamination constitute the dominant sedimentary structures. The consistently fine-grained nature of the slackwater deposits distinguishes them from the slope deposits which are composed of sand and pebble-sized grus fragments derived from the weathering of granite outcrops. Slope deposits overlying certain slackwater sedimentation units contribute to separating the flood events. In the middle set, the slackwater deposits are rhythmically bedded and contain 13 sediment couplets 20 to 30 cm thick. These also show upward fining of grain size. Such sand-and-silt rhythmite deposits show evidence for their deposition by separate floods, e.g. the distinct reddish colour and increase in the content of organic matter of silt layers. Most of these units are structureless although parallel lamination, presumably formed as the result of migration of plane beds, may occur in the sand layers. The upper set of the sequence is dominated by silty sand and sandy silt textures with individual flood units of 12 to 60 cm thick. Detailed stratigraphy shows at least 11 slackwater sedimentation units. Stratigraphical breaks in the upper part are more difficult to determine due to minor changes in texture and massive structure. Stone lines, broken root marks and rapid changes in sediment induration were used as the main criteria for flood unit separation. River mollusc and terrestrial gastropod shells are scattered or concentrated in discontinuous layers. Radiocarbon datings (by accelerator mass spectrometry) of shell material from flood units of the base of this upper set indicated absolute ages of 8490 ± 80 and 8300 ± 80 ^{14}C yr BP, respectively. Several pits were dug along the El Puente del Arzobispo gorge in the highest slackwater bench. The number of slackwater sedimentation units varies from four to 12 at sites 4 and 5 (Figure 23.3). The middle and lower sets of the flood units are probably absent in these sites.

The stratigraphy of intermediate slackwater bench sediments (Bench 2) was described in three sites (1, 2 and 3) where thickness ranges from 1 to 2 m. These sites contain between four and five slackwater sedimentation units composed of medium to fine sand and silty sand. Massive structure is most common in these slackwater sequences although parallel lamination and trough cross-bedding may also occur.

The lower slackwater bench (Bench 3) records floods of relatively high frequency and is periodically flushed out and replaced. This lower slackwater bench deposit ranges from 2 to 5 m in thickness with seven to 19 slackwater sedimentation units of 15 to 75 cm, and is the most ubiquitous in the El Puente del Arzobispo gorge (Figure 23.3). These slackwater sediment units fine upwards from coarse sand to silt with decreasing energy conditions (Figure 23.5). Sedimentary structures range from large-scale planar or trough cross-bedding and ripples to massive or parallel lamination. Stratigraphical breaks separating individual floods are frequently represented by fine drapes of silt and clay. The lower contacts of flood beds are usually sharp and no fluted or tool marks were detected suggesting that the flow-produced sedimentary sequence was initially non-erosional. These highly tractive sedimentary structures and sediment sizes indicate higher energy conditions related to proximity to the eddy shear line. In sites 1, 2 and 3 (Figure 23.3) the slackwater Bench 3 overlies an archaeological site with abundant remains of pottery and bones. Two pottery pieces were dated by a thermally stimulated luminescence technique as 378 ± 55 and 354 ± 47 yr BP. The archaeological site predates the slackwater sedimentary sequence, although sediments are probably of a much earlier age. Charcoal flecks are

Figure 23.5   Stratigraphy of the slackwater flood deposits in Bench 3. Note the multiple flood units
with contacts frequently represented by fine drapes of silt and clay. Sedimentary structures range
from large-scale planar or trough cross-bedding and ripples to massive or parallel lamination

abundant in the flood units although radiocarbon dating indicates that they are older
than the archaeological site, which is evidence that the charcoal was reworked. In site 9,
the stratigraphy was interpreted as evidence for 12 slackwater flood units. In the third
uppermost flood unit an aluminium paper fleck with 88% aluminium oxide content
(determined by scanning electron microscopy with energy disperse X-rays) suggested a
maximum age of around AD 1975 for these flood events, probably corresponding to the AD
1978 or 1979 floods.

Slackwater sedimentology is far from being fully understood, probably since its application to palaeoflood hydraulic reconstruction mostly requires the identification of
stratigraphical breaks and contacts between flood units. The sedimentary structures
preserved in these slackwater sediment units may be used for the indirect assessment of
local hydraulic conditions. From a strictly sedimentological point of view, seven different
slackwater flood deposit sequence types can be recognised (Figure 23.6). In each type of
sequence, the sedimentary structures of the slackwater sediment units permit the indirect

FACIES	SEQUENCE TYPE	BED OR SET THICKNESS	THICKNESS (cm)	TYPICAL LOG Cl-Si Sand	INTERNAL STRUCTURE
HETEROLITHIC FACIES ASSOCIATION	7	Min.: 10 cm Max.: 16 cm	10 / 0	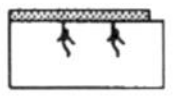	Massive
	6	Min.: 8 cm Max.: 40 cm	10 / 0	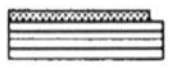	Massive Parallel lamination
	5	Min.: 12 cm Max.: 40 cm	40 / 30 / 20 / 10 / 0	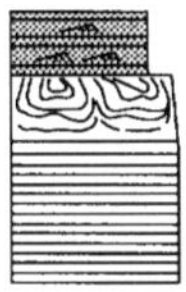	Parallel lamination Small scale cross-bedding Convolute bedding Parallel lamination
	4	Min.: 55 cm Max.: 80 cm	40 / 30 / 20 / 10 / 0	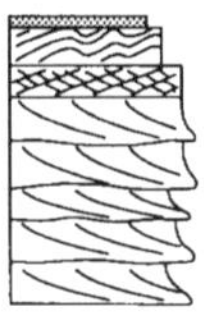	Massive Laminae in phase } Climbing ripples Laminae in drift } Large scale planar cross-bedding
SAND FACIES	3	Min.: 10 cm Max.: 40 cm	10 / 0		Parallel lamination Large scale trough cross-bedding Parallel lamination
	2	Min.: 20 cm Max.: 87 cm	40 / 30 / 20 / 10 / 0	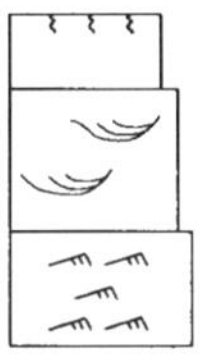	Massive or parallel lamination Large scale trough cross-bedding Small scale planar cross-bedding
	1	Min.: 25 cm Max.: 51 cm	50 / 40 / 30 / 20 / 10 / 0	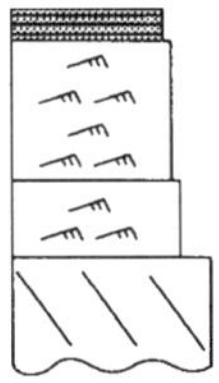	Parallel lamination Small scale planar cross-bedding Large scale planar cross-bedding

Figure 23.6 Slackwater flood deposits sequence types recognised in the stratigraphy of individual flood units

evaluation of local hydraulic conditions such as flow velocity and water depth. In addition, slackwater sediment thickness can be used as an approximate index of relative magnitude of palaeofloods in a particular site, but should not be used to compare those of different sites (Kochel and Baker, 1988).

Three main sand facies associations were found (Figure 23.6). The first consists of large-scale planar cross-bedded coarse sand, overlain by small-scale planar cross-bedding and parallel-laminated fine and very fine sand. The flow conditions, according to flume experiments performed by several authors using a water depth of 0.3 to 1 m, indicate a decrease in flow velocity from 45–50 cm s^{-1} to 10–30 cm s^{-1} with a fairly high deposition rate accumulating units of 25 to 51 cm in thickness. The second facies association consists of small-scale planar cross-bedded medium sand at times eroded by large-scale trough cross-bedded medium to fine sand and massive or parallel-laminated very fine sand. This sequence shows an increase in the local energy conditions into the second facies, probably due to a flood surge followed by a sudden decrease in flow velocity during the waning flood stage. In a slackwater bench, these facies are associated with floods of the highest magnitude. The third sand sequence type of slackwater deposit consists of medium to fine sand with parallel lamination overlain by thin sets of large-scale trough cross-bedded very fine sand with parallel lamination, deposited during the waning flood stage.

The slackwater heterolithic facies association comprises four sequence types characterised by overlain fine textures (clay and silt-sized particles) with parallel or massive structures (Figure 23.6). Most energetic eddy flow conditions give rise to large-scale planar cross-bedded medium sand in sets of 7 to 10 cm thickness, medium to very fine sand forming climbing ripples with laminae in phase and in drift, indicating high sedimentation rates, and massive silt deposited from suspension. Planar cross-bedding stratification is thought to occur due to the migration of two-dimensional dune bedfoms. The energy flow conditions decreased from the bottom to the top of the sequence with a fairly high sediment discharge. A second type of heterolithic slackwater facies association consists of medium to fine sand with parallel lamination and upward-increasing sediment supply giving rise to fluid escape structures, density inversion and silt with small-scale cross-bedding and parallel lamination. The third and fourth heterolithic facies associations are composed of very fine sand which is either parallel laminated or of massive structure, capped with silt and clay textures deposited from suspension. These sequence types are associated with the lowest local energy flow conditions in areas of water stagnation.

PALAEOFLOOD HYDROLOGY AND HYDRAULICS

The estimation of discharge associated with geological evidence was performed using the US Army Corps of Engineers HEC-2, Water Surface Profile computer program (Feldman, 1981; Hydrologic Engineering Center, 1985). The computational procedure is based on the one-dimensional energy equation, derived from the Bernoulli equation for steady, gradually varied flow. The procedure accounts for the energy expended by flow between discrete cross-sections, i.e. channel roughness (Manning's n), and for losses due to channel expansion and contraction. The palaeohydrological reconstruction is based on the calculation of the step-backwater profile which best matches geological evidence of the flow stage. Fifty-four cross-sections of the study area were measured using an electrical distance meter (EDM) surveyor (Figure 23.3). Model calibration is critical for the modelled discharge results. Daily mean peak discharges of the 1996 winter floods and the water surface profile obtained from high-water marks and flood deposits were used to 'tune up' the model. Manning's n values of 0.045 and 0.06 over the valley floor and margins, respectively, were assigned. The sensitivity test, performed using different expansion and contraction coefficients (energy losses), is of minimal importance to the modelled discharge results. For a 25% variation in roughness values, an error of 4% was introduced into the discharge results. It was assumed for the calculations that flow was subcritical along the modelled reach.

The step-backwater calculations and slackwater deposits show a minimum peak discharge for the highest bench of 3200 m^3 s^{-1}. A stage–rating relationship was developed to estimate the discharge required to deposit several of the other flood units (Figure 23.7). The upper section, comprising 11 flood units, is associated with minimum discharges of 1500 to 3200 m^3 s^{-1}, the rhythmites (13 flood units) with minimum discharges of 1100 to 1500 m^3 s^{-1}, and the lower section (six flood units) with minimum discharges between 700 and 1100 m^3 s^{-1}.

There is no evidence of floods with discharges attaining 7320 m^3 s^{-1} as recorded for the 1947 flood by the Talavera gauge station (Figure 23.2). The water surface elevation of this flood was, in fact, well documented. Marks on the walls of buildings at El Puente del Arzobispo confirmed the estimated discharge of about 2500 m^3 s^{-1}. In the study area, this flood is represented by a 5–30 cm thick slackwater deposit unit that overlies an A-horizon soil near the bench scarp.

The slackwater Bench 2, comprising between four and five flood units, was associated with an estimated minimum discharge of 1300 to 1000 m^3 s^{-1}. The lower inset Bench 3 is present in almost all of the sites within the study area. The upper slackwater sedimenta-

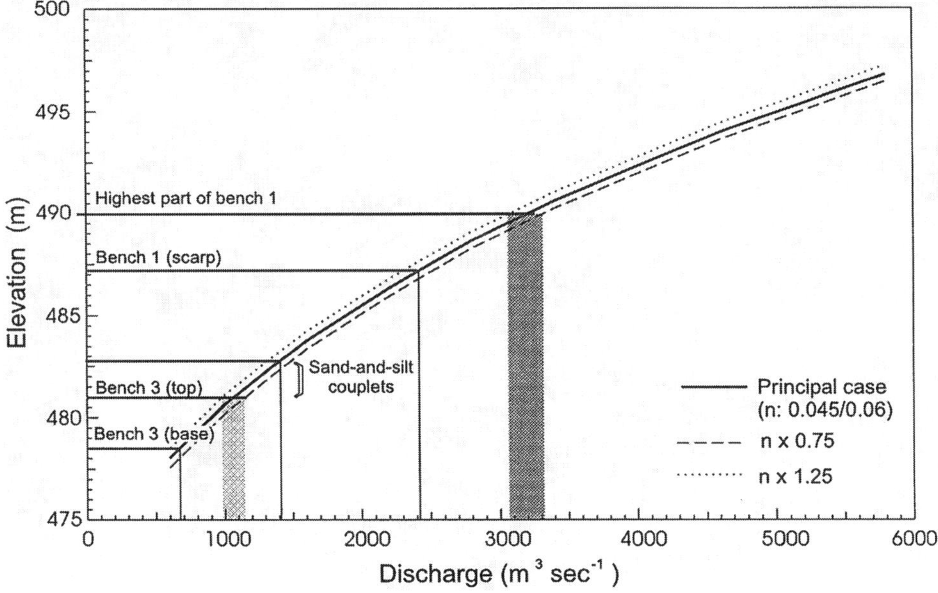

Figure 23.7   Discharge rating curve for cross-section 48. The elevations of the bottom and the top of Benches 1 and 3 are shown together with their estimated discharges

tion units of this bench are mainly composed of recent deposits and probably represent the 1978, 1979, 1984 and 1989 floods recorded by the Talavera gauge station with discharges between 900 and 1150 m^3 s^{-1}. A modelled discharge of 600 and 1100 m^3 s^{-1} brackets this lower inset.

PALAEOFLOOD HYDROLOGY: FROM A SCIENTIFIC DISCIPLINE TO A FLOOD ESTIMATION METHOD

Palaeoflood hydrology shows great potential for understanding the spatio-temporal variation of extreme events under different climatic and environmental conditions. The study of the past flooding of the Tagus river based on historical or geological records provided significant data concerning peak discharges in this respect. The floods of the Tagus are related with the development of meridional flows and atmospheric blocking situations which give rise to persistent rainfall for several weeks. Historical data reflect an increase in the number and the magnitude of floods occurring during the periods AD 1150–1290, AD 1400–1500 and AD 1850–1910. The first of these periods corresponds to the climatic transition at the end of the Medieval Warm Period or Little Climatic Optimum (from AD 900–1000 to AD 1200–1300; Hughes and Diaz, 1994). Floods took place in late autumn and winter and by analogy to present-day floods would be induced by S and SW synoptic situations. The second period is characterised by wet and warm conditions (Font, 1988) which precede the severe, cold climate of the Little Ice Age (AD 1600 to AD 1800–1900; Grove, 1988). The third period coincides with great variability in temperature from widespread cold temperatures (AD 1820) to warm (AD 1885), then returning to cold (AD 1915). Some of the flood periods correspond to high fluvial activity in regions south of the Alps (see Chapter 4), although they seem to show certain discordance with the records of northern and central Europe.

Registers of slackwater deposits suggest that floods of magnitude similar to or greater than the 1947 flood recorded by the Talavera gauge station have occurred at least nine times in past millennia. Application of such information suggests that these palaeofloods provide discharge data associated with extreme floods corresponding to high recurrence periods and permits the correction of errors detected in discharge values recorded by gauge stations of the area. Thus, the flooding in 1947, with an officially registered discharge of 7319 m^3 s^{-1}, was assigned a value in the order of 2500 m^3 s^{-1} in the present model.

In times such as the present, when catastrophic floods have generated national debate on the adequate structural and non-structural measures needed to prevent and mitigate flood-associated hazards, palaeohydrological criteria provide a feasible solution for the planning of flood zones. However, solutions based on the comprehension, interpretation and explanation of past episodes of flooding seem to foment the eternal controversy between the different ways of understanding floods (see Baker, 1994).

As technology advances, humans try to subject the behaviour of natural phenomena to their strict control. Their greatest error probably stems from their own gullibility. Attempts to counterbalance the risks of floods include the building of dams, channelling and other structural measures in general, and warning systems or non-structural solutions. There is no doubt that the greatest preventative measure is based on good planning in areas susceptible to flooding. Extreme floods are the response to cycles which may fall outside the scope of the temporary range of instrumental registers of the gauge station network, which are generally limited to a few decades. Geomorphological and palaeohydrological investigations provide the tools and the methodology needed for the estimation of discharge of large magnitude floods occurring over large periods of time. Such palaeohydrological studies should be taken into account in each measure taken to classify a flood-prone area in terms of risk potential. Similarly, the evaluation of the impact and damage caused by past events will permit the determination of the potential damage of future floods under current or future conditions. Despite the potential of palaeohydrological and geomorphological criteria for the prevention of hydrological risks, they have not yet been adopted as general tools in planning studies. The following courses of action are proposed:

(1) interdisciplinary investigations aimed towards the evaluation of damage and the control of flood risk;
(2) efforts to systemise palaeohydrological guidelines to be used in planning studies, involving different geomorphological settings such as alluvial plains, bedrock canyons, alluvial fans, mountain streams, etc.;
(3) The training of professionals in applied hydrology and geomorphology to systematically provide palaeohydrological data for the evaluation and prevention of flood risk.

ACKNOWLEDGEMENTS

This research project was financed by Spanish Committee for Science and Technology (CICYT) grants CLI95–1748 and HID96–0383. The authors would like to gratefully acknowledge the assistance given by Satur de Alba (CCMA – CSIC) throughout the topographical survey.

REFERENCES

Baker, V.R., 1973. *Paleohydrology and sedimentology of Lake Missoula flooding in eastern Washington*, Geological Society of America, Special Paper 144, 1–79.
Baker, V.R., 1978. Large-scale erosional and depositional features of the Channeled Scabland. In: V.R. Baker and D. Nummendal (eds), *The Channeled Scabland*, NASA, Washington DC, 81–115.
Baker, V.R., 1994. Geomorphological understanding of floods. *Geomorphology*, **10**, 139–156.
Baker, V.R., Kochel, R.C., Patton, P.C. and Pickup, G., 1983. Paleohydrologic analysis of Holocene flood slack-water sediments. In *International Conference on Fluvial Sedimentology*, Glasgow, 1981, Blackwell, 229–239.
Benito, G., 1997. Energy expenditure and geomorphic work of the cataclysmic Missoula flooding in the Columbia River gorge, USA. *Earth Surface Processes and Landforms*, **22**, 457–472.
Benito, G., Machado, M.J. and Pérez-González, A., 1996. Climate change and flood sensitivity in Spain. In J. Branson, A.G. Brown and K.J. Gregory (eds), *Global Continental Changes: the Context of Palaeohydrology*, Geological Society, London, Special Publication 115, 85–98.
Boshoff, P., Kovacs, Z., Van Bladeren, D. and Zawada, P.K., 1993. Potential benefits from palaeoflood investigation in South Africa. *South African Civil Engineer*, **35**, 25–26.
Bretz, J.H., 1929. Valley deposits immediately east of the Channeled Scabland of Washington. *Journal of Geology*, **37**, 393–427, 505–541.
Capel, J., 1981. *Los Climas de España*. Col. Ciencias Geográficas. Oikos-Tau, Barcelona.
Chatters, J.C. and Hoover, K.A., 1986. Changing later Holocene flooding frequencies on the Columbia River, Washington. *Quaternary Research*, **26**, 309–320.

Ely, L.L. and Baker, V.R., 1985. Reconstructing paleoflood hydrology with slackwater deposits: Verde river, Arizona. *Physical Geography*, **6**, 103–126.

Ely, L.L., Enzel, Y., Baker, V.R. and Cayan, D.R., 1993. A 5000-year record of extreme floods and climate change in the southwestern United States. *Science*, **262**, 410–412.

Enzel, Y., 1992. Flood frequency of the Mojave River and the formation of the late Holocene playa lakes, southern California USA. *The Holocene*, **2**, 11–18.

Feldman, A.D., 1981. HEC models for water resources system simulations. Theory and experiences. *Advances in Hydrosciences*, **12**, 297–423.

Font, I., 1988. *Historia del clima en España. Cambios climáticos y sus causas*, Instituto Nacional de Meteorología, Madrid.

Greenbaum, N. and Frumkin, A., 1996. Surface and cave paleofloods: evidence for climatic fluctuation in the Northern Negev (Nhal Zin and Mount Sedon), Israel. In G. Benito, A. Pérez-González, M.J. Machado and S. de Alba (eds), *Palaeohydrology and Modelling of Environmental Change*, Second International Meeting on Global Continental Palaeohydrology, Toledo, Spain, 34.

Grove, J.M., 1988. *The Little Ice Age*, Methuen, London.

Hughes, M.K. and Diaz, H.F. (eds), 1994. *The Medieval Warm Period*, Kluwer Academic Press, Dordrecht.

Hydrologic Engineering Center, 1985. *HEC-2 Water Surface Profile: Program User's Manual*, US Army Corps of Engineers, Davis, California.

Jarret, R.D., 1990. Paleohydrologic techniques used to define the spatial ocurrence of floods. *Geomorphology*, **3**, 181–195.

Jarret, R.D., 1991. *Paleohydrology and its value in analyzing floods and droughts*, US Geological Survey, Water Supply Paper 2375, 105–116.

Kale, V.S., Ely, L.L., Enzel, Y. and Baker, V.R., 1994. Geomorphic and hydrologic aspects of monsoon floods in the Narmada and Tapi rivers in central India. *Geomorphology*, **10**, 157–168.

Kale, V.S., Ely, L.L., Enzel, Y. and Baker, V.R., 1996. Paleo and historical flood hydrology, Indian Peninsula. In J. Branson, A.G. Brown and K.J. Gregory (eds), *Global Continental Changes: the Context of Palaeohydrology*, Geological Society, London, Special Publication 115, 155–163.

Knox, J.C., 1985. Responses of floods to Holocene climatic change in the upper Mississippi Valley. *Quaternary Research*, **23**, 287–300.

Knox, J.C., 1995. Fluvial systems since 20000 years BP. In K.J. Gregory, L. Starkel and V.R. Baker (eds), *Global Continental Palaeohydrology*, Wiley, Chichester, 86–108.

Kochel, R.C., 1981. *Interpretation of flood paleohydrology using slackwater deposits, lower Pecos and Devils rivers, SW Texas*, University Microfilms International, Ann Arbor, Michigan.

Kochel, R.C. and Baker, V.R., 1982. Paleoflood hydrology. *Science*, **215**, 353–361.

Kochel, R.C. and Baker, V.R., 1988. Paleoflood analysis using slackwater deposits. In V.R. Baker, R.C. Kochel and P.C. Patton (eds), *Flood Geomorphology*, Wiley, New York, 357–376.

Kochel, R.C., Baker, V.R. and Patton, P.C., 1982. Paleohydrology of Southwestern Texas. *Water Resources Research*, **18**, 1165–1183.

Martínez-Goytre, J., House, P.K. and Baker, V.R., 1994. Spatial variability of small-basin paleoflood magnitudes for a southeastern Arizona mountain-range. *Water Resources Research*, **30**, 1491–1501.

O'Connor, J., 1993. *Hydrology, Hydraulics and Geomorphology of the Bonneville Flood*, Geological Society of America, Special Paper 274, 1–83.

O'Connor, J.E. and Webb, R.H., 1986. Paleohydrology of pool-and-riffle pattern development: Boulder Creek, Utah. *Geological Society of America Bulletin*, **97**, 410–420.

Partridge, J. and Baker, V.R., 1987. Palaeoflood hydrology of the Salt River, Arizona. *Earth Surface Processes and Landforms*, **12**, 109–125.

Pickup, G., Allan, G. and Baker, V.R., 1988. History, paleochannels and paleofloods of the Finke river, Central Australia. In *Fluvial Geomorphology of Australia*, Academic Press, London, 177–200.

Webb, R.H., O'Connor, J.E. and Baker, V.R., 1988. Paleohydrologic reconstruction of flood frequency on the Escalante River, South-Central Utah. In R.V. Baker, R.C. Kochel and P.C. Patton (eds), *Flood Geomorphology*, Wiley, New York, 403–418.

Wells, L.E., 1990. Holocene history of the El Niño phenomenon as recorded in flood sediments of northern coastal Peru. *Geology*, **18**, 1134–1137.
Yevjevich, V., 1994. Floods and Society. In G. Rossi, N. Harmancioglu and V. Yevjevich (eds), *Coping with Floods*, NATO ASI Series, Series E Applied Sciences, 257, 3–9.
Zawada, P.K., 1997. Paleoflood hydrology: method and application in flood-prone southern Africa. *South Africa Journal of Science*, **93**, 111–132.

CHAPTER 24

Regional Palaeoflood Databases Applied to Flood Hazards and Palaeoclimate Analysis

ANDRÉS DÍEZ-HERRERO

Departamento de Medio Ambiente, Universidad Europea de Madrid, Spain

GERARDO BENITO

CSIC-Centro de Ciencias Medioambientales, Madrid, Spain

AND

LUIS LAÍN-HUERTA

Instituto Tecnológico Geominero de España, Madrid, Spain

INTRODUCTION

The aim of palaeohydrological flood analysis is to provide information on extraordinary events which are outside the timescale covered by systematic gauge records (Baker, 1989). This includes the study of past floods, both those for which historical records exist, and those which took place in prehistoric times or in areas remote from human observation. The study of spatial and temporal flood distribution and the palaeohydrological analysis of flood magnitude and frequency, based on stratigraphic and historical records, provides a highly effective tool for risk analysis and long-term planning (Baker, 1989; Ely and Enzel, 1992; Benito et al., 1996a) and for defining periods with a high frequency of climatic conditions favourable to flood generation (Hirschboeck, 1988; Ely et al., 1993). Owing to the close interaction between climate and the hydrological cycle, the study of the spatial and temporal variability of extraordinary events is essential for identifying anomalous patterns of atmospheric circulation on different scales. The organisation and systematisation of hydrological information on past floods enables a detailed regional analysis to be made of extraordinary events and their related hydroclimatic scenarios.

Since the beginning of the 20th century, hydrological services in most countries have compiled systematic hydrological databases based on discharge records. Advances in information technology have undoubtedly contributed to the development of hydrological databases which can manage considerable amounts of information in a simple, immediate and reliable manner within a system which may include tools for hydrological calculation and modelling (Quintas, 1996). Palaeohydrological databases, however, are a new development, hitherto limited by the scarcity of palaeohydrological data. This information is restricted to certain regions, and its main drawback is the considerable diversity of the methods used in hydrological reconstruction and the chronological uncertainties which hinder analysis and comparison. It is therefore necessary to define the type of data to be used in palaeohydrological analysis and the temporal and spatial scale which will enable us to provide databases whose degree of development and usefulness is comparable with that of contemporary hydrological databases.

GLOBAL PALAEOHYDROLOGICAL DATABASES

Differences in the amount of data and the methods used to calculate the hydrological parameters of hydrological and palaeohydrological databases has meant that the two types of database have been conceived and developed very differently. Contemporary hydrological databases are based on regions, with the drainage basin (catchment) as the superunit. The main palaeohydrological databases developed to date are global in nature and were chiefly designed for global analysis in terms of the calibration, validation and interpretation of models such as the general circulation models, and for predicting future hydrological changes. At present, different databases dealing with palaeodata information such as palaeoenvironmental data (WDC-A, 1992) and palaeolake level data (Street-Perrott and Harrison, 1982) have been implemented. In terms of specifically global fluvial palaeohydrological data sets, two comprehensive databases have been designed in recent years: (1) the Global Continental Palaeohydrology (GLOCOPH) database, and (2) the Global Palaeoflood Database.

(1) The GLOCOPH database is managed by the GeoData Institute at Southampton University (UK) and was developed to facilitate the compilation and distribution of fluvial palaeohydrological data compiled prior to, during and after the Global Continental Palaeohydrology Project (Branson, 1995; Branson et al., 1995). The aim of this database is to provide a resource for the study of palaeohydrological change in the long term, improving access to datasets on primary and secondary environmental parameters of interest for modelling fluvial palaeohydrological processes and their controls. The data include details of channel morphology, drainage basin features, sedimentology, and the calculation of hydraulic and hydrological parameters such as discharge, flow velocity, sediment load and groundwater flow, derived from the primary data (Branson, 1995). In the GLOCOPH database the data are stored hierarchically and have first been grouped in sets of individual records relating to data for an individual site provided by a researcher or taken from publications relating to the same author (researcher's table, bibliography table). Metadata about each dataset are provided in the dataset table which gives information on the spatial and temporal coverage of the records in the datasets, the techniques used during compilation, a brief description and related key words. The raw data tables contain primary field information and the hydrological reconstruction is presented in the derived data table. The data can be modelled and analysed using a model table, which provides routines for applying several hydrological models to the data for calculating discharges, flow velocity, etc. using the primary data.

(2) The Global Paleoflood Database (Hirschboeck et al., 1996) is designed to compile

different types of information on palaeofloods and include them in a flexible but structured database allowing analysis and regional and global comparisons. The ultimate aim of the project is to supply a forum for communication and storage in order to foster and stimulate research into hydrological matters through the growing demand for information on palaeofloods. The project originated at the Arizona Laboratory for Paleohydrological and Hydroclimatological Analysis (ALPHA), which for several decades has been compiling information on palaeofloods in the southwest of the USA, an area which has become a pilot region due to the numerous palaeoflood hydrology studies conducted using various techniques. The records specify the nature of the flood (rain, snow, dam break, etc.), the techniques used to define the palaeoflood (through palaeostage indicators, flow competence, hydraulic geometry, historical observations and botanical data), the methods for calculating the discharge (slope–area, Manning equation, step-backwater modelling, empirical regression, etc.). In addition, a form is included for a description of the site, with data on the geometry of the drainage basin, the nearest gauge and precipitation stations and the characteristics of the maximum likely discharge and precipitation.

These global databases are well established although further development may be increased by integrating reference and access to regional database systems. A regional database allows different resolutions of information as well as independent collation and management at a regional level. This could equally well be maintained within a global system, although the updating and maintenance may be best handled by the data generators; the custodial and wider access and exchange role of global databases suggests a cooperative link between regional and global systems. It is therefore proposed to increase the use of regional palaeohydrological databases, using standard palaeohydrological reconstruction techniques enabling comparison and global analysis of the data in the databases referred to above. The main advantages and contributions of regional databases are detailed below.

(1) The palaeohydrological data generated by small research groups are compiled, analysed and interpreted at the regional level, providing results in the short term due to their application to risk analysis and hydrological planning.
(2) In view of the intrinsic spatial component of palaeohydrological data, the use of geographical information systems (GIS) is justified at the regional level to acquire and analyse information and present results. This GIS makes it possible to include maps and derived data by means of topologic relations, the application of filters and algebraic operations with the information coverages.
(3) Quantitative and qualitative studies of palaeohydrological data and their connection with graphic applications and numerical analysis tools allow the calibration of hydrological models and the simulation of hydrological variations for different climatic and environmental scenarios.

IMPLEMENTATION OF REGIONAL PALAEOHYDROLOGICAL DATABASES: THE PALAEOTAGUS DATABASE

One of the first proposals for the creation of a database to compile palaeohydrological information at regional level was made by Gregory and Lewin (1987) for the Severn and Wye river basins (UK). In view of the huge amount of palaeohydrological data generated in prior studies and collected in a complete work (Gregory et al., 1987), they decided to structure them in several figures and tables. The main objective was to provide the basis of present-day river management and allow secular climatic and runoff trends to be identified. In addition, there are relevant examples of compilations of data of past floods, chiefly historical, which in the case of the People's Republic of China was part of a national survey of historical flood marks (Chen et al., 1974; Hua, 1985). In other European countries, for example in Spain, an inventory was made of the main historical floods (Spanish Civil Protection; Comisión Técnica de Inundaciones, 1985) to determine the main points of conflict for the purpose of risk analysis.

In all these cases, it remains necessary to systematise palaeohydrological information using computerised databases which efficiently organise data and link them with both graphic data with a geographical reference (thematic maps), and with hydrological models for the regional analysis of different hydroclimatic scenarios in the past.

The Palaeotagus database was established in 1996 to compile, analyse and interpret the palaeohydrological data generated within a geographical region (the Tagus river basin, central Spain) by the National Climate Programme research project CLI95-1748. This relational database is supported by a standardised and compatible database management system which not only allows information to be stored but enables it to be studied quantitatively and qualitatively using graphic applications and numerical analysis tools.

AIMS AND STRUCTURE

The general aims of the regional Palaeotagus database are diverse and relate to both its immediate regional application and its future integration with global databases:

(1) to compile and store effectively all the diverse palaeohydrological information in the bibliography or generated in subsequent studies;
(2) to analyse palaeohydrological data, using their spatial and temporal variability and the quantitative values of certain parameters (discharge, flood stage, etc.);
(3) to interpret the results of the analysis from a twofold viewpoint: the understanding of regional hydroclimatic development and the definition of parameters applied to flood study;
(4) to facilitate the integration of this palaeoclimatic information and these palaeoclimatic interpretations into the most commonly used databases worldwide.

To achieve these aims, the structure of the Palaeotagus database is based on storage using information tables and a hierarchy of data. Since there are two different palaeohydrological datasets (historic floods and palaeostage-based discharge data using slackwater deposits), two independent but interrelated databases have been created (Historic and Geologic).

In the historic floods database (Historic), the primary data table contains information on the year, month and day(s) of the flood, stream, site, discharge, causes and damage inflicted by the flood (Figure 24.1). The information about the damage provides an indication of the severity of the flood for events for which water stage data are not available. Damage has been categorised into four groups: (1) loss of human lives; (2) infrastructures; (3) irrigation and drainage networks; (4) river bed floodplains and

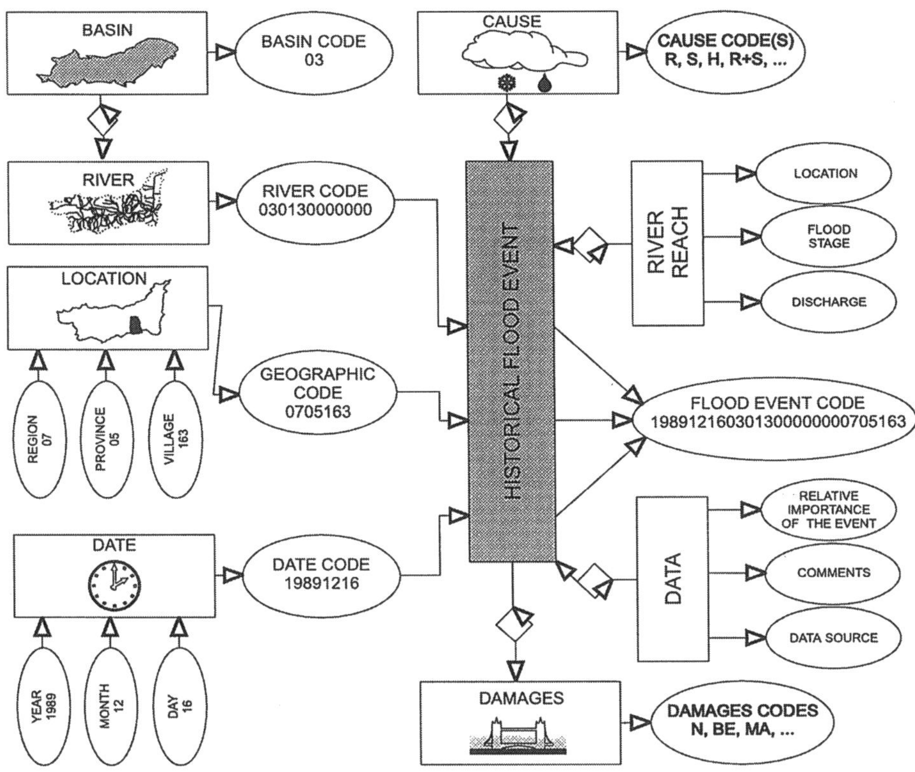

Figure 24.1 Diagrammatic representation of the conceptual model and the formal structure of the historic floods database (Historic) included in the Palaeotagus database. The rectangular boxes, ellipses and rhombuses represent the entities, attributes and relations among the associations, respectively

geomorphological changes. For each type of damage, three categories have been considered: no damage, minor damage and severe damage. These fields have been used in combination to generate other fields which will be used as coders in the database, such as *flood event* (year, month, day, stream and location) and *date* (year, month, day). To facilitate subsequent spatial and temporal flood analysis, events which affected different locations and streams on the same date have been considered as different events. In addition, the primary data have been used to generate a series of reference tables relating to locations, streams, causes and damages. The location and stream codes are linked to georeferenced points, arcs and polygons in the graphic coverages (Table 24.1).

Flood data obtained from palaeostage indicators (Geologic) have been stored in a hierarchical structure, similar to that of GLOCOPH (Branson, 1995). This makes it possible to group together datasets in subdrainage basins and subdivide them progressively into specific locations, places and records. By way of an example, we can cite the set of individual data provided by Benito et al. (1996b) in their study of palaeofloods using slackwater flood deposits (Figure 24.2), in the Tagus basin, located at the El Puente del Arzobispo gorge with several stratigraphic profiles and multiple flood units.

At present, the degree of developement of the Historic and Geologic databases is uneven. The Historic database has actually been developed and the data set comprises

Table 24.1   Example of a card extracted from the historic floods database (Historic) integrated in the regional Palaeotagus database. The data table, reference table (for code transfer) and the code combination table are shown

*** DATA TABLE**

FIELD	CONTENTS
YEAR	1876
MONTH	12
DAY	06
RIVER CODE	030100000000
GEOGRAPHIC CODE	1010008
RIVER REACH LOCATION	Roman Bridge
FLOOD STAGE	33.37 m
DISCHARGE	15,800 m^3/s
RELATIVE IMPORTANCE OF THE EVENT	The Tagus river reached a flood stage of 33.37 m above the base level. This stage was 1 m higher than the one reached in 1856, and only 5 m below the uppermost brick portion of the Alcántara Roman Bridge. A discharge of 15,800 m^3/s was estimated by Water Resource Office. The hydraulic computations made by G. Benito show us values of the peak discharge near to 18,000 m^3/s.
CAUSES	R
DAMAGES A	I2
DAMAGES B1	NH
DAMAGES B2	
DAMAGES B3	UH
DAMAGES C1	
DAMAGES C2	AH
DAMAGES C3	IH
DAMAGES D	BE
COMMENTS	On 7 December 1876, the hydrograph recorded in the Villa Velha de Rodão gauge station (Portugal) shows a peak discharge of 15,000 m^3/s.
DATA SOURCE	Comisión Técnica de Inundaciones, 1985

*** REFERENCE TABLES**

*** TABLE R1 (INE).**

FIELD	CONTENTS
GEOGRAPHIC CODE	1010008
GEOGRAPHIC NAME	Alcántara

*** TABLE R3.**

FIELD	CONTENTS
CAUSE CODE	R
CAUSE DESCRIPTION	River flood by rain

*** TABLE R2 (CDR).**

FIELD	CONTENTS
RIVER CODE	030100000000
RIVER NAME	Tagus River

*** COMBINATION TABLE**

FIELD NAME	CONTENTS
FLOOD EVENT	187612060301000000001010008
DATE	18761206

*** TABLE R4.**

RECORD NUMBER	FIELD CONTENTS	
	DAMAGE CODE	DAMAGE
1	I2	Injuries/1–2 casualties
2	NH	Destruction of bridge/road/railway
3	UH	Damage to buildings
4	AH	Major damage to agricultural fields and cattle
5	IH	Major damage to industries
6	BE	Bank erosion/channel incision

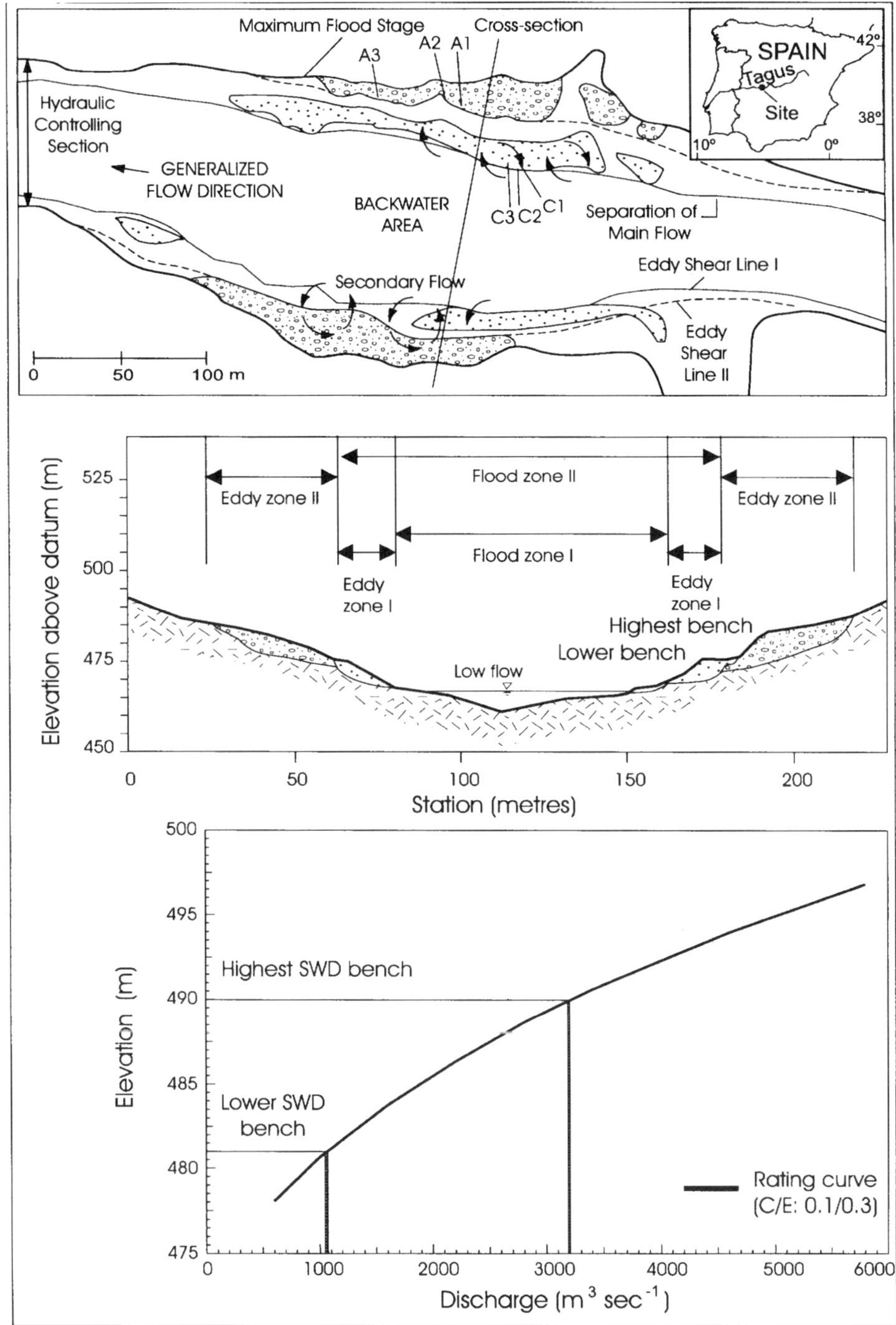

Figure 24.2   Output of a slackwater deposit (SWD) site at the El Puente del Arzobispo gorge included in the geologic flood record (Geologic) of the Palaeotagus database. A1–A3 and C1–C3 refer to the location of stratigraphic profiles

about 400 records. The Geologic database will be developed eventually when the number of flood deposit sites along the Tagus river basin increases in the near future.

DATA INTEGRATION AND ANALYSIS ON GIS

A significant feature of Palaeotagus is that the database is connected to a GIS installed on ARC/INFO v. 7.03 for Workstation (ESRI) software which is widely used worldwide. The vector coverages incorporated in the GIS are:

- boundary of the Tagus basin;
- autonomous community, provincial and municipal boundaries; polygons labelled with the Spanish National Statistics Institute (Instituto Nacional de Estadística, INE) numbers;
- isohypses from provincial scale 1:200 000 maps published by the National Geographical Institute (Instituto Geográfico Nacional, IGN); arcs encoded with their related altimetric values in metres above sea level;
- main drainage network of the Tagus basin; arcs encoded with the Decimal River Code (Código Decimal de Ríos, CDR) numbers;
- network of meteorological stations of the National Weather Institute (Instituto Nacional de Meteorología, INM); points encoded with their related numbers;
- network of gauge stations of Tagus River Water Resources Authority (Confederación Hidrográfica del Tajo, CHT); points encoded with their related number;
- geological map reclassified according to the hydrogeological properties of the lithology;
- soil map, with classes distributed according to hydrological parameters;
- land use map, reclassified according to the proposal of the Soil Conservation Service (SCS, 1972);
- potential vegetation map (vegetation series);
- isomaxima of precipitation in 24 h for different return periods;
- detailed maps including isohypses and siting of deposits (polygons) or landmarks (arcs and/or points).

The raster information incorporated in the GIS is:

- digital terrain model of the Tagus basin; pixel size 80×80 m (Centro de Estudios y Experimentación de Obras Públicas, CEDEX).
- images (photographs, diagrams and/or drawings of locations, sites or records of interest).

The alphanumeric information in the data tables was imported to the Arc/Info system from the Access database. In addition, it was necessary to integrate new reference tables relating to the alphanumeric equivalents of the encoding of the vector coverages (CEDEX, INE and CDR codes), and tables with supplementary information (statistical parameters of INM and CHT stations). Other graphic material to be integrated in the GIS are the graphics, functions and representations obtained from the analysis of the individual datasets and contained in the publications.

Once the information has been entered in the GIS, the first uses of the system focus on visualisation and selective searches for information (Figure 24.3). It is thus possible to

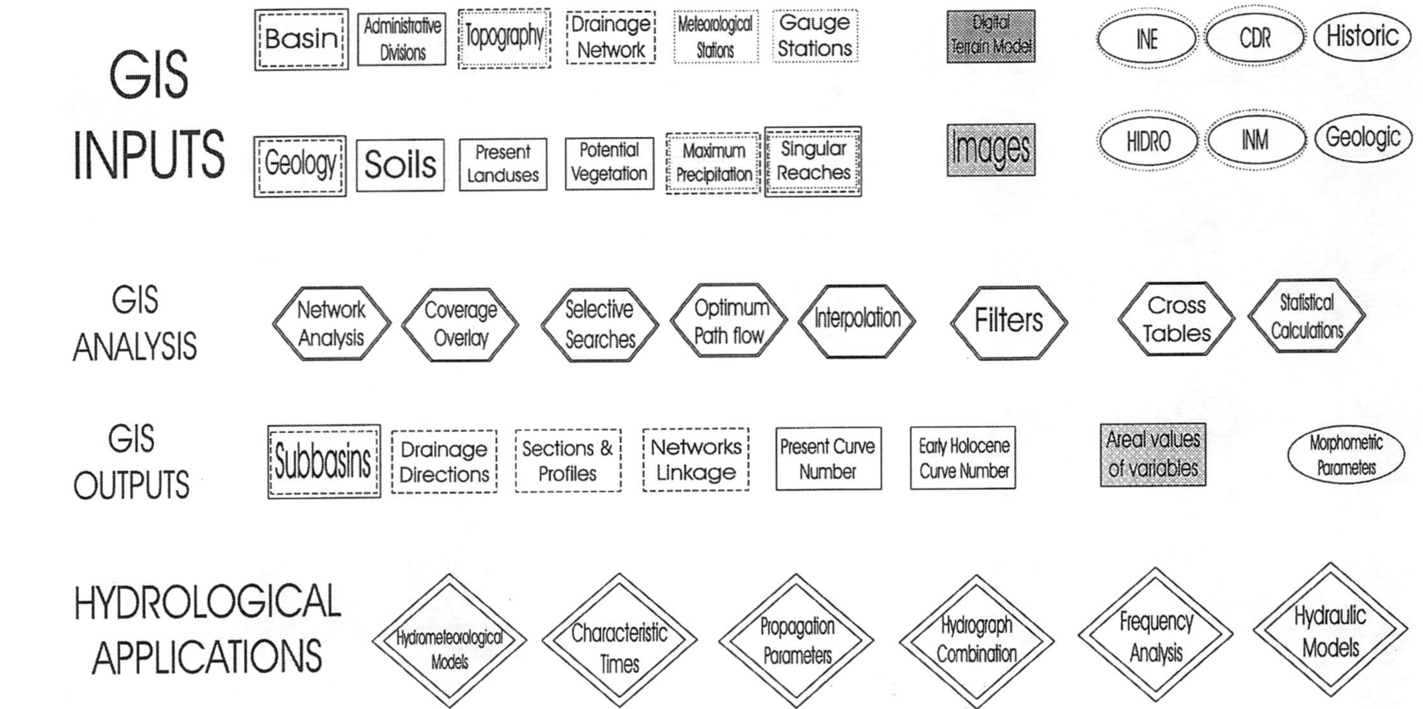

Figure 24.3 Organisation of the graphic coverages available in the Palaeotagus database and functions of the geographical information system applied to the palaeohydrological analysis

obtain different synthesis maps, detailed maps and compositions of the mapping with other study diagrams and graphics; selective searches can be used to obtain areal distributions of indicators according to their characteristics at different time intervals, or by geographical situation. Both types of analysis make it possible to draw conclusions regarding the spatial and temporal distribution of the flood events and the establishment of groups of streams or sub-basins with uniform behaviour, and to calculate the classic statistical parameters for each group according to different database variables.

At a more advanced level, the GIS makes it possible to obtain mapping and data through topologic relations, the application of filters and algebraic operations with the information coverages. The following techniques are noteworthy due to their application to the study of flood parameters.

(1) Boundary of drainage basins and sub-basins and preferential directions of drainage or of surface runoff concentration can be determined using the digital terrain model.

(2) Automatic estimation of different morphometric parameters of the basin (surface area, elongation, etc.) or of the streams (length, gradient, etc.); determination of timing parameters (concentration, time to peak, etc.) using classic formulae (US Corps of Engineers, etc.), and the necessary parameters in the flood wave propagation studies (Muskingum or Puls methods).

(3) Delineation of longitudinal and transverse profiles based on different directions or following the course of a particular fluvial stream, which can be used to reconstruct the extent of flooded areas during the peak flood discharge.

(4) Visualisation of the interconnections between streams and related analyses with regard to flood routing applicable to the propagation and combination of flood hydrographs.

(5) Interpolation of specific hydrological parameters based on a calculation of the spatial distribution of meteorological data from weather stations (precipitation in 24 h) and their statistics (means, rates of variation, etc.).

(6) Solution of the curve number or runoff threshold (mm) of an area (complete basin or sub-basin) by superimposing the gradient map (obtained from the digital terrain model), the land use map and the soil map, all reclassified according to the proposal of the Soil Conservation Service (SCS, 1972). By overlaying these maps (see Ferrer et al., 1995; Díez and Pedraza, 1997) and applying the SCS tables, we obtain a new raster mapping with the spatial distribution of the curve number which is essential in hydrometeorological flood calculations. This value, which refers to the current situation, can be modified towards hypothetical past situations (e.g. runoff threshold 10 000 years ago) by applying the potential vegetation map instead of the land use map (Figure 24.4).

(7) Integration of hydraulic models into the GIS, making it possible to supply directly to the database important parameters such as the circulating flood and velocity, using the digital terrain model. Two effective applications in this field were the integration

---

Figure 24.4 (opposite) Case study of the Palaeotagus GIS implementation applied to the determination of the curve number for the El Burguillo sub-basin (Alberche river, Tagus basin) both for the present vegetation and land use conditions and for a scenario previous to the land use change (Early Holocene). The rainfall–runoff models (e.g. HEC-1) account for palaeohydrological conclusions as a result of the analysis of the hydrographs obtained for present and past scenarios

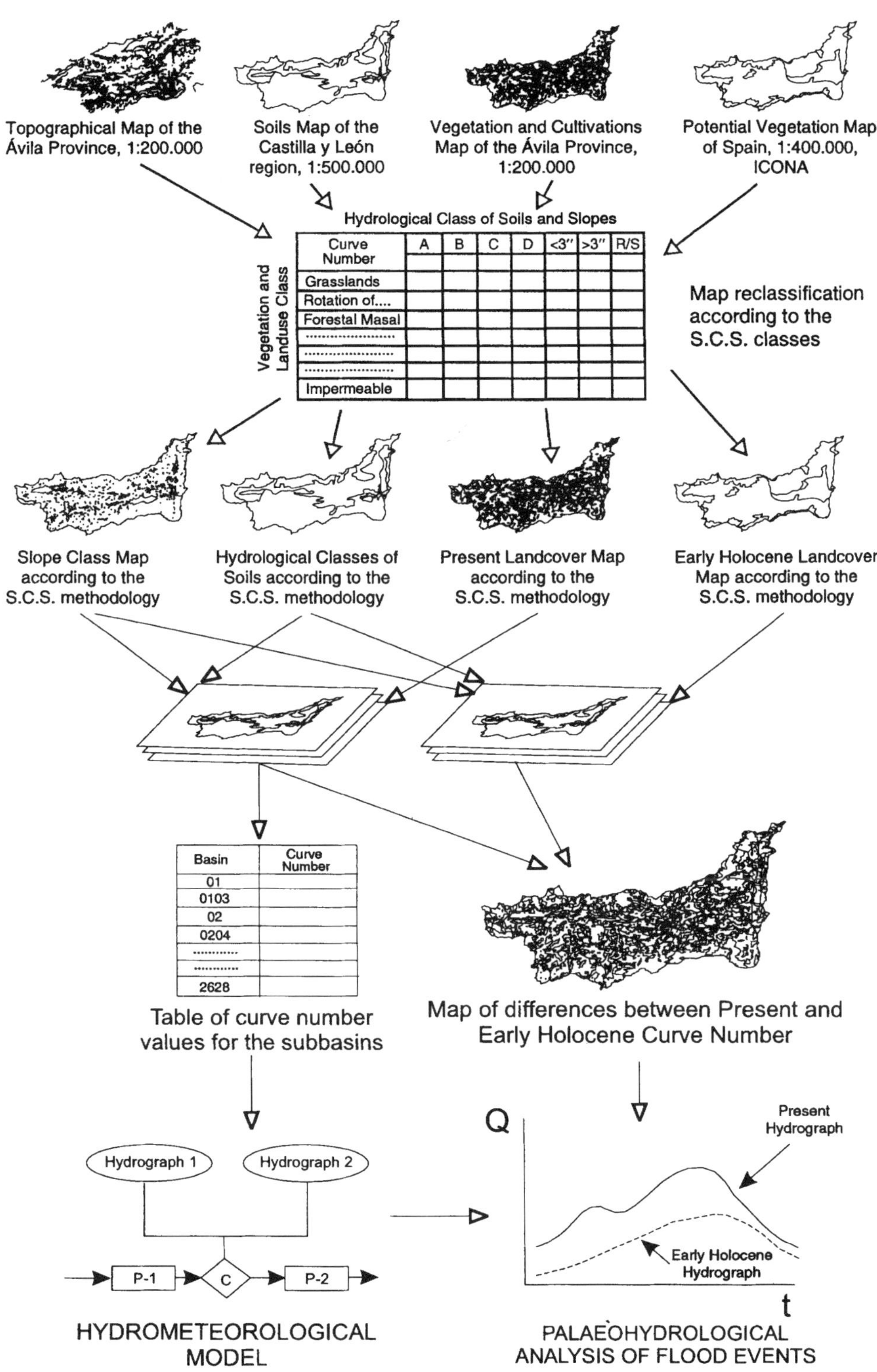
Topographical Map of the Ávila Province, 1:200.000
Soils Map of the Castilla y León region, 1:500.000
Vegetation and Cultivations Map of the Ávila Province, 1:200.000
Potential Vegetation Map of Spain, 1:400.000, ICONA
Hydrological Class of Soils and Slopes
Vegetation and Landuse Class
Curve Number
A
B
C
D
<3"
>3"
R/S
Grasslands
Rotation of....
Forestal Masal
Impermeable
Map reclassification according to the S.C.S. classes
Slope Class Map according to the S.C.S. methodology
Hydrological Classes of Soils according to the S.C.S. methodology
Present Landcover Map according to the S.C.S. methodology
Early Holocene Landcover Map according to the S.C.S. methodology
Basin
Curve Number
01
0103
02
0204
2628
Table of curve number values for the subbasins
Map of differences between Present and Early Holocene Curve Number
Hydrograph 1
Hydrograph 2
P-1
C
P-2
HYDROMETEOROLOGICAL MODEL
Q
Present Hydrograph
Early Holocene Hydrograph
t
PALAEOHYDROLOGICAL ANALYSIS OF FLOOD EVENTS

of Arc/Info with the HEC-2 model, and of Grass with two-dimensional flow models in the floodplain.

(8) Integration of the discharge data on historical floods and palaeofloods into the flood frequency analyses, using maximum likelihood estimates (Stedinger and Cohn, 1986).

The original maps and information or those derived from the GIS analysis can be visualised or printed using multiple final editing tools. An application with a menu system (user-friendly interface) could be intalled on Visual Basic or similar, which, through the use of an ArcView (ESRI) medium, would make handling of the spatial database easier for users not familiar with GIS.

CONNECTION TO GLOBAL DATABASES

The integration of the information contained in Palaeotagus into global databases requires a similar degree of hierarchical structure of the information and the compatibility of the management systems of the two databases.

The structure and nomenclature of the GLOCOPH fields, with their multiple tables, ensure a high degree of division of the information, which is ideal both for compiling and organising information and for carrying out selective searches. The structure of the flood database in the Palaeotagus geological record takes over the GLOCOPH tables whose techniques and methodologies (slackwater deposits) or results (hydroclimatic, discharges and stages) are common to both. A simple computer application extracts the necessary information in the fields common to GLOCOPH and, after adapting it to the Global Paleoflood Database codes, fills in the related record in the global database.

Also with the aim of integrating the information, the Access Database Management System, which is widely used and compatible with other systems (including GLOCOPH's Oracle) has been selected for the computerised implementation of Palaeotagus.

CONCLUSION

The design and development of palaeohydrological databases at regional level enables information to be organised and analysed prior to its integration into global databases, facilitating the standardisation and validation of results. Furthermore, coupling such databases with tools such as geographical information systems considerably increases the possibilities of spatial and temporal analysis, implementation of hydrological models and graphic output, facilitating palaeohydrological interpretation.

ACKNOWLEDGEMENTS

The authors are grateful to Francisco Flores Montoya (Confederación Hidrográfica del Tajo) and to Luis Quintas Ripoll (Centro de Estudios Hidrográficos del CEDEX) for supplying GIS information on the Tagus river basin. This research was supported by the Spanish Committee for Science and Technology (CICYT), Spanish National Climate Programme grant CLI95-1748, and Water Resources Programme grant HID96-0383. The review and comments of the original manuscript by Dr J. Branson (GeoData Institute, University of Southampton) are also gratefully acknowledged.

REFERENCES

Baker, V.R., 1989. Magnitude and frequency of palaeofloods. In K. Beven and P. Carling (eds), *Floods: Hydrological, Sedimentological and Geomorphological Implications*, Wiley, Chichester, 171–183.

Benito, G., Machado, M.J. and Pérez-González, A., 1996a. Climate change and flood sensitivity in Spain. In J. Branson, A.G. Brown and K.J. Gregory (eds), *Global Continental Changes: the Context of Palaeohydrology*, Geological Society, London, Special Publication, No. 115, 85–98.

Benito, G., Machado, M.J., Pérez González, A. and Sopeña, A., 1996b. Paleoflood analysis of the Tagus River in the El Puente del Arzobispo Gorge (Central Spain). In G. Benito, A. Pérez González, M.J. Machado and S. de Alba (eds), *Paleohydrology in Spain*, Centro de Ciencias Medioambientales, CSIC, Madrid, 5–16.

Branson, J., 1995. *The GLOCOPH database. Data structure information*, Geodata Institute, University of Southampton.

Branson, J., Clark, M.J. and Gregory, K.J., 1995. A database for global continental palaeohydrology: technology or scientific creativity? In K.J. Gregory, L. Starkel and V.R. Baker (eds), *Global Continental Palaeohydrology*, Wiley, Chichester, 303–321.

Chen Jia-Qi, Ye Yong-Yi and Tan Wei-Yan, 1974. The important role of historical flood data in the estimation of spillway design floods. *Scientia Sinica*, **18**(5), 669–680.

Comisión Técnica de Inundaciones, 1985. *Estudio de inundaciones históricas: mapa de riesgos potenciales*, Comisión Nacional de Protección Civil.

Díez, A. and Pedraza, J., 1997. Cálculo hidrometeorológico de caudales de avenida para la subcuenca de El Burguillo (río Alberche, Cuenca del Tajo). *Geogaceta*, **21**, 93–96.

Ely, L.L. and Enzel, Y., 1992. Paleoflood records and risk assessment: Examples from the Colorado River Basin. In J. Ganoulis (ed.), *Water Resources Engineering and Risk Assessment*, NATO Advanced Study Institute, Series G. Ecological Sciences, no. 29, Springer-Verlag, Berlin, 105–112.

Ely, L.L., Enzel, Y., Baker, V.R. and Cayan, D.R., 1993. A 5000-year record of extreme floods and climate change in the southwestern United States. *Science*, **262**, 410–412.

Ferrer, M., Rodríguez, J. and Estrela, T., 1995. Utilización de un Sistema de Información Geográfica en la obtención del umbral de escorrentía. *Ingeniería del Agua*, **2**(4), 43–58.

Gregory, K.J. and Lewin, J., 1987. Conclusions: palaeohydrological synthesis and application. In K.J. Gregory, J. Lewin and J.B. Thornes (eds), *Palaeohydrology in Practice: A River Basin Analysis*, Wiley, London, 341–360.

Gregory, K.J., Lewin, J. and Thornes, J. B. (eds), 1987. *Palaeohydrology in Practice: A River Basin Analysis*, Wiley, London.

Hirschboeck, K.K., 1988. Flood hydroclimatology. In V.R. Baker, P.C. Patton and R.C. Kochel (eds), *Flood Geomorphology*, Wiley, New York, 27–49.

Hirschboeck, K.K., Wood, M.L., Fenbiao, N. and Baker, V.R., 1996. The Global Paleoflood Database Project, *Second International Meeting on Global Continental Palaeohydrology (GLOCOPH'96)*, Toledo, Spain, 37.

Hua Shi-Qian, 1985. A general survey of flood frequency analysis in China. In *U.S.–China Bilateral Symposium on the Analysis of Extraordinary Flood Events*, Nanjing, China.

Quintas, L., 1996. La base de datos hidrológicos 'HIDRO' del CEDEX. *Ingeniería Civil*, **104**, 117–126.

SCS, 1972. *National Engineering Handbook*, Section 4, Soil Conservation Service, US Department of Agriculture, Washington.

Stedinger, J.R. and Cohn, T.A., 1986. Flood frequency analysis with historical and paleoflood information. *Water Resources Research*, **22**, 785–793.

Street-Perrott, F.A. and Harrison, S.P., 1982. Temporal variations in lake levels since 30,000 yr BP – An index of the global hydrological cycle. In J.E. Hansen and T. Takahashi (eds), *Climate Processes and Climate Sensitivity*, American Geophysical Union, 118–130.

WDC-A, 1992. A new focus, the paleoclimate record. *World Data Center-A for Paleoclimatology Newsletter*, **3**(2), 1.

Index

Note: Page numbers in *italics* refer to Figures; those in **bold** refer to Tables

Index compiled by Annette Musker